Practical Management for the Digital Age

Practical Management for the Digital Age

An Introduction for Engineers, Scientists, and Other Disciplines

Martin Baumers and John Dominy

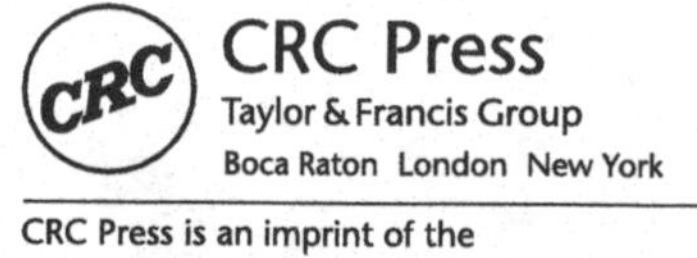

CRC Press is an imprint of the
Taylor & Francis Group, an informa business

First edition published 2022
by CRC Press
6000 Broken Sound Parkway NW, Suite 300, Boca Raton, FL 33487-2742

and by CRC Press
2 Park Square, Milton Park, Abingdon, Oxon, OX14 4RN

Library of Congress Cataloging-in-Publication Data

Names: Baumers, Martin, author. | Dominy, John (Professor of engineering), author.
Title: Practical management for the digital age : an introduction for engineers, scientists, and related disciplines / Martin Baumers and John Dominy.
Description: First edition. | Boca Raton : CRC Press, 2022. | Includes bibliographical references and index.
Identifiers: LCCN 2021028884 (print) | LCCN 2021028885 (ebook) | ISBN 9781032041865 (hardback) | ISBN 9781032120713 (paperback) | ISBN 9781003222903 (ebook) | ISBN 9781032132129 (ebook other)
Subjects: LCSH: Management. | Project management.
Classification: LCC HD31.2 .B38 2022 (print) | LCC HD31.2 (ebook) | DDC 658--dc23
LC record available at https://lccn.loc.gov/2021028884
LC ebook record available at https://lccn.loc.gov/2021028885

ISBN: 978-1-032-04186-5 (hbk)
ISBN: 978-1-032-12071-3 (pbk)
ISBN: 978-1-003-22290-3 (ebk)
ISBN: 978-1-032-13212-9 (E-book+)

DOI: 10.1201/9781003222903

Typeset in Times
by Deanta Global Publishing Services, Chennai, India

To Tina and Billie, maybe you'll read this one day

To my wife Megan, for her help and support during this project

Contents

PART II The Activity of Managing the Business

PART III *Practical Management Techniques*

Preface

We have written this book with the ambition of giving the reader a systematic and contemporary introduction to important management principles and techniques. The approach is systematic in two ways: first, it provides the reader with a broad framework of relevant concepts, with one definition building upon another as the book progresses. Second, the book explains – wherever possible – how management concepts and techniques flow from rationality-based models of human behavior. We have refrained from attempting to produce a contemporary management textbook that summarizes the latest fashions and trends in management thinking. Instead, our book aims to be contemporary by teasing out the all-changing implications of the information technology revolution. As we hope this book shows, the current reshaping of industry and commerce cannot be grasped without an understanding of the digital technologies behind these developments. In the same vein, this book highlights how the wider effects and impacts of business activity, such as its environmental footprint, are interwoven with management decision-making. This allows the reader to draw connections to the other great topic of our time, climate change.

This book is aimed primarily at undergraduate and postgraduate-taught students who have not yet had the chance to learn about management and commerce. We anticipate that most readers will be students undertaking courses in engineering, science, computer science, medicine, pharmacy, and the social sciences that feature an introductory taught element of management studies. Though not primarily intended for students on degree courses leading to qualifications in management, business, accounting, or finance, these students will find this book worthwhile. In particular, we think that MBA students without formal training in management or business will benefit from this book. Beyond the student audience, our book is aimed at management practitioners seeking to brush up on basic management principles and techniques and those interested in the broad impacts of information technology.

Especially with students in mind, we hope that this book will furnish the reader with actionable knowledge of managerial principles and techniques that will prove invaluable in many situations beyond graduation. As soon as graduates achieve a modicum of success in their early careers, they will likely be rewarded with a promotion to a role that requires them to manage. The contents of this book will help budding decision-makers engage with managerial situations confidently. Perhaps not unlike a course in self-defense, our intention is to provide a solid foundation of concepts and methods to fall back on when needed. As new managers will inevitably discover, many of their manager colleagues operate without a reliable safety net of fundamental management knowledge underneath.

BORN DIGITAL

Today's world is undergoing a profound transformation through the emergence of increasingly pervasive and powerful digital technologies. Technologist Neil

Gershenfeld once quipped that "we've had a digital revolution, but we don't need to keep having it".* What Gershenfeld meant is that the transformation caused by information technology is now well underway. Any attempt to halt or reverse it is misguided and managers and decision-makers need to systematically and unflinchingly draw conclusions for the road ahead.

Our book's approach to information technology, and technology more generally, is inspired by the teachings of philosopher Martin Heidegger, who urged that technology shouldn't be thought of as a set of tools that can be adopted if deemed beneficial. More appropriately, technology should be seen as an enduring and inescapable force that grips us and changes us over time in ways that don't necessarily serve our own interests. In the competitive context of business, this means that managers, now more than ever before, must understand the implications of new technologies and adapt their management practices. Ignoring this imperative will likely spell doom for any business.

Fortunately, over the last decade or so, a clear appreciation of the mechanisms through which information technology is reshaping business has emerged in the management literature. Unfortunately, this insight has not yet percolated through to introductory management texts. The fact that digital technology is now interwoven almost everywhere in business is addressed throughout this book, including a dedicated chapter entitled "The Economics of Digitalization and Automation" (Chapter 4). Rather importantly, and perhaps ambitiously, this chapter includes a treatment of what information technology actually is.

BALANCING COMPACTNESS, ACCESSIBILITY, AND RIGOR

Heeding writer Alan Watts' sentiment that, as a successful man in business, he hated business because it is an arid and colorless occupation,† we have written this book in the hope that a modern course in introductory management can be delivered with minimal tedium. To achieve this, our first objective was to produce a compact and self-contained text that can comfortably be read cover-to-cover within a few weeks. Our second objective was to write the book in a style that is accessible to readers with little or no prior knowledge of the concepts of management and business. Our third objective was to produce an academically rigorous, accurate, and consistent treatment of a subject that draws on a wide field rife with competing definitions, methodological variety, conceptual fuzziness, and inconsistent use of terminologies. Naturally, our objectives ran against each other – we hope that the compromise reached is a happy one.

PEDAGOGICAL APPROACH: LETTING BUSINESS BE BUSINESS

The pedagogical approach along which this book is structured draws on our own experiences of learning and teaching management in an engineering department.

* Gershenfeld (2006). The beckoning promise of personal fabrication. TED2006 Conference.

† This statement forms part of Eastern philosopher and writer Alan Watts' lecture "How to attract what you love". Different recordings of this lecture are available online.

Management is undoubtedly a huge field of many disciplines, professional orientations, and traditions. We found that introductory management texts often contain quite specific assumptions and motivations that are not always explained clearly and explicitly enough to learners without prior knowledge. For example, the basic theory of the firm taught in undergraduate economics is built entirely on the idea that the objective of business activity is to extract the greatest possible profit. Other schools of thought, however, such as the operational philosophy of Lean or the service-dominant logic in marketing, stress that the objective of management is to provide genuine value to customers. Again, other disciplines involved with management such as accounting are preoccupied with the accurate and truthful representation of facts. Because fundamentally different ideas about why managers behave the way they do are baked into the basic theories and methods in various disciplines, it can be difficult and sometimes frustrating for beginners to draw important connections.

To address this issue, we have deliberately chosen a perspective concentrating on the management of private, for-profit businesses. More specifically, we view businesses through the lens of the *private venture*. This perspective sees businesses as having a life cycle consisting of discernible stages and a finite lifespan. Over this life cycle, businesses are expected to generate a sufficiently large stream of financial benefits to compensate those who have invested money into it or otherwise contributed effort. This perspective also highlights that businesses require planning and resources to set up and are subject to considerable risk and adversity.

Portrayals in the media often seem to suggest that it is the objective of business to make a cultural contribution, provide a lifestyle to those involved, or act as a stage for the self-expression of celebrity entrepreneurs. Not so. In this book, we unromantically emphasize that businesses are, and should primarily be, means to the ends of achieving their owners' financial objectives. For this reason, the activity of managing is marked by a strongly instrumental character directed at achieving financial goals. Letting business be business in this way benefits new students of management because it provides a clear logical framework in which they can locate various theories and activities.

We are conscious that by portraying management in this way we are unable to reflect some management practices in non-profit organizations and public institutions. However, we feel that the clarity in our approach outweighs its limitations especially for readers at the beginning of their management journey. Crucially, in this book, we emphasize that viewing businesses as means to generate financial gains for owners does not absolve managers from their duties of ethical decision-making and caring for others, including future generations.

Acknowledgments

A substantial number of colleagues and peers have helped develop this book. We would like to thank (in no particular order) John Edwards, Helen Rogers, Ender Özcan, Mike Clifford, Mike Walsh, Elizabeth Clark, Zsofia Toth, Martin Robinson, Jack Jones, Matthias Holweg, Jon McKechnie, Stephen Detsch, Brendan Ryan, and Michal Konturek for their reviews at various stages of this project. We would like to express our gratitude to the students on the Engineering Management 1 and 2 courses at the Faculty of Engineering at the University of Nottingham for "test-driving" this material – always enthusiastically and sometimes to breaking point. We also thank the team at CRC Press, Joe Clements, and Lisa Wilford for their patient support. Additionally, we would like to thank Vijay Bose and the team at Deanta for their professionalism and support in producing this book. Martin would like to thank his wife and family for allowing him to work on this book in long evenings during the corona pandemic. John would like to thank Dr John Johnson who was his mentor for many years and without whom he would not have set off in this direction.

Author Biographies

Martin Baumers is Associate Professor of Additive Manufacturing Management at the University of Nottingham. With a background in economics, Martin worked in the metals industry before becoming an academic. Martin completed his PhD at Loughborough University in 2012 on the management and operation of additive manufacturing technology, commonly known as 3D printing. As an academic researcher, Martin has written over 20 research articles, appeared at more than 15 conferences, and written three book chapters. At Nottingham, Martin teaches a range of management-related subjects mainly to engineering students. Martin is fascinated by the implications of digital technologies for business, especially in manufacturing. This is Martin's first book.

John Dominy is Honorary Professor in the Composites Group at the University of Nottingham. John has worked in the engineering industry since graduating from the University of Middlesex (then Middlesex Polytechnic) in 1974. John was awarded a PhD in 1980 for work at Rolls Royce on the lubrication of high-speed roller bearings. He has experience in management in very large (aerospace), medium (motorsport), and micro (specialist composites) businesses. Since 2003, John has been teaching in the Faculty of Engineering at Nottingham, mainly in the areas of management and engineering design. Although now semi-retired, John takes a keen interest in teaching what he has learned over the course of a career in engineering management to undergraduate students.

About This Book

The field of management is vast, so reading (and writing!) this book is unavoidably an exercise in dipping various toes into various ponds. Figure 0.1 outlines the learning journey that readers will embark on when working through this book. To provide some initial orientation, the figure names important concepts and ideas introduced along the way. The book begins with an introduction of the context of management by providing a brief history of management and then establishes what a private business

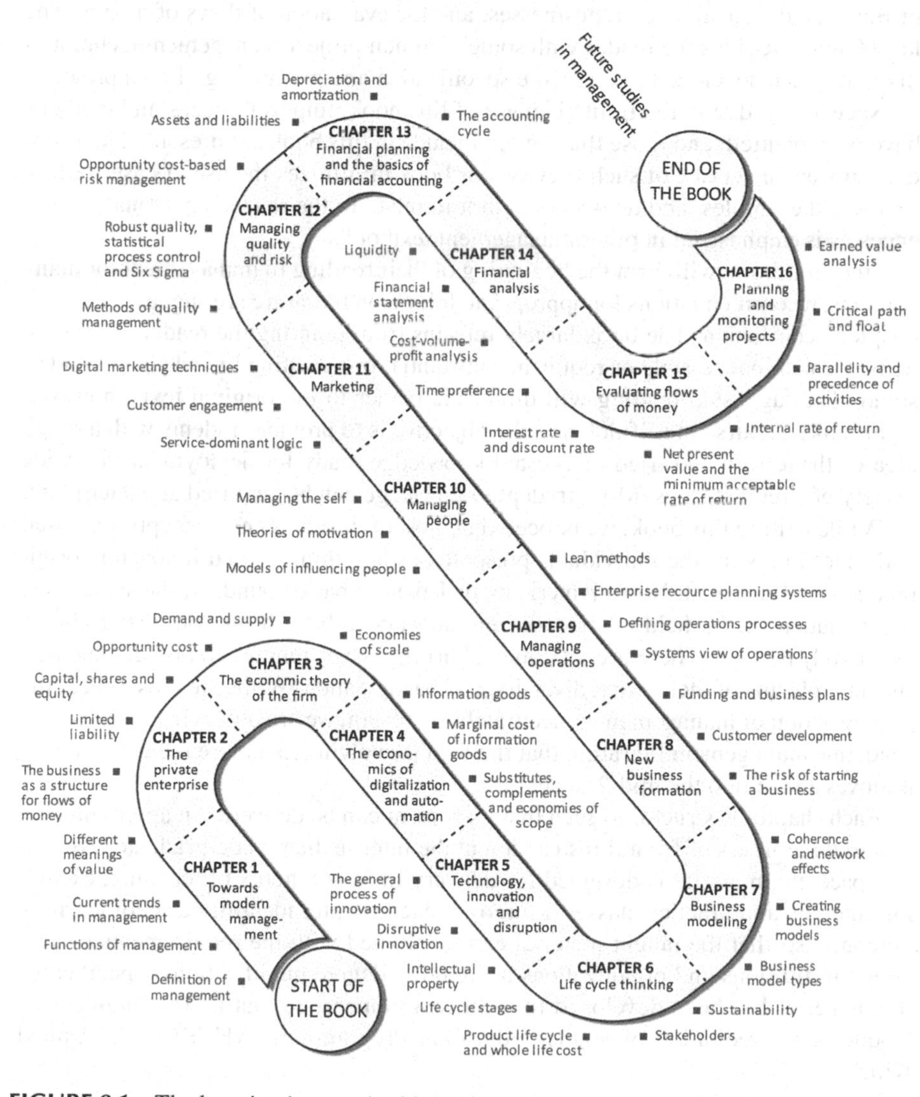

FIGURE 0.1 The learning journey in this book

actually is. This is followed by two chapters related to economics, the first of which introduces the rational behavior of a business and the second of which explores the general economic impact of information technology. The subsequent two chapters introduce the broad importance of technological change as well as the need to take a life cycle view to understand the full impact of managerial decisions. The following chapters of the book introduce the reader to various interlinked activities of management, including business modeling, starting new businesses, managing operations, managing people, marketing, as well as managing quality and risk. The final part of this book intends to equip the reader with some basic, yet extremely widely used practical techniques. These include basic financial planning processes, the analysis of financial data available on businesses, and the evaluation of flows of money. The last chapter provides the reader with some common project management techniques that have been developed to allow the specification and monitoring of own projects.

Necessarily, due to the limited length of this book, important ideas and methods have been omitted, and those that we did include in this book are presented in a condensed way. In service of such brevity, our book minimizes the use of case studies, real-world examples, and does not attempt to relate to the reader's personal experiences as is emphasized in other management textbooks.

Since this book will form the beginning of their reading in management for many students, recommendations for appropriate follow-on literature are suggested in each chapter's conclusion. The book largely refrains from pointing the reader to original texts deemed inaccessible or requiring substantial pre-existing knowledge to understand. The suggested reading will direct the reader to the original texts, however. Again, like a course in self-defense, the objective is to provide students with a rough idea of the terrain and tried-and-tested knowledge ready for deployment in a wide variety of circumstances. More in-depth knowledge can be acquired at a later point.

While writing this book, we noticed that most of the ideas and concepts discussed and indeed most of the individuals presented as key thinkers and innovators originate from a European, North American, or Japanese background. At the same time, the evolution of the field of management appears to have been dominated almost exclusively by men. These are, of course, burning issues relating to broader and persistent problems of inclusivity, diversity, and participation. While we must reflect the past evolution of management as accurately as we can, we are observing in the recent academic management literature that the field is evolving in a more inclusive way as it moves ahead into the mid-21st century.

Each chapter is structured such that its content can be delivered in approximately two to three hours of frontal instruction at the intermediate undergraduate level. At this pace, the material is designed for delivery in 32–48 hours of teaching, excluding tutorials and practice classes. Moreover, each chapter identifies a set of learning outcomes so that the taught material can be aligned with the requirements of academic institutions and organizations accrediting degree qualifications. Specifically, this material has been developed to meet requirements set out by the Engineering Council's Accreditation of Higher Education Programmes (AHEP) in the United Kingdom.

Because we must digitally practice what we preach, especially at the time of writing during the global coronavirus pandemic, this book has been developed with remote electronic assessment in mind. Several software systems are available for e-assessment, most of which emphasize convergent lines of questioning, meaning that students are expected to produce unambiguously correct answers to questions. The review questions contained in each chapter have been drawn from a pool of such questions and should aid the creation of additional questions. To guide the reader, each question identifies a question type, such as *multiple choice* or *calculation*. However, the material covered is also assessable by more traditional means. The correct answers to the review questions are available for download at the publisher's website.

Due to the vastness of management as a field of activity and the available literature, we have had to make many subjective judgments on what to include in this book. These judgments were made on the basis of the perceived relevance to our target audience. If we have unjustifiably excluded important aspects, this is solely our fault. Equally, any factual and logical errors contained in this book are due to us.

The guiding idea behind this book was to write the introduction to management that we would have wanted to read at the start of our own journeys in management. Some early reviewers have made comments in this direction, such as "I absolutely wish I would have had this book in school", which encouraged us greatly. As discussed in this book, the powerful logic of platforms now dictates that comments and reviews from our readers (i.e., you!) are decisive for the success of a project like ours. So, if you feel that our book has been of assistance or stirred some other – hopefully positive – emotions, we would be very grateful for a review on a platform such as Google Books or Amazon. In any case, we hope that this book turns out to be useful.

Part I

The Context of Management

1 Toward Modern Management

OBJECTIVES AND LEARNING OUTCOMES

The objective of this chapter is to establish the basis of this book by defining what management is and to introduce the reader to how management has emerged as a distinct practice and field of inquiry. The chapter presents a brief history of management, showing that the automobile industry in the 20th century was the venue for crucial developments in the field of management. In doing so, it presents the origins of mass-production and outlines developments in Japan ultimately leading to the Lean approach, which is extremely influential in contemporary management. The chapter finishes with a brief overview of current trends in management. Specific learning outcomes include:

- the ability to usefully define management and identify the five general functions of management;
- knowledge of precursors of modern management and understanding how innovation in management follows technological development;
- understanding the evolution of mass-production in the 20th century and the role of managerial innovation in this;
- knowledge of Lean processes and their impact on management as a response to mass-production, including a definition of Lean;
- knowledge of major trends that are addressed by management and current themes in management;
- understanding important terms, synonyms, and accepted acronyms; and
- appreciation of important thinkers in the history of management.

FIRST THINGS FIRST: WHAT IS MANAGEMENT?

Management is a wide field. It is both a practical endeavor as well as an academic discipline. It comprises many different activities and can be applied in many different areas. Perhaps best thought of as a body of knowledge that should be put into action, management is rife with competing definitions, inconsistent use of terminology, a wide variety of methods, and conceptual fuzziness. While these aspects make management an interesting and rich area for reflection and academic study, they can also pose challenges for learners at the beginning of their management journey. The ambition of this book is to equip readers with the knowledge needed to cut through this complexity.

Many aspects and concepts have been added to the field of management in an ongoing process, especially throughout the 20th century. This chapter, therefore,

DOI: 10.1201/9781003222903-2

puts first things first* and provides a brief history of management. This is useful since the simplicity and clarity of the early management innovations make them particularly instructive for novice readers. In this spirit, the book opens with a traditional definition of management and presents an – even more traditional – first characterization of management as consisting of five general functions.

The word *management* is commonly used in two senses. In the first sense, it describes the general human activity of managing, referring to the process of taking responsibility for the purpose, progress, and outcome of events. Used this way, the word management captures human *agency*, which can be defined as a universal human ability to make a difference in the world through intentional and goal-oriented action. Management can, therefore, be described as the pursuit of goals with the support of other people and the resources available. This makes clear that management is a common activity in which we all engage almost every day.

The second sense of the word management reflects a distinct role that people, especially in organizations, can take on. To understand this meaning, it is necessary to first understand that management activity, as defined in the previous paragraph, is separate from the actual work to be done. Only if these two activities are separate, is it possible for a person to assume the role of a manager and in most cases, this will entail that the effort of other people is to be managed. Following this line of thought, this book uses a modified version of a traditional definition of management:†

IMPORTANT DEFINITION

Management is that group of functions in an organization which concerns itself with the direction of various activities to attain the organization's objectives. In doing so, management deals with the active direction of human effort.

FAMOUS THINKER

This two-part definition shows that the role of management is about directing activities in order to reach the goals of an organization. It also shows that the element of human effort is central. In line with this definition, anyone performing this role, which amounts to being in charge of an organizational unit (however small) is a manager and should be referred to as one.

Henri Fayol (1841–1925)

A first impression of the scope of management activities can be obtained by considering Henri Fayol's model of the five functions of management, first published in 1916. Fayol was a mining engineer and is seen as one of the founders of the field of management as we understand it today. His model views management as formed of five general functions that allow the goals of an organization to be achieved.

* As the reader will see in Chapter 10, even the seemingly obvious logic of "first things first" has been absorbed into the management literature (Covey, 1989).

† Adapted from Spriegel and Lansburgh (1947).

1. *Planning*

   A course of future action must be planned to achieve a goal or desired state. Planning includes the development of a logical sequence of stages, timing, allocating activities to people, and determining the resources needed for execution. Both short-term and long-term planning are required.
2. *Organizing*

   Building on a plan, managers make available the resources required for execution, including any materials, equipment, funding, and people. Organizing also requires assigning responsibilities to people, establishing who reports to whom, and how the elements of the organization relate to each other.
3. *Commanding*

   To execute the plan, managers lead people to achieve the goals of the organization. Directing people in this way requires communication skills, the capacity to motivate people, and an ability to balance the needs of those involved in the organization with the requirements of the tasks at hand.
4. *Coordinating*

   Managers coordinate the required activities for successful execution. The underlying structure of communication between people provides the basis for this. Managers must ensure that the timing of activities within an organization is aligned and that individuals have the authority they require to fulfill their tasks.
5. *Controlling*

   Managers assess whether there are actual or potential deviations from the plan, thereby ensuring the attainment of the organization's goals. Controlling includes the managing of information (especially in the form of feedback), the evaluation of performance, and the initiation of corrective action.

The significance of Fayol's model of the general functions of management, summarized in Figure 1.1, lies in its applicability to all kinds of organizations and that it describes behavior that managers ought to adopt in order to reach the organization's goals. Fayol's ideas form a fundamental part of how we think about management in the present.

Setting managers apart from other members in organizations creates a split between the actual work to be done and the activities of managers. Naturally, this can carry the problem that some managers become distant from the actual work performed within the organization. This distancing between managers and others is a well-recognized and long-standing issue and can undermine a manager's ability to direct and control what is going on. This aspect is particularly important in more recent management methodologies, as discussed throughout this book.

The remainder of this chapter outlines the evolution of professional management, especially during the 20th century. This evolution has led to identifiable fields, or branches, within management. These branches include, but are not limited to, financial management (see Chapter 13), human resources management (see Chapter 10), information technology management (see Chapter 4), marketing management (see Chapter 11), operations management (see Chapter 9), and business models (see Chapter 7).

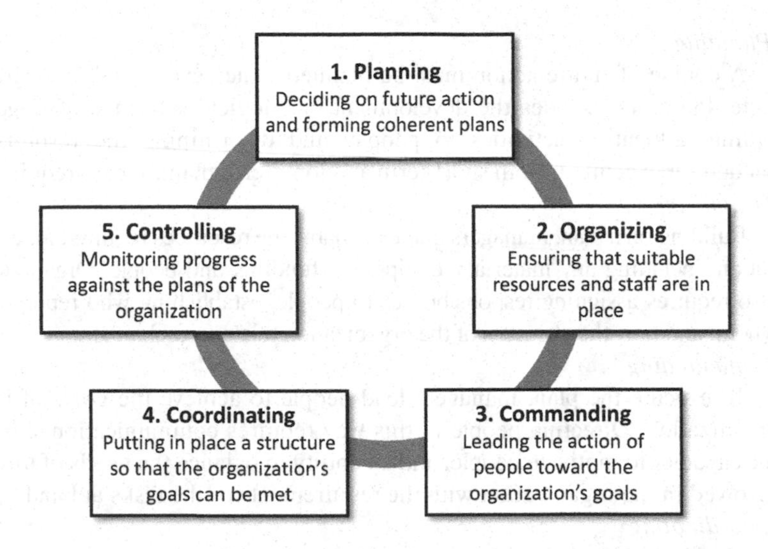

FIGURE 1.1 The five general functions of management

In the present, however, it is becoming increasingly clear that effective practical management often involves combinations of these branches. For this reason, modern managers tend to think in terms of processes and activities that are shaped by the problem at hand rather than in terms of distinct techniques or fields of specialization.

EARLY FORMS OF MANAGEMENT

During the 1760s, the story of management as we understand the term today begins with the industrial revolution. The industrial revolution itself was made possible by a broad transition in technology from hand production to machines that were driven by steam and waterpower. The textile industry in the United Kingdom pioneered the adoption of these new technologies, and so many related technological and organizational innovations occurred in this setting.

In this period in the United Kingdom, many workers who had been self-employed and operated simple machinery at home to produce textile products in rural areas migrated to large industrial centers. In these locations, workers were employed in new facilities that became known as *factories.* Factories were owned and controlled by another emerging – and much less numerous – group of people, the *factory owners.* New forms of mechanization and centralization within factories meant that the bulk of the new work required little skill and the workers were paid very low salaries. This resulted in significant poverty among the factory workers which, in turn, set in motion long-term political changes. This was reinforced by the overbearing power of the factory owners, expressing itself in the close control of the factory workers and highly regimented patterns of work in which the individual factory worker counted very little. Overall, life as a factory worker in the 18th and 19th centuries was generally not a happy existence.

To characterize the managerial attitude of the factory owners, it is important to note that wages among factory workers were so low that it did not matter significantly if the activities of each worker were carried out in an optimal way. If a factory owner wanted to increase production quantity, they simply employed more people. This also meant that there was little incentive to improve the abilities of individual workers. Where product quality mattered or specialist skills were required, in the manufacture of machine tools, for example, factory owners still relied on highly skilled and well-paid craftspeople. Such specialists occasionally came together under one roof but retained their independence and were generally self-employed. It is important to note that, especially during the early stages of the industrial revolution, limited transport networks meant that the newly formed industrial structures remained strictly local.

MANAGEMENT INNOVATION FOLLOWING TECHNOLOGICAL INNOVATION

In the early 19th century, technological developments centered on steam and steel. The introduction of canals, railways, and steamships made new forms of transport possible but initially did not change the way in which work was organized. The volume of products transported by railway and shipping remained small and there was still little incentive to increase the effectiveness of the factory workers. This changed in the second half of the 19th century with additional technological advances, most notably the telegraph. The telegraph enabled the formation of larger organizations with multiple locations because real-time communication made it possible to coordinate activities between localized operations and company headquarters. At this point, increasingly large quantities of goods began to be moved over long distances using the available transport infrastructure. This also meant that goods could be transported for sale in new geographical markets.

FAMOUS THINKER

Frederick W. Taylor (1856–1915)

The technological innovations driving the industrial revolution were accompanied by a rapidly spreading scientific view of the world. This included the spread of a range of ideas centered on the importance of reason and the role of physical evidence as the source of knowledge. It is important to note that this rationality-based outlook was not applied in all areas immediately. Surprisingly, the importance of a scientific, evidence-based way of thinking for the control of organizations was not recognized until much later, becoming widely accepted only after 1910.

The revolutionary idea of *scientific management*, which aims to improve industrial efficiency by establishing a science of factory production, is closely associated with Frederick W. Taylor, who was a mechanical engineer and a contemporary of Henri Fayol. Taylor promoted five principles that deviated strongly from how factory owners had structured the work in their operations in the past. These principles were:

- use of scientific methods to find the best way to perform activities, rather than rule-of-thumb methods;

- choice of the best available worker to perform an activity, making sure that they are capable of performing the task successfully;
- training and teaching workers to be able to perform activities, as defined by managers;
- payments to motivate workers to precisely execute the defined methods; and
- establishing a clear division of responsibilities so that managers plan and control activities and the workers execute the tasks.

Taylor's approach to scientific management was extremely influential over the course of the 20th century, unlocking large improvements in the effectiveness of factory work. His ideas also led to important innovations supporting production, such as accounting processes and methods for stock control. To this day, Taylor's methods continue to be used in many modern-day factories across the globe, especially in industries that employ relatively low-skilled workers, such as textile production and the assembly of electronics.

Later in the 20th century, the implementation of Taylor's ideas came to be widely criticized for imposing a significant human cost on factory workers. Following Taylor's approach tends to make workers highly exchangeable, which has a negative impact on individual job security. Additionally, the requirement to adhere to tightly defined processes at all times and to perform limited, repetitive activities under close supervision makes workplaces organized in this way very dull. Throughout the history of the factory, these practices have caused mental and emotional problems in many factory workers and should be viewed critically for this reason.

The span of time between the technological innovations at the start of the industrial revolution and the widespread adoption of a scientific approach to management based on Taylor's ideas is remarkably long. It is now widely accepted that the invention of new organizational and managerial methods (and other aspects of administration) lags behind the arrival of new technologies. This is known as the organizational lag model* and describes a general inability of management methods to adapt in step with rapidly evolving technologies. This carries two important implications for managers:

- change in management methods is frequently, but not always, driven by constantly evolving technologies; and
- highly effective organizations will be faster to adopt new management methods than low-performing organizations.

To remain competitive, managers must therefore maintain an awareness of ongoing technological developments and understand their impact on business and management methods over time. As a core topic of this book, this idea applies to information technology in particular. Chapters 4 and 5 concentrate on these aspects.

---

* Damanpour and Evan (1984).

FIRST STEPS TOWARD MODERN MASS-PRODUCTION

Important elements and prerequisites of modern management methods emerged in specific locations and industries during the industrial revolution. One important place for such developments was the city of Portsmouth in the United Kingdom during the age of sail. This era lasted to the mid-19th century and was marked by sailing ships dominating international trade and naval warfare. To operate the rigging on each of these ships, high-quality pulley blocks were required in large numbers. To address this manufacturing challenge, new forms of organizing production and methods of standardization were introduced in the early years of the 19th century to ensure the delivery and quality of these essential components. These ideas proved to be extremely influential in many other industries and established the idea that standardization is an essential pillar of high-volume manufacturing.

However, the period during which arguably the most significant progress in management methods occurred so far was in the early years of the 20th century following the arrival of the automobile. The car was one of the first commercially manufactured products requiring precision engineering while – unlike pulley blocks on sailing ships – having the potential consumer appeal to be demanded in extremely large numbers by ordinary people.

Interestingly, the arrival of the car in the 1890s did not have an effect on manufacturing and management methods immediately. During this time, cars were produced in relatively small factories and workshops employing craft production methods. These operations still relied on the skills of specialized craftspeople. In this early part of the history of the automobile, no two cars were exactly the same; seemingly identical parts were not usually expected to be interchangeable. The prevalence of handwork also meant that there was no benefit in forming large organizations in order to increase the number of cars made. These circumstances led to early cars being expensive to buy, difficult to operate, and unreliable. In consequence, early cars generally suited the requirements of wealthy enthusiasts rather than being useful for everyday transportation.

FAMOUS THINKER

Henry Ford (1863–1947)

Sensing an opportunity to overcome these limitations, a major wave of managerial innovation was unleashed in the early 20th century in the United States by the inventor and entrepreneur Henry Ford, founder of the Ford Motor company. The focus of Ford's activities was to systematically lower manufacturing costs in automobile production. After several earlier products, Ford introduced the highly influential Ford Model T in 1908. Its innovative design and low price fully established the car as a utilitarian consumer product that could be afforded by the masses. The Model T's design objectives included low manufacturing cost (permitting a low price that could be afforded by people on average incomes), design with ease of manufacturing in mind, and maintainability by the owner.

Ford achieved these goals through a series of remarkable steps. Unlike many other manufacturers, production at Ford worked to a standard measurement gauge within the factory and used new materials allowing the heat treatment of parts without causing significant distortion. This allowed Ford to reduce the number of expensive skilled workers in factories. Significant organizational changes soon followed. In 1903, a Ford car was still assembled by one employee, known as a *fitter*. In this approach, the car remained in a single location during assembly as the fitter moved to collect the components needed from the factory's stock. With the introduction of the famous Model T in 1908, Ford had *de-skilled* the assembly work by breaking it down into very small elements such as mounting a wheel, each carried out by a single worker. This further reduced Ford's need for more skilled, and therefore expensive, fitters. At this point, the car still stayed in one place while the workforce came to it.

In 1913, the *continuous flow assembly line* was introduced at Ford, inspired by the way meatpacking operations in slaughterhouses worked. Both an organizational and technological innovation, the chief purpose of an assembly line, also known as the *production line* or *manufacturing line*, is to reduce production time. If the item to be assembled comes to the worker, then that person will not waste time walking around in the factory. If the worker does the same task repeatedly, then the required tools can remain in hand so no additional time is lost retrieving tools. Because time-saving accumulated over the multitude of steps required to assemble a car, the overall impact on production time, and hence cost was game-changing. As discussed in Chapter 9, line operations systems constitute a relevant and important type of operational processes to this day.

Using these methods, Ford halved the average time needed to assemble a car and almost completely de-skilled work in the factory. This resulted in dramatic increases in the number of cars built, reaching a volume of 2 million per year by 1923,* an unprecedented number at the time. Additionally, the need to coordinate and maintain a steady flow of parts to the manufacturing line gave rise to new jobs and specializations such as industrial engineers and production engineers.

Ford's factories were highly self-contained, meaning that most components needed by Ford were produced within the factory, extending even to the production of steel and glass in the 1930s. This was partly motivated by a business need but also due to Henry Ford's personal distrust of other organizations and people. By the mid-1930s, Henry Ford had expanded his business significantly and owned rubber plantations, iron mines, ships, and railways. This approach is known today as *vertical integration* and forms an important business model, as discussed in Chapter 7. During this period, Ford also attempted entries into other industries which were unsuccessful. At this point, Ford had become one of the first truly international companies, with factories in the United States, the United Kingdom, Germany, and France.

However, the managerial and technological innovations achieved by Ford led to a number of problems. An early obstacle faced by Ford was the poor motivation within the workforce. While Ford's factory employees were well-paid by the standards of contemporary farmworkers, they were still relatively low-paid workers. Typically, factory workers were migrants who joined Ford for a period of time to save up money

* Gross (1997).

and then return to their homes elsewhere in the country. This meant that Ford's workforce suffered from a *high staff turnover*, which forms an important metric in human resources management reflecting the attitudes of workers (introduced in Chapter 10).

Responding to this problem, Ford doubled the wages of the factory workers to the famous $5 per day. This improved staff morale and reduced the desire among factory workers to return to the relative poverty of their rural homes. However, while the working conditions and pay at Ford were acceptable in the short term, they were often still not attractive for an entire career. In consequence, many workers found the conditions at Ford unsatisfactory and agitated for a better life. This dissatisfaction contributed to the unionization of the motor industry in the United States. In turn, this led Henry Ford to impose an increasingly restrictive set of rules for his staff, which constrained the growth of his organization.

Competing car manufacturers quickly recognized the power of Ford's technological and organizational innovations and began also adopting these methods. In 1927, the Model T, still remarkably unchanged, went out of production. By the mid-1930s, Ford had reached the limits of its founder's approach, and other manufacturers overtook Ford in terms of market share.

SUPPORTING MANAGERIAL INNOVATION

Alfred Sloan (1875–1966)

FAMOUS THINKER

Ford's main competitor at the time, General Motors (GM), had been founded in 1908 by William Durant. Durant was a tycoon who had purchased a number of companies such as Buick and Cadillac but struggled to maintain control of his organization. In the early years of GM's existence, its business units were not coordinated effectively and were unable to react to changes in market conditions. These problems were resolved after 1920 when engineer and businessman Alfred Sloan took over the leadership of GM. Sloan was an extremely talented manager, developing numerous highly influential and successful management techniques.

Sloan understood the importance of brands for consumer appeal and organized GM's product range according to this. This approach resulted in a brand lineup ranging from offerings for price-sensitive customers (Chevrolet) to luxury brands (Cadillac). This was done so that individual business units would not compete for the same customer and so GM's customer audience could remain loyal to GM throughout their lives. Sloan established the units within GM as *profit centers* and monitored them closely, replacing managers as and when necessary. A profit center can be defined as follows:

IMPORTANT DEFINITION

A profit center is a segment of an organization that achieves inflows of money and incurs costs. A profit center has a positive effect on the financial performance of the organization.

Accepting the need for external sources of money to fund the activities of GM, Sloan gained support from major banks and financial institutions. This enabled Sloan to carry out a systematic program of expansion in the business. His innovations also allowed GM to effectively manage overseas factories.

Where Ford had introduced the professions of the industrial engineer and the production engineer, Sloan went further, recognizing the need for new roles such as financial managers and marketing specialists. Additionally, while each unit within GM was responsible for a different product, GM pioneered a strategy of part commonality so that certain components, such as pumps and generators, could be used extensively across the entire organization. This further reduced costs.

With his keen understanding of the aspirations and needs of people, Sloan made GM's product offering more attractive to customers by introducing a range of additional innovations:

- to create a continuous demand for the new automobiles, Sloan revised the styling of each model every year;
- GM began offering optional extras in cars, such as air conditioning units and automatic transmissions, so that customers could adapt their car to their preferences; and
- recognizing that many customers would not want to save to be able to purchase a car, Sloan began offering loans to customers, thereby innovating the idea of combining financial services with the production of consumer goods.

Together, Sloan's visionary ideas were highly successful. In the 1930s, GM took over leadership of the global car industry from Ford and became one of the largest businesses in human history.

With the addition of the innovations at GM in the 1940s and largely as a result of developments in the car industry in the first half of the 20th century, the elements for mature mass-production as we understand the term today were in place.

Mass-production combines Ford's factory practice, Sloan's management techniques, and the workers' union movement supporting the interests of factory workers. This combination of technological, managerial, and social innovation proved extremely influential in many other industries and, through gradual rises in worker salaries, led to significant improvements in the living conditions of hundreds of millions of people across the globe. The role of these innovations in lifting a large number of people out of poverty around the world is easily forgotten when criticizing mass-production approaches from today's perspective.

Overseas makers such as Morris in the United Kingdom and Citroën in France had recognized the advantages of mass-production. However, with economic depression and the lasting impact of World War II in their home countries, these manufacturers had been unable to adopt such methods immediately. While the rise of mass-production continued into the 1970s and 1980s in the West, the growth of the trade unions and their consequent gain in influence caused major disputes and disruption. This was fueled, to a large extent, by the perceived lack of prospects for factory workers.

THE GROWTH OF LEAN INTO A PRACTICAL MANAGEMENT PHILOSOPHY

As described in this chapter, the development of new management methods in the early 20th century was focused on the automobile industry in the United States. Following this period of significant and largely undisturbed growth, major challenges to the US car industry in the 1950s and later were the direct result of new foreign competition made possible by global trade. Developments in Japan played a particularly important role in this.

Kiichiro Toyoda, the founder of the Toyota Motor Corporation, had been trying to manufacture cars for consumers in Japan since the 1920s but had been constrained by the economic and political circumstances in Japan at the time. As Japan set out to develop its own car industry following World War II, companies such as Toyota found that they were unable to copy the Western mass-production model. Reasons for this failure included a lack of financial resources, the unacceptability of Ford's practices to Japanese culture, and restrictive labor laws introduced by the United States as an occupying power.

Taiichi Ohno (1912–1990)

FAMOUS THINKER

Following a number of visits to the United States, particularly to the Ford plant in Detroit, Taiichi Ohno, a senior engineer at Toyota, came to believe that there were inherent flaws in the mass-production system and that it would be possible to solve these issues together with Toyota's own problems. The revolutionary approach to operations that Ohno invented and eventually perfected is known today as *Lean*. Lean differs from the original concept of mass-production in many respects. Instead of using high-cost systems relying on specialized tooling such as large presses, more flexible and interchangeable tools were prioritized. Additionally, factory workers were given the responsibility of changing and maintaining their tooling and granted substantially more autonomy in their day-to-day activities. Together, these steps allowed the cost-effective production of smaller batches of products, significantly cutting the need to hold and store large volumes of intermediate items, such as materials and components, known as *work in progress*. Chapter 9 introduces Lean as a broad approach to operations and summarizes a range of important Lean ideas and methods.

As realized in the West in the 1980s when Lean successfully challenged mass production in several industries, the impact of Lean on the field of management would be substantial. Rather than promoting work in rigid hierarchies with fixed responsibilities, Lean is based on a culture of working in small teams. In factories that follow Lean, teams of factory workers share the responsibility not only for production but also for maintaining quality, housekeeping, and minor tool repair. Ohno emphasized that substantial gradual improvements can come from small changes, many of which would be identified by the factory workers rather than managers. The method of systematically pursuing continuous incremental improvement, which was stressed by Taylor also, became known under the Japanese expression *Kaizen*.

As a general approach to management, thereby going beyond a set of techniques for the optimization of manufacturing operations, Lean can be characterized in different ways. A useful way of thinking about Lean is as a *philosophy* based on a group of principles directed toward an ongoing transformation. The following offers an accepted current definition:*

IMPORTANT DEFINITION

Lean is an on-the-job learning method based on two pillars: continuous improvement, which means continuously challenging oneself and learning by taking small steps, and respect, which means making the best effort to understand the obstacles each person encounters, supporting their development, and making the best possible use of their abilities.

The role of a worker in a Lean factory differs strongly from one in a factory operating in the style of traditional mass-production. An important difference is that time is allocated to allow team members to discuss their responsibilities and work on ways of improving their activities. Realizing that everyone in an organization carries a share of the responsibility for its success, even junior workers are given the right to stop production if necessary. This is done so teams are in the position to resolve problems as they arise, rather than correcting problems after products have been completed. This approach helps avoid substantial additional costs resulting from the storage and rework of faulty products, which usually plague large-scale traditional mass-production operations. By implementing a rigorous approach to identify the root causes of failure, the adoption of Lean and related methods, such as Six Sigma (introduced in Chapter 12), most quality problems were eventually eliminated in many industries. Ultimately, this resulted in higher profitability, better products, and greater success in the global marketplace.

Gradually, Lean principles and ideas have spread throughout most functions in businesses, including office work. As a testament to the power of these ideas, Lean has made the traditional mass-production approach obsolete in many industries. By the end of the 20th century, the Japanese motor industry had come to dominate the global market in terms of quality and customer acceptance.

CURRENT TRENDS IN MANAGEMENT

At present, many organizations are facing a complex and competitive business environment. Whether or not this environment has become more challenging than in the past is unclear. What is certain, however, is that the management competencies needed to thrive in the business world of today are changing. This is, as outlined in this chapter, at least in part due to ongoing and rapid technological changes requiring new management methods.

---

* Adapted from Ballé and Jones (2015).

TABLE 1.1
Megatrends Changing the Way Business Is Done

Technological change	New technologies, in particular new information technologies, are rapidly changing the behavior and expectations of customers as well as the tools and methods available to organizations. Chapter 4 is devoted to the unfolding impact of new information technologies on management. Chapter 5 more generally discusses innovation processes.
Progressing urbanization	An increasing share of the world's population is living in cities. Significant challenges are posed in particular by the rapid growth of mid-sized cities in developing countries.
Global warming and scarcity of resources	Growth in the demand for energy, food, and water is currently at an unsustainable level. The same applies to carbon emissions. The resulting increase in average temperatures across the globe may lead to catastrophic consequences if left unaddressed. Chapter 6 presents an integrated approach to assessing environmental and economic impacts.
Rebalancing of the economic performance of countries and regions	Some emerging economies that were growing strongly in the recent past are now stagnating. Organizations investing in these emerging economies will need new methods to manage this uncertainty.
Demographic and social changes	It is expected that unsustainable population growth will occur in some developing countries in the foreseeable future. In developed countries, an aging population will pose different challenges.

Changes of a global magnitude are viewed by managers as part of a range of major trends, often labeled *megatrends* due to their worldwide significance. Apart from ongoing technological change, megatrends include rapid urbanization, global warming and the increasing scarcity of some natural resources, changes in the economic performance of different countries and regions, and ongoing demographic and social changes. All megatrends require managerial responses and create the opportunity for commercial innovation. Table 1.1 briefly summarizes each of these megatrends.

As described in this chapter, management ideas have evolved in an ongoing way in the past, often doing so following technological breakthroughs. This process will undoubtedly continue in the future. Many new management techniques will be developed over the coming decades to address the outlined megatrends and exploit new commercial opportunities.

Current developments in leadership and management that aim to increase the success of organizations focus on a number of topics. These include:

1. *Adopting new organizational structures*

   As organizations adopt flatter structures, meaning that the number of hierarchical levels is reduced (see Chapter 2), new management styles will be needed to facilitate collaboration and stimulate the interaction between different functions and roles. In particular, there is an emphasis on involving workers in management decisions, known as *workplace democracy*, with the aim of distributing some management functions among workers. At the same time, new forms of organization are emerging that increasingly involve external partners, such as contractors, for many different roles. This phenomenon is known as the *gig economy*. Such organizational configurations are introduced in detail in Chapter 7.

2. *Promoting the development of workers as individuals*

   Since technological progress generates an ongoing stream of changes, workers and managers can no longer assume that the skills they have acquired during their years in full time education will remain relevant for the entirety of their career. It is now accepted that lifelong learning will be required to cope with ongoing change. This is particularly urgent in developed countries where a significant share of the workforce is currently moving into retirement. A concerted management effort is needed to ensure that the transition of management roles to following generations occurs smoothly and without the loss of expertise.

3. *Striving for gender balance and inclusivity*

   The increasing representation of women and minorities in senior management roles has frequently been shown to lead to higher organizational performance.* Successful management practice will increasingly capitalize on this insight by addressing, among other things, the lack of role models, dealing with the unavoidable presence of unconscious bias, and unequal salary levels. Both an issue of equity and organizational performance, the topics of gender balance and inclusivity are receiving a high degree of managerial attention in the early 21st century.

CONCLUSION

This chapter has introduced management as directing an organization's activities to attain goals. In particular, this involves leading and mobilizing others. This chapter has discussed how management techniques and methods change and evolve following technological changes. This core idea is introduced as the organizational lag model, which has been particularly visible in the evolution of the automotive industry in the 20th century. The chapter has further described the emergence of mass-production as an extremely potent combination of technological, managerial, and social innovations as a core idea. It has showed how the shortcomings of mass-production ultimately led to the emergence of Lean as a broadly applicable management philosophy. The final section of this chapter has discussed dominant global issues to which management is responding at the present and outlined directions in which current management practice is evolving.

Most management textbooks provide useful definitions of management and introduce the reader to the history of management. A particularly thorough and detailed textbook is *Management – An Introduction* by David Boddy (2017). For a first-hand account of the evolution of the automotive industry, Alfred Sloan's *My Years with General Motors* (1964) is insightful. An interesting book that popularized Lean for a

* Gaudiano, 2020.

professional audience in the West is *The Machine That Changed the World* by James Womack and colleagues (1990).

REVIEW QUESTIONS

1. Complete the following definition of management.
   (Question type: Fill in the blanks)
   "Management is that ____________ in an organization which concerns itself with the direction of ____________ to attain the organization's ____________. Management deals with the active direction of ____________".

2. Complete the following figure of the general functions of management by inserting the correct labels.
   (Question type: Labeling)

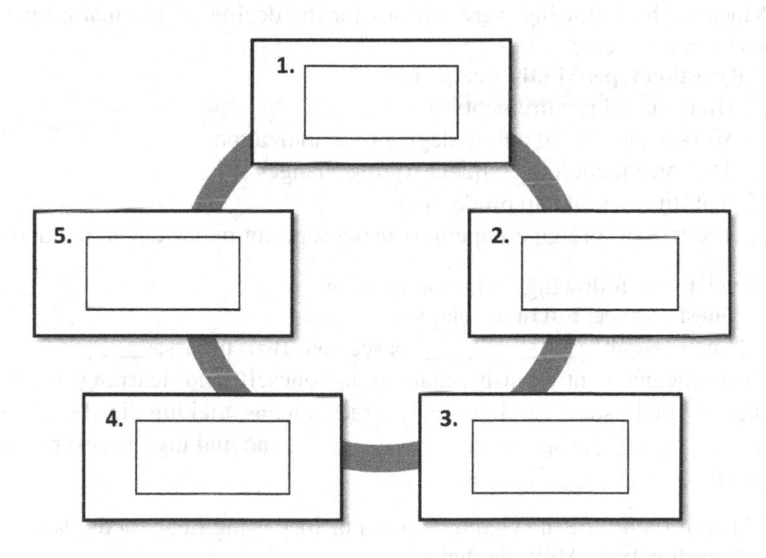

FIGURE 1.2 Insert the appropriate labels

3. Which of the following is a characteristic of craft manufacture?
   (Question type: Multiple choice)
   ○ Workers gathered in large groups working for a central employer
   ○ Workers, each with one simple task, working locally
   ○ General agricultural workers making craft products
   ○ Specialist craftspeople generally working alone or in small, local groups

4. Which of the following is not an innovation associated with Henry Ford?
   (Question type: Multiple choice)
   ○ Petrol engines
   ○ Emphasis on interchangeable components

○ The moving manufacturing line
○ The "$5 a day" pay structure

5. Managerial innovation usually occurs simultaneously with technological innovation.
   (Question type: True/false)
   Is this statement true or false?
   ○ True ○ False

6. Which of the following was an innovation of Alfred Sloan?
   (Question type: Multiple choice)
   ○ Empowerment of manufacturing line workers
   ○ Introduction of production engineers
   ○ Profit centers
   ○ Tool and die changes performed in a single minute

7. Which of the following were reasons for the decline of the mature mass-production system?
   (Question type: Multiple response)
   ☐ High capital requirements
   ☐ Worker unrest and a high degree of unionization
   ☐ The introduction of frequent styling changes
   ☐ Inability to scale up production
   ☐ The rise of foreign competitors using superior management methods

8. Complete the following definition of Lean.
   (Question type: Fill in the blanks)
   "Lean is an ____________ based on two pillars: ____________, which means continuously challenging oneself and learning by continuous small steps, and respect, which means making the best effort to ____________, supporting ____________, and making the best possible use of ____________".

9. Which of the following characterizes a factory using Lean methods?
   (Question type: Multiple choice)
   ○ Large groups controlled directly by an operations director
   ○ A high degree of unionization
   ○ Groups changing their tasks weekly
   ○ Staff working in small groups under a team leader

10. Which of the following are seen as megatrends changing the way business is done?
   (Question type: Multiple response)
   ☐ Increasing globalization
   ☐ Innovation in financial products
   ☐ Rapid urbanization
   ☐ Consumerism
   ☐ Demographic changes

REFERENCES AND FURTHER READING

Ballé, M. and Jones, D., 2015. The future of work: 10 signs you respect me as an employee. *Fast Company.* Available at: https://www.fastcompany.com/3036623/10-signs-you-respect-me-as-an-employee.

Boddy, D., 2017. *Management: An introduction.* 7th ed. New York: Pearson Education.

Covey, S.R., 1989. *The 7 habits of highly effective people.* New York: Simon and Schuster.

Damanpour, F. and Evan, W.M., 1984. Organizational innovation and performance: The problem of "organizational lag". *Administrative Science Quarterly*, 29(3), pp.392–409.

Fayol, H., 1916. *Administration Industrielle et Générale*, Dunod (Trans.). London: Sir Isaac Pitman and Sons.

Gaudiano, P., "How Inclusion Improves Diversity And Company Performance, Forbes, www.forbes.com/sites/paologaudiano/2020/07/13/how-inclusion-improves-diversity-and-company-performance. Accessed 14 September 2021.

Gross, D., 1997. *Forbes: Greatest business stories of all time.* Hoboken: John Wiley & Sons.

Sloan, A.P., 1964. McDonald, J. (ed.), *My years with general motors.* Garden City, NY, US: Doubleday.

Spriegel, W.R. and Lansburgh, R.H., 1947. *Industrial management.* Hoboken: John Wiley.

Womack, J.P., Jones, D.T. and Roos, D., 1990. *The machine that changed the world: The story of lean production.* New York: Rawson Associates.

2 The Private Enterprise

OBJECTIVES AND LEARNING OUTCOMES

The objective of this chapter is to introduce the idea of the private enterprise, commonly referred to as a "business" or "company", as the basic unit of organization in the field of management. After providing a general definition, this chapter presents a range of co-existing perspectives on businesses that managers frequently take. These vantage points include viewing businesses as structures for flows of money, as sets of interrelated activities, as legal forms, and as organizations of people. The chapter further introduces the objectives commonly associated with managing and owning businesses. Specific learning outcomes include:

- understanding what a business is and how it can be defined, and that there are several synonyms in general usage;
- understanding the business as a framework for different regular flows of money;
- understanding how the business forms a set of activities that make up a value chain;
- an appreciation of the business as a legal structure, including a definition of the share as an important unit of ownership;
- understanding the business as an organizational structure containing a hierarchy;
- understanding the objectives of a business, including a definition of the term "capital";
- an appreciation of the owners' objectives and understanding the difference between investors and speculators;
- understanding important terms, synonyms, and accepted acronyms; and
- appreciation of key management thinkers.

THE PRIVATE ENTERPRISE AS THE FUNCTIONAL UNIT OF MANAGEMENT

Chapter 1 introduced the general field of management by simply referring to the running of organizations. If an organization is owned and managed by individuals that freely decide to do so and these individuals use their ideas and the things they own to produce a financial gain for themselves, it constitutes a *private enterprise*. Often used interchangeably* with the terms *business, company*, or *firm*, the private enterprise is the basic functional unit and building block of a free economy. It can be defined as follows:

* The usage of the terms "private enterprise", "company", "business", and "firm" in this book reflects the general usage of these terms in management practice. Legal professionals, accountants, and public officials may use these terms in more specific ways.

DOI: 10.1201/9781003222903-3

A private enterprise is an organization owned and managed by individuals that freely decide to do so and for their own financial benefit. In many contexts, "business", "company", and "firm" are used as synonyms.

Following this definition, the individuals founding a private enterprise are empowered to act in their own self-interest and are motivated by the prospect of creating wealth in the form of a financial gain. This financial gain is known as *profit*. The freedom from interference and outside control, at least in principle, sets private enterprises apart from government organizations or public institutions. Known collectively as the *private sector*, private enterprises employ the majority of the workforce in most countries.

Many private enterprises – referred to mainly as "businesses" in this book – are managed by their owners. Other, especially larger, businesses employ professional managers. Generally, businesses can be owned by one or more individuals through what is known as *shared ownership*. In most countries, businesses are obliged to pay taxes to fund public expenditure. Despite operating without state interference in principle, businesses are subject to regulation that enforces certain standards and conduct. This can include, for example, adherence to labor laws, ethical business practices, or certain standards of reporting financial results. However, within these limitations, businesses are free to engage in any activity and to pursue any project considered beneficial by the managers and owners.

As evident from this book, management practitioners and businesspeople frequently talk about *value* as something to be pursued by businesses. This suggests that value, normally defined as either "the amount of money that can be received for something" or "the importance or worth of something for someone"* plays a central role in management. However, the idea of value can be invoked in many ways and contexts – in fact, discussions of the different meanings of value go back to the teachings of Aristotle in classical antiquity. This underscores that value itself is a complex and multi-layered concept that is surprisingly hard to define. The question of what kind of value is meant in a given management situation and to whom it arises is a common source of confusion and conflict. To address this issue, this book highlights to the reader where a specific understanding of value is required to grasp the meaning of the presented material.

Businesses are generally seen to be valuable in the sense that they are of worth to their owners and could be sold to alternative owners in exchange for a payment. This implies that the running of a business involves the management of valuable items used or owned by the business or its owners. Such items are frequently referred to as the *resources* of the business. There are three basic resources in every business: *financial resources* (such as money, often referred to by managers as *cash*), *capital resources* (the objects and equipment used, the term "capital" is defined later in this chapter), and *human resources* (the people employed, also known as *employees* or *staff*).

A fundamental distinction is often made between manufacturing and service businesses. Manufacturing businesses bring in materials, add value through a process controlled by the business, and sell their products at a profit. Service businesses provide services for which they are paid. This book takes the position that it is not always possible to distinguish clearly between manufacturing and service businesses

* Both definitions from Cambridge Dictionary (2020).

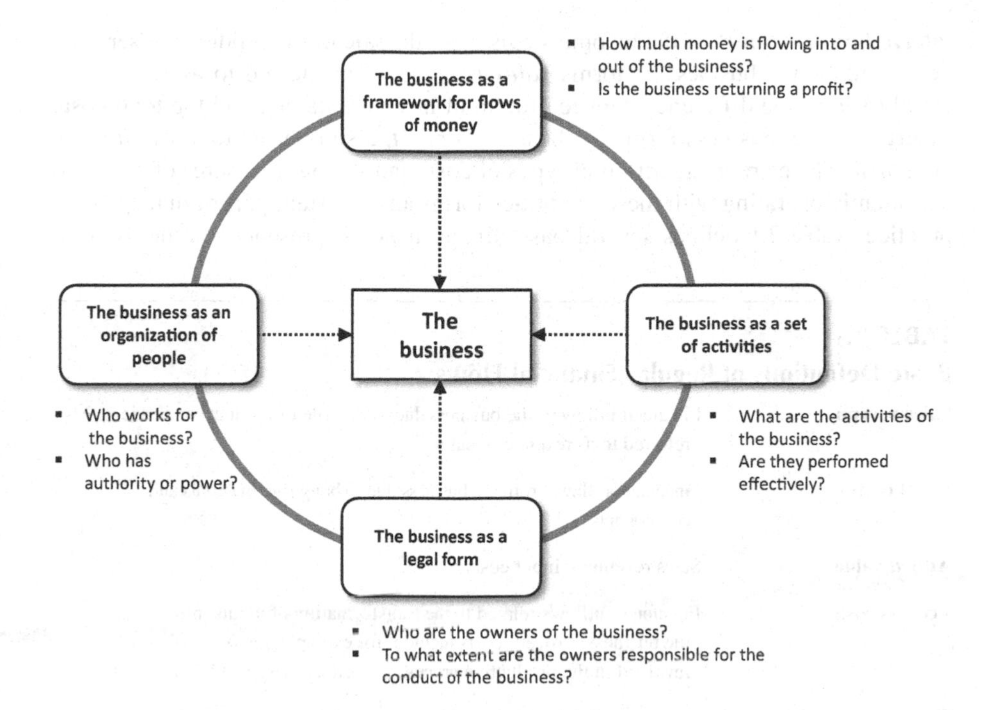

FIGURE 2.1 Different perspectives on a business

since businesses often engage in both sides, or manufacture products that cannot be clearly distinguished from services, or vice versa.

As will be shown in the following sections of this chapter, businesses are complex in that they have multiple and quite diverse characteristics. To manage a business effectively, managers must understand these different aspects and know when to view the business from a suitable perspective. Figure 2.1 summarizes important perspectives, or vantage points, that managers frequently adopt. The figure also indicates some major questions associated with each. The following four sections provide introductions to each perspective. The remainder of the chapter is concerned with the objectives of managing or owning a business.

THE BUSINESS AS A FRAMEWORK FOR DIFFERENT FLOWS OF MONEY

Perhaps the most important way to capture the general logic of a business is by viewing it as a framework for different flows of money, or cash, occurring on a regular basis. Adopting this view, an outflow of cash, from the perspective of the business, is mostly referred to as a *cost*[*] or an *expense*, whereas an inflow of cash is mostly

[*] A useful and very general way to think about costs is that they represent the resources that are sacrificed to achieve benefits (see, for example, Atrill and McLaney, 2018).

referred to as *income*[*]. If the income stems from the sale of the products or services generated by the business, it forms *sales revenue*, also referred to as *revenue* or simply *sales*. The difference between the total income obtained and the total costs incurred by the business forms the business's *profit*, also referred to as *net income* or *return*. There are many different types of costs and various measures of profit, so confidently operating with these quantities forms an important part of management practice. Table 2.1 defines several basic financial flows, presents commonly used

TABLE 2.1
Basic Definitions of Regular Financial Flows

Sales revenue	Financial inflow to the business due to the sale of products or services. Also referred to as revenue or sales
Input costs	Financial outflows from the business due to bought-in materials and components
Added value	Sales revenue – input costs
Process costs	Financial outflows related to the transformation of inputs into outputs by the business. Process costs include, for example, the costs of labor involved in the productive process as well as energy and consumables
Cost of goods sold (COGS)	Input costs + process costs. Also known as *cost of sales*
Operating costs	Financial outflows related to the operation of the business, reflecting the costs of resources employed to maintain its existence over time, such as labor costs, marketing, human resources management, and cleaners. Especially if process costs are excluded, operating costs are referred to as *overheads*
Total costs	Input costs + process costs + operating costs + other costs
Gross profit	Sales revenue – COGS. Also known as *gross income*
Operating profit	Sales revenue – COGS – operating costs. Also known as *operating income*
Net profit	Sales revenue – COGS – operating costs – other costs. Also known as *net income, net earnings*, or *bottom line*
Earnings before interest, taxes, depreciation, and amortization (EBITDA)	Net profit + interest payments + tax payments + depreciation + amortization. The concept of interest is presented in detail in Chapter 15. The concepts of depreciation and amortization are introduced in Chapter 13
Earnings before interest and taxes (EBIT)	Net profit + interest payments + tax payments

[*] It is important to note that those concerned with taxation, such as tax advisers and tax administrators, often operate with a more narrow definition of "income", generally reserving the term to payments to actual people in the form of salaries or pensions. This alternative usage frequently causes confusion between managers and non-managers.

synonyms, and shows how additional simple financial terms can be constructed from these. Note that it is assumed that sales revenue forms the only source of income to the business.

The presented definitions identify *input costs* and *process costs* as distinct categories. This distinction is made to highlight that different types of costs arise in the productive process of the business, especially in manufacturing businesses. The costs incurred for purchases of the materials processed in the business's activities are also known as *direct material costs*. It should be noted, however, that there are competing, alternative definitions for such costs. Perhaps more reflective of manufacturing activities rather than service provision, the presented framework suggests that purchasing the required material forms an important part of the business's activity. Moreover, the framework defines *added value* as the difference between total sales revenue and input costs.

The framework of financial flows can also be expressed in graphical form, as shown in Figure 2.2. Alternatively, this structure can be depicted as a sequence of cumulative cash positions, as done in Figure 2.3. Again, it should be noted that sales revenue forms the only financial inflow to the business in this simple model.

THE BUSINESS AS A SET OF ACTIVITIES

It is common for managers to view a business as a collection of processes that can be identified, measured, and controlled. As suggested in the previous section, such

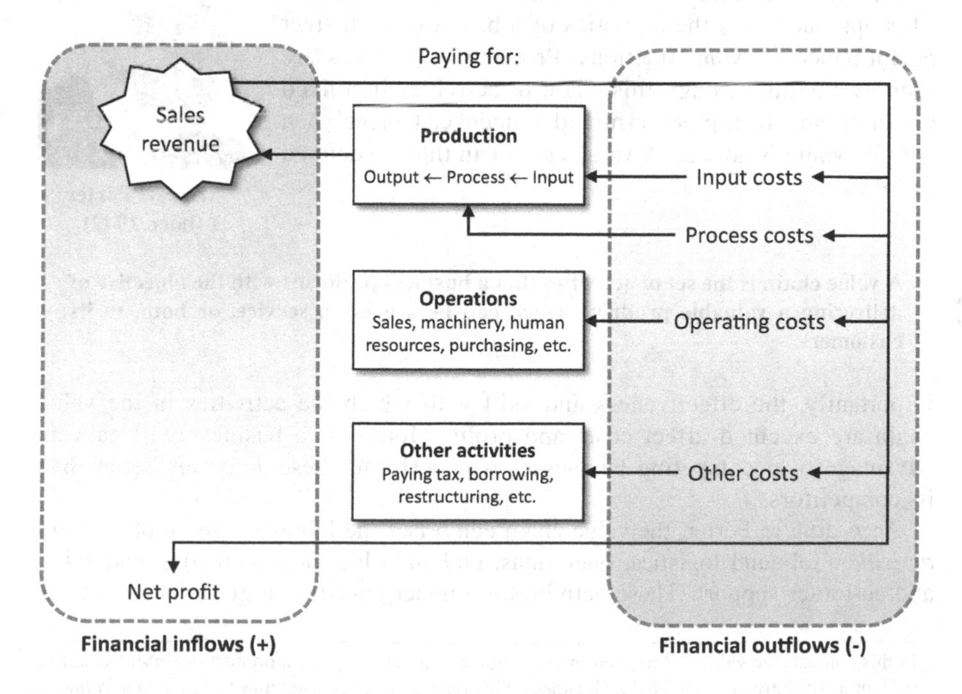

FIGURE 2.2 The business as a framework for financial flows

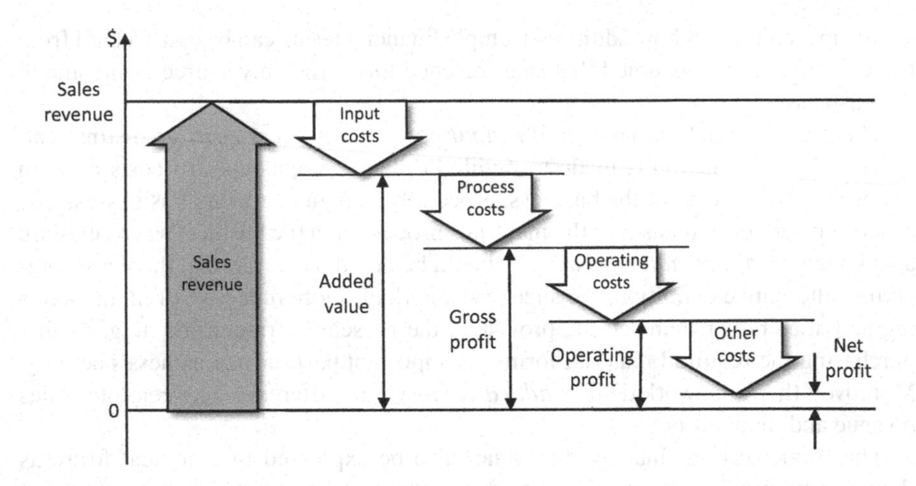

FIGURE 2.3 The business as a sequence of cash positions

processes can be seen as ongoing transformations in which inputs are converted into outputs. This idea forms the basis of the topic of operations management, which is introduced in Chapter 9.

To increase the acceptance of this idea, management scholar Michael Porter developed the concept of the *value chain* in the 1980s, which has since then become an extremely influential way of thinking about businesses. This approach sees the activities of a business as distinct performance-relevant functions. Products or services are seen as moving through this chain of activities in a fixed direction and in a predetermined sequence. During each activity value is added.* A value chain can thus be defined as follows:

Michael Porter (born 1947)

A value chain is the set of activities that a business performs with the objective of delivering a valuable product, which can be a good, a service, or both, to its customers.

Importantly, the effectiveness and skill with which the activities in the value chain are executed affect costs and profit. Moreover, a business will gain an advantage over competing businesses if it performs these activities better than its competitors.

According to Porter, the value chain comprises the following group of *primary activities*: inbound logistics, operations, outbound logistics, marketing and sales, and customer support. These activities are underpinned by a group of *secondary*

* In this context, the value added is an increase in the amount of money a product or service could be sold for as they are generated by the business. This form of value is also known as *exchange value*.

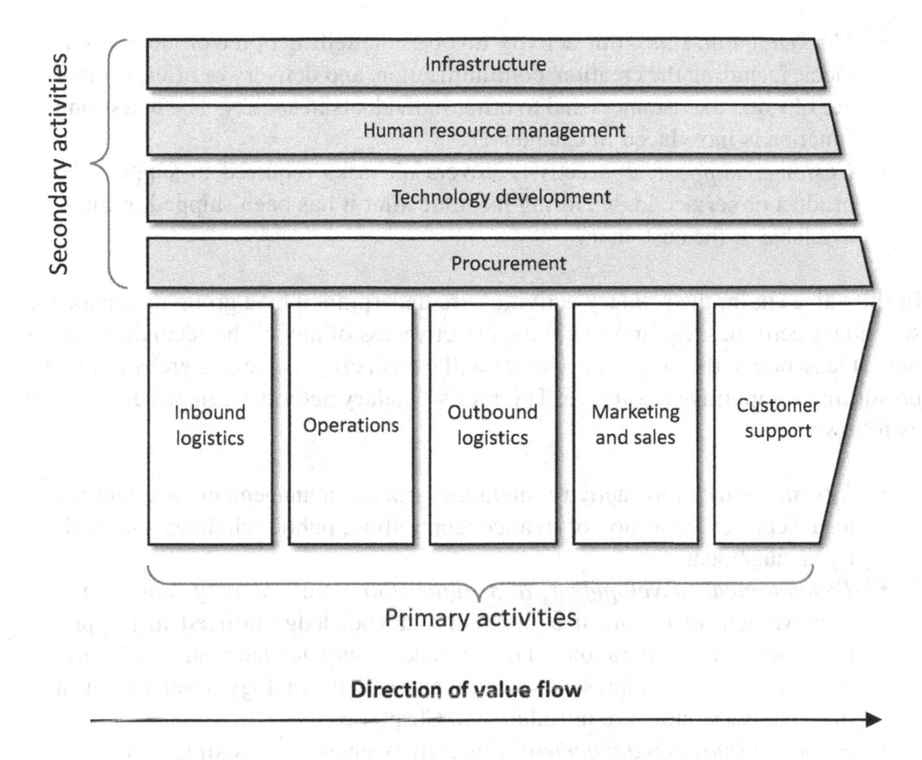

FIGURE 2.4 The value chain, adapted from Porter (1985)

activities comprised of firm infrastructure, human resource management, technology development, and procurement. The standard model of the value chain, slightly adapted, is shown in Figure 2.4.

In this model, all five primary activities are required for the business to operate as a whole. If the business is able to perform any one of the primary activities better than competing businesses, it can build what is known as a *competitive advantage*, which will secure its profitability. The five primary activities can be briefly outlined as follows:

- *Inbound logistics*: this activity involves the transfer of raw materials, components, or other inputs into the facilities of the business, such as a warehouse or a factory.
- *Operations*: in the value chain, the term "operations" denotes the activity that transforms the inputs, in forms such as raw materials, labor, and energy, into the outputs of the business, in the form of goods, services, or both. Note the differing usage of the term "operations" in the model of a business as a framework for flows of money, where it denotes activities required to maintain the business over time (see, for example, Figure 2.2). The operations function is introduced in Chapter 9.
- *Outbound logistics*: this activity encompasses the storage, movement, and transfer of finished products to the customers. For service provision, this activity refers to making the services available to customers.

- *Marketing and sales*: this activity involves the selling of the product or services, including the creation, communication, and delivery of offerings that are of value to customers and to other individuals in society. The marketing function is introduced in Chapter 11.
- *Customer support*: this activity covers all tasks required to support the product or service in delivering its value after it has been shipped or made available to the customer.

In the value chain, the primary activities are underpinned by a group of supporting secondary activities. By improving the effectiveness of any of the secondary activities, at least one of the primary activities will gain in effectiveness, thereby indirectly promoting competitive advantage. The four secondary activities can be summarized as follows:

- *Infrastructure*: this activity includes general management, accounting, legal services, provision of finance, controlling, public relations, and quality management.
- *Technological development and innovation*: this activity covers the improvement of technical equipment and knowledge utilized in the primary activity of operations. This includes computer hardware, software, and operating procedures. Innovation and new technology development in the business context are introduced in Chapter 5.
- *Human resources management*: this activity encompasses all tasks carried out to recruit, hire, train, develop, pay, and, where necessary, dismiss or lay off employees. The human resources management function is introduced in Chapter 10.
- *Procurement*: this activity pertains to the purchase of goods and services from outside providers.

THE BUSINESS AS A LEGAL FORM

Every business must select a legal structure to follow, referred to as the *legal form* of ownership. Usually, this decision is made by the founders before the business begins to operate. Occasionally, however, businesses change their legal form later on. The choice of a legal form to operate under is a very important decision with many consequences. It determines how resources are structured, how management roles are administered, how taxes are paid, and how financial information is reported. It is important to note that legal forms vary from country to country. Nevertheless, there are two broad classes of legal forms for businesses that can be found in most jurisdictions around the globe, albeit under different names: *incorporated* and *unincorporated* legal forms.

Incorporated legal forms are based on the concept of *legal personhood*. This means that, in the eyes of the law and under certain circumstances, the business can do things normally reserved for human beings. This includes the ability to enter into contracts, incur debt, own property, sue and be sued, and so on. The choice of an incorporated legal form, thus, confers the status of being a legal person on a business. This process is referred to as *incorporation*.

The key advantage of having legal personhood is that owners of such a structure cannot be held to account fully for what the company does – this special protection is referred to as *limited liability*. At the same time, and somewhat oddly, the owners are allowed to privately reap the benefits of the business in the form of profit. Despite not being a common topic in everyday life, the importance of legal personhood is profound. Without legal personhood, and the separation of the risk of owning a business from its rewards, market-based economies could not exist in their current form, especially with respect to the trading of shares in the financial markets. For this reason, legal personhood is occasionally described as capitalism's most important institution.

The idea of legal personhood is absent from a business adopting an unincorporated legal form. In unincorporated businesses, the owners are not separated legally and financially from what is going on in the business, and therefore do not enjoy limited liability. In other words, they are fully responsible for any money owed or damages created by the business and any income generated by the business arises as their own personal income. As a topic touched upon repeatedly in this book, such risks and their avoidance are of deep practical importance to managers.

Across the globe, there are many different incorporated and unincorporated legal forms, each with distinct characteristics, rights, and obligations. The remainder of this section briefly presents the four most important legal forms. It should be noted that this introduction adopts terminologies based on English Law.

Sole Trader

The most straightforward way for an individual to set up and run a business is assuming the legal form of a *sole trader*, also known as a *proprietorship*. As a sole trader, ownership and control of the business rest with a single individual. The business can have employees, however, and operate under a business name. As an unincorporated legal form, operating as a sole trader is inherently risky since the owner is not separated legally from the business and is, therefore, fully liable for any debts, damages, or contractual obligations the business may incur. In practice, this will mean that if the business produces significant losses, perhaps due to unfortunate circumstances, the owner will be responsible for these losses with their personal wealth. The main advantage of this legal form is that the requirement for paperwork and administrative duties is minimal. Apart from unlimited liability, an important additional disadvantage is that it is considered difficult to raise funds from third parties as a sole trader, for example, by borrowing money from a bank.

Partnership

A *partnership* is formed when two or more people join forces to form a business as a group. Normally, this involves drawing up a legally binding contract referred to as a *partnership agreement*, determining various aspects such as the amount of money or the equipment made available to the business by each partner. Importantly, this agreement will also specify how the profits or losses will be shared among the partners. Because the partnership is unincorporated, it is not a separate legal person. This means the partners will bear all consequences of their decisions and may lose their personal wealth. Due to this risk, the day-to-day activities of partnerships are normally managed by the partners. As with a sole trader, operating as a partnership will limit the ability to borrow money or obtain funding from third parties to grow the business.

Private Limited Company

The *private limited company* is the most common legal form used for running a business in many countries. As an incorporated form, private limited companies are entities with separate legal personhood. This means that such businesses can do things such as enter into contracts in their own name or take on debt. Unlike a real person, however, at least in the modern age, a private limited company is owned by people. This entails that the private limited company's finances are separate from the owners' finances – which means that *creditors*, individuals to whom the private limited company owes a financial debt, may not pursue the owners' personal wealth to settle the debt.*

Private limited companies must have at least one owner who owns a fraction of the company called a *share*. Collectively, all shares are known as the *equity* of a business. Apart from conferring ownership and the associated responsibilities, owning shares normally gives shareholders the ability to vote on important decisions affecting the business.† Shares can, thus, be defined as follows:

IMPORTANT DEFINITION

A share is a unit of ownership in an incorporated business. Owning a share entitles its holder to a portion of the profits of the business and confers a range of responsibilities. A shareholder normally has the right to vote on certain issues pertaining to the business.

The management of a private limited company is normally considered separate from its ownership and is the responsibility of a director or board of directors who are obliged to act in the interest of the business and its owners. Net profits generated by a private limited company can be given to the shareholders as payments called *dividends* or retained by the business. A private limited company can collect additional

* Despite the principle of limited liability, owners may contractually agree to be held liable in some form for the debt their business incurs.

† Whether or not ownership of shares confers voting rights depends on the type of shares held. As a general rule, ordinary shares confer voting rights whereas preference shares do not.

funds from new or existing shareholders by selling shares in the form of *ownership stakes* in the business. However, directors may also own shares and be partly paid in shares.

The structure of private limited companies gives businesses a far greater ability to borrow money than unincorporated legal forms. This is because incorporated businesses can own valuable items that act as security against which the business can borrow. However, incorporated businesses are generally subject to stricter rules and legal requirements than sole traders or partnerships.

Public Limited Company

Companies with limited liability that are permitted to offer shares for sale to the wider public are referred to in some jurisdictions as *public limited companies*. This naming convention is somewhat confusing since such businesses are private businesses. Due to the greater degree of separation between the managers and shareholders in public limited companies, this legal form attracts strict regulation and reporting requirements, especially if the shares of such a business are traded on a stock exchange. Most large and international businesses are public limited companies.

THE BUSINESS AS AN ORGANIZATION OF PEOPLE

Apart from being a framework for flows of cash, a set of activities, and a legal form, a business is also an organization comprising people to achieve a collective goal. From this perspective, an initial point of interest is the general relationship between the age and the size of a business in terms of the number of people it employs. Unsurprisingly, newer businesses tend to be smaller while more mature businesses with a history of successful trading tend to be larger. Figure 2.5 summarizes this relationship and identifies a range of categories of business size.*

Most businesses exhibit some form of *hierarchical organizational structure*, also known as a *vertical structure of management*. The underlying idea of an organizational hierarchy is that every person or group, except a single individual at the top, is subordinate to the authority of another person or group. Normally, the hierarchy in a business has an individual person or board of directors at its top with additional levels of power or authority below. Individuals within hierarchical organizations normally interact mainly with the tiers immediately above and immediately below. The purpose of this structure is to enable effective communication because it limits unnecessary interactions between members of the organization. At the same time, limited information flow also forms the main disadvantage of hierarchical organizations, slowing down change processes and adaptation to changing circumstances.

As indicated in Chapter 1, there is a long-standing trend towards less hierarchical organizations with *flatter* organizational forms, known as *horizontal corporations*.† Businesses adopting a flat organizational form typically exhibit a range

* Organisation for Economic Co-operation and Development (2017).
† Castells (1996).

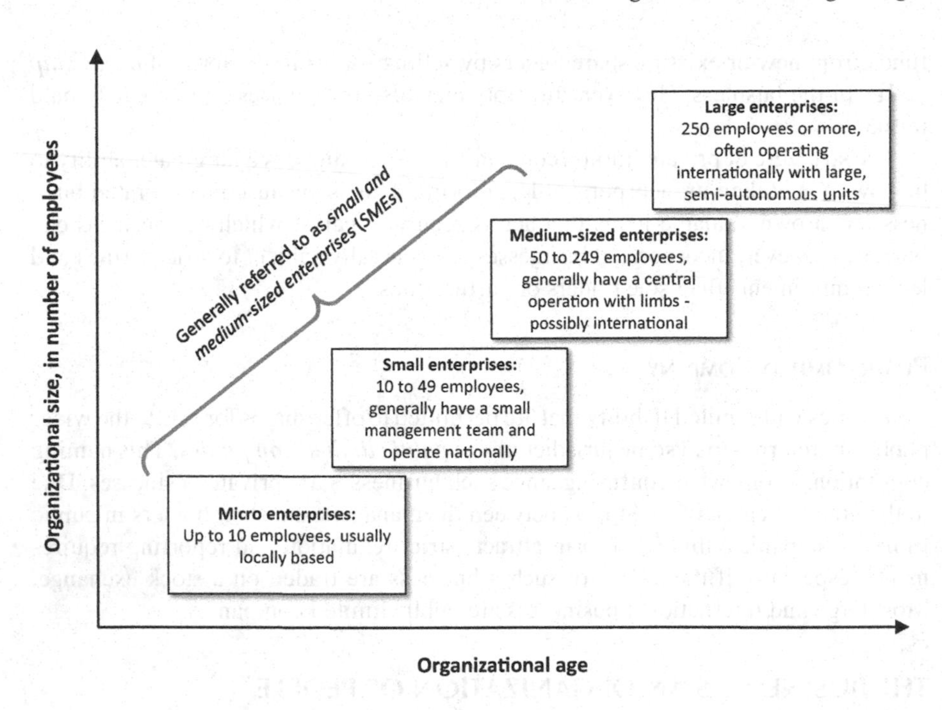

FIGURE 2.5 Categories of business size and relationship with age

of characteristics that set them apart from more vertical organizations. These characteristics include being organized around processes instead of tasks, having fewer hierarchical levels, managing teams of employees, measuring performance directly in terms of customer satisfaction, allocating rewards on the basis of team performance, having as many contacts with customers and suppliers as possible, and showing a strong focus on ongoing training and retraining at all levels throughout the business. In businesses with very pronounced hierarchies with many tiers, known as *tall organizational structures*, these characteristics can be entirely absent for some roles.

There are many ways of representing the hierarchy within a business; the most common way is to draw it as an *organizational chart*. Figure 2.6 presents a stylized example of an organizational chart showing a typical engineering business. Some additional explanations have been added to the figure.

The main purpose of organizational charts is to identify lines of reporting. As shown in Figure 2.6, more senior staff are located higher in the diagram and will generally hold greater responsibility and authority. Sometimes a member of staff may have a secondary reporting line. In the example provided, the research and development manager additionally reports to the chief marketing officer to ensure that new developments are consistent with market requirements. This will usually be indicated by a dotted line in organizational charts. Moreover, some larger businesses with very distinct products may choose a structure that subdivides the organization along the different product lines or service types offered.

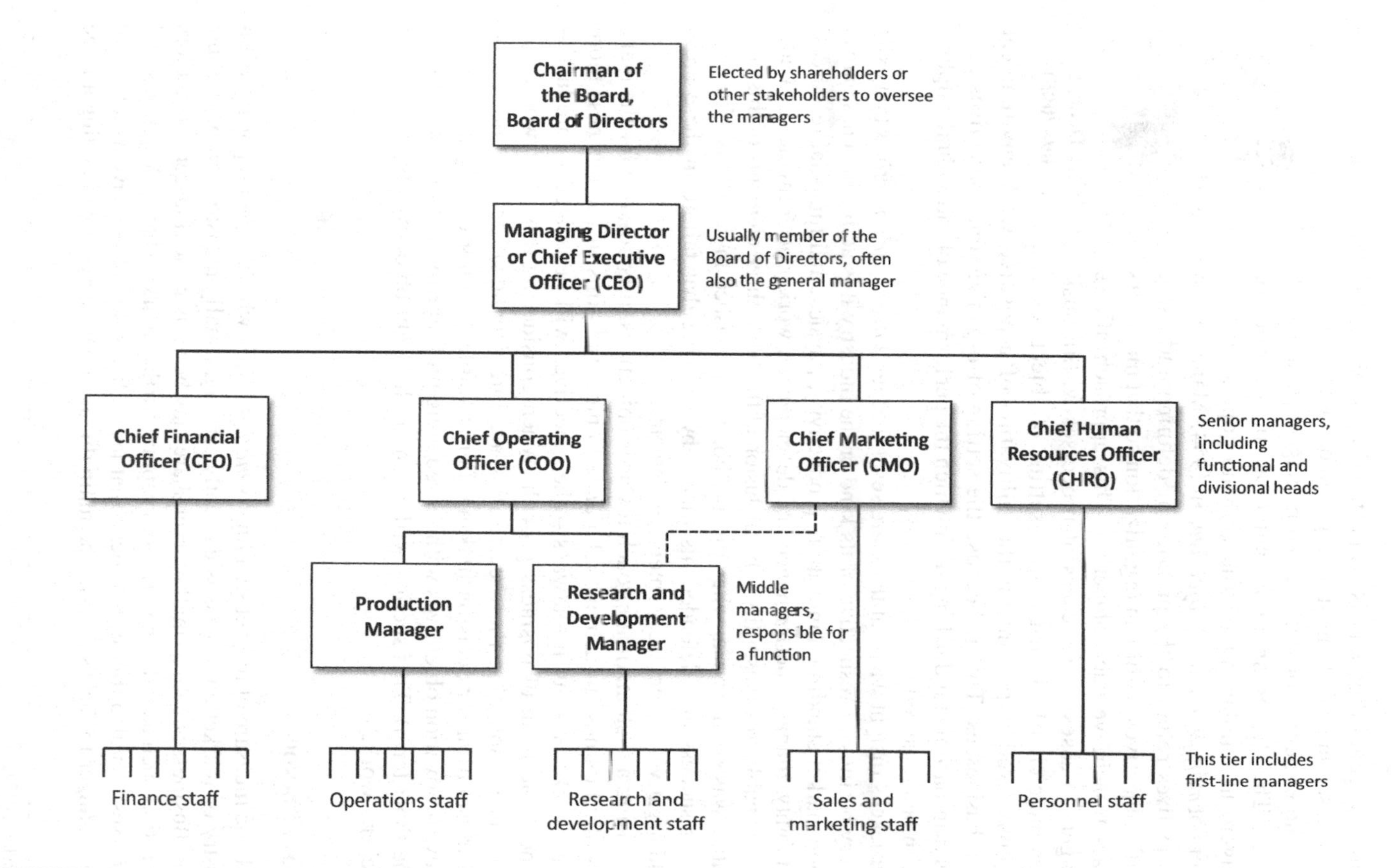

FIGURE 2.6 A typical organizational structure of a company

THE OBJECTIVES OF THE BUSINESS

FAMOUS THINKER

Peter Drucker (1909–2005)

Despite the general emphasis on profitability in the world of business, there is a long-standing line of thought focusing on more specific objectives that managers should work towards. This forms an important topic in the work of management scholar Peter Drucker, who popularized the concept of *management by objectives* in 1954.* Management by objectives refers to the practice of defining and using specific objectives within an organization so that managers can act to achieve each objective. This approach allows managers to assess the success of processes within businesses more clearly. Clarity in the setting of objectives is also considered important for the cultivation of a positive work environment within businesses. This is because the setting of organizational objectives provides transparent standards against which the performance of individual employees can be measured.

Drucker's insight was that there are several general *areas for objectives* that exist in every business, irrespective of its kind or the industry it is active in. These areas include market standing, innovation, productivity, physical and financial resources, profitability, manager performance and development, worker performance and attitude, and public responsibility. The inclusion of the latter three elements of this enumeration was controversial when this method was published in the 1950s. However, Drucker successfully defended his theory by showing that failure in these areas would negatively affect the former five elements.

By using a systematic approach towards objectives, it is possible for the managers of a business to achieve a balance to meet the aims of the business more generally. Often – but not always – the objectives will be related to productivity and some form of business growth. When considering this framework, it is important to note that these categories are interdependent and should not be used in isolation for decision-making. For example, it is not usually possible to achieve innovation objectives without reducing profitability in some way (at least in the short term). This section will examine the eight objectives highlighted in Drucker's work.

MARKET STANDING

The basic measure of market standing is *market share*, which is the proportion of an industry or market a business is currently serving, usually measured in percentage terms. Businesses with a small market share may be vulnerable if larger competitors take action such as reducing prices to capture further market share. However, small businesses can also thrive in a niche that larger firms do not perceive as attractive or as a danger to their success. The main threat to businesses commanding a large

* Drucker (1954).

market share is that they typically have a much higher profile and run the risk of attracting unwelcome government scrutiny and intervention, possibly also in foreign markets. Additionally, it is not uncommon for the management of large and successful businesses to lose view of potential dangers to their market standing and to become complacent. Moreover, businesses commanding significant market share may, simply by being very large, lose the ability to respond quickly to rapid changes in technology or market expectations.

Innovation

In every business there are two broad kinds of innovation. The first kind is innovation that leads to new products or services, or new variants thereof, for sale to customers. The second kind is innovation in the processes and activities available to supply to the customers. This distinction is discussed in detail in Chapter 5. There may be different reasons to engage in innovation of either type. Some businesses innovate out of necessity or to safeguard the survival of the business. Other businesses take a more strategic or proactive approach and innovate to exploit specific market opportunities or to address an identified customer need. There is a consensus, however, that in the long run, innovation is necessary to maintain or gain market share.

As is evident in the evolution of the automotive industry discussed in Chapter 1, innovation need not be technical. It can be commercial or organizational in nature, leading to more competitive business practices. In order to be relevant, it is often argued that innovation should be closely focused on the existing activities of the business.

Also referred to as *research and development* (*R&D*) in the private sector, innovation results in a short-term cost against which there must be an expectation of a future financial return. R&D can, therefore, be thought of as an investment into the future subject to the risks of technical failure and rejection by the market. It is also notable that R&D expenses tend to be among the first costs to be cut when a business encounters financial difficulties. The dilemma faced by management in such situations is having to choose between cost savings in the short term and the ability to compete technologically or financially in a recovering market.

Productivity

Productivity is an apparently intuitive concept that can be hard to define and difficult to measure in practice. Metrics of productivity usually express the ratio of an output quantity over an input quantity. It is possible to construct many alternative metrics of productivity within a business, each capturing a specific facet. This means that picking an appropriate measure of productivity forms a crucial part of the assessment of managerial performance and for comparisons between competing businesses.

Frequently used measures of productivity focus on how effectively a business exploits its resources, including people, buildings, equipment, and financial reserves. Table 2.2 summarizes the basic concept of productivity and defines two frequently

TABLE 2.2
Measures of Productivity Used in Management

General measure of productivity	$\frac{\text{Quantity of an output}}{\text{Quantity of an input}}$
Labor sales productivity	$\frac{\text{Sales revenue}}{\text{Number of employees}}$
Labor profit productivity	$\frac{\text{Operating profit}}{\text{Number of employees}}$

used measures, *labor sales productivity* and *labor profit productivity*. Labor sales productivity expresses the relationship between the revenue generated by the business and its overall number of employees, reflecting the general effectiveness of its workforce. Labor profit productivity is a more specific metric based on the relationship between operating profit and the overall number of employees, showing how employees on average contribute to profitability. Naturally, increasing all kinds of productivity helps businesses reduce costs and hence become more profitable and competitive.

Physical and Financial Resources

Whether a business can meet its objectives often hinges on the availability of physical and financial resources. Such resources include valuable objects such as buildings, factories, machinery, and equipment as well as money. Modern versions of Drucker's framework of objectives also include the number of employees, their skills as well as ownership of technology and ideas in this category.

Financial resources are generally viewed as the most important type of resource. However, a lack of other resources, such as technology or employee skills, can be catastrophic. Financial resources can be raised through a number of avenues, including the use of retained profits or other cash reserves, loans from a bank, or funding obtained from the emission of new shares.

At this point, it is useful to introduce the term *capital*. Capital can have different meanings in different contexts of management such as accounting, economics, law, and even in the social sciences. The term is derived from the Latin word *caput*, meaning "head". When used in the general management context, the term "capital" is best understood as describing an important and valuable object. Note that in accounting, capital sometimes denotes the profit retained in a business. For the purposes of this book, capital is defined and used as follows, if not indicated otherwise:

Capital is any kind of financial or valuable object, physical or intangible, that is owned by a business or an individual and is useful in generating profit or progressing development otherwise.

Profitability

As one of the main reasons for the existence of a business, profit is of central importance to managers. In the comparison of different businesses, however, it can be misleading to compare net profit alone. For a more meaningful picture, it is important to understand how net profit relates to the scale of the activities of a business. For example, if one business reports the same profit in absolute terms as another business, yet is twice the size in terms of revenue, the former business would be seen as less profitable than the latter. Different financial ratios can be used to assess different aspects of profitability. The construction of these metrics and their use is discussed in detail in Chapter 14.

Healthy profits allow businesses to maintain and grow their capital, progress research, and reward owners through the payment of dividends. A balance between these aspects is necessary in order to ensure growth while meeting the expectations of the shareholders. A further point of contention in private businesses is whether managers should focus on profitability in the long term or in the short term. There is no right or wrong answer to this question. Rather, the profit objectives will depend on the ownership structure and the interests and preferences of the involved partners or shareholders.

It is possible to generalize that middle-sized businesses are often at the greatest risk of low profitability. This is because they have neither the low overheads of a small business nor access to the returns to scale and benefits of low-cost labor available to large international businesses with locations in multiple countries and access to global networks of suppliers.

Manager Performance and Development

The performance of a manager or management team is highly important for business success and, thus, forms an important area for objectives. Management performance can be assessed relatively easily. Since competing businesses operate within the same market and with access to similar resources, a common measure of management effectiveness is the manager's ability to grow market share. Increasing market share is normally taken to indicate that managers are making the most of the available resources.

For the development of manager performance, it is relevant that traditional management career paths lead up through various levels in the organizational hierarchy over time. In junior management roles, such as first-line managers, effective managers will require specialist skills to supervise the employees doing their work. Additionally, managers at this level will operate according to relatively short time frames. In more senior management roles, the situation is reversed: successful senior managers will usually require an excellent generalist skill set and be able to plan far into the future. The assessment of managers must, hence, take into account their level of seniority.

Worker Performance and Attitudes

Worker attitude is an often-satirized topic; it is often joked that "job satisfaction is the same as stealing from the company". Nevertheless, the importance of creating job satisfaction as a management objective is not always acknowledged. It is generally assumed that employees with a higher level of job satisfaction are more motivated and will contribute more to business performance. As opposed to the performance of managers, the performance of non-management staff and their attitudes are notoriously difficult to quantify. The topic of worker satisfaction forms an important topic in the management of people and is discussed in detail in Chapter 10.

Generally, methods based on motivating staff through reward and punishment, known as *stick and carrot*, are seen as outmoded. Job satisfaction and personal well-being among staff are increasingly perceived to be essential elements for high productivity. Moreover, in a competitive labor market, workers have the option to move to other businesses if they feel their jobs are unsatisfactory. Over recent decades, these insights have contributed to the general recognition among managers that businesses progress better if workers are empowered to influence their job and working conditions. The rise of Lean, introduced in Chapter 1 and further discussed in Chapter 9, is a testament to this development.

A further underlying issue for the measurement of job satisfaction is the categorization of staff into workers and managers. This is problematic since workers are increasingly involved in decision-making and share responsibility for the conditions they work in. This blurs the traditional division between workers and managers according to which the workforce executes work and managers direct work.

Public Responsibility

While profitability in the short term and long term are likely to be of major concern for most businesses, many will wish to be seen by wider society as acting responsibly. This is reinforced by the insight that public attitudes, for example among potential customers, are important for business performance. While many aspects of the environment a business operates in are subject to legislation, such as pollution and working conditions, businesses will often want to be perceived as achieving more than the legal minimum in these aspects.

In pursuit of such objectives, small businesses with local customer bases may wish to contribute towards local issues, such as supporting their community services or engaging in local charitable initiatives. Large businesses will often become involved in a much wider range of issues. They may contribute to education, engage with professional institutions and activities, and partake in local, national, and international initiatives to address environmental issues, for example.

Such considerations have led to the development of frameworks for self-regulation such as *corporate social responsibility*, also known as *corporate conscience*, *corporate citizenship*, or *responsible business*. These approaches aim to act as a set of mechanisms integrated into the business by which managers evaluate and ensure compliance with ethical standards, the spirit of the law, and national or international norms.

THE OBJECTIVES OF OWNERS

In businesses without limited liability, the managers and the owners are, in the vast majority of cases, the same people. This implies that there is normally no difference between the owners' and the managers' objectives. For small limited companies, it is often the case that the directors are also major shareholders, creating close alignment between the interests of managers and shareholders.

For larger incorporated businesses, however, the situation is more complex. Here, the most senior managers in the business, the board of directors, are elected by the shareholders to oversee the other managers in the business. In such businesses, particularly those listed on the stock market, there can be a greater distance between shareholders and managers. While managers are obliged to make decisions and take actions on behalf of the owners, in practice, there may be considerable leeway to serve their own interests. This will, of course, run against the interests of the owners. An exacerbating factor is that managers usually have better information regarding the business they run. Together, these factors create a situation in which owners cannot always ensure that the managers are acting as desired. This aspect has attracted significant interest in the management literature and is known as the *principal-agent problem* between corporate managers and shareholders.*

In practice, various methods are available to the owners to align the interests of the managers with their own, apart from performance measurement and the threat of refusing to re-elect directors. These methods form part of the system of rules and processes that exists in a business to control the activities of managers at various levels, known as *corporate governance*. Common ways of influencing the conduct of managers include offering shares or the option to buy shares to managers as well as financial incentives in the form of *bonuses* or *performance-related pay*. These aspects are introduced in Chapter 10.

It is important to note that there can be differences in the objectives even among the owners of the business. While these differences may be minor, such as the prioritization of short-term profits or long term profits, there may be situations in which the interests of owners are fully opposed. This happens in particular, if *speculators* are involved.

Any shareholder in a limited company who seeks to profit through the long-term ownership of shares through capital gains or dividends or through interest payments (introduced in detail in Chapter 15) can be characterized as an *investor*. However, it is important to note that not all shareholders are investors. Shareholders to whom the characteristics of the business are largely irrelevant, and instead attempt to profit from short-term fluctuations in the market value of shares, are known as speculators. The objectives of speculators and managers are not aligned and may even be in opposition to each other.

Speculators who attempt to benefit from increases in the share price or the payment of dividends by buying shares are said to have a *long position*. In this case, their interests are largely aligned with the interests of other owners, managers, and also those of the employees of the business. However, speculators may also take the opposite stance and adopt a *short position* through a process known as *short selling*. The goal of short

* For a brief introduction to the principal-agent problem, see Boddy (2017).

selling is to benefit from decreases in the value of the shares of a business. Short selling requires the services of specialist financial companies who make available shares to rent in exchange for a fee. The process of short selling is summarized by the following sequence:

1. a speculator obtains temporary possession of a share by renting it from an owner, normally a specialist financial services company, for a fee;
2. the share is sold in the market by the speculator at market price;
3. the speculator waits until the market price falls;
4. the share is bought back at a new, lower market price; and
5. the speculator returns the share to its permanent owner and keeps the difference between the sales price and the re-purchasing price.

Short selling carries the reputation of being risky since the share price is not guaranteed to fall as planned. If the share price rises strongly and the short seller is forced to buy back the share for a multiple of its original price, the financial loss can be severe. This situation is known as a *short squeeze*. Despite this risk, short selling is, in fact, a routine activity in the investment community and financial products based on short selling are widely available to non-professional investors in many countries.

A further situation in which the objectives of shareholders and managers can deviate strongly is when a *hostile takeover* of a business is attempted. The aim of a hostile takeover is to allow an outside bidder to acquire ownership of a business whose managers are unwilling to agree to the acquisition. There are three general ways in which a hostile takeover can be pursued: *tender offer*, *creeping tender offer*, and *proxy fight*. Table 2.3 briefly characterizes each.

CONCLUSION

This chapter has defined the private enterprise as the basic functional unit of management. It has highlighted basic features of businesses from four different perspectives:

TABLE 2.3
Ways in Which a Hostile Takeover Can Be Attempted

Tender offer	A tender offer is a public takeover bid in which existing shareholders are invited to sell their shares to an acquirer. This bid is normally significantly above market value and conditional on enough shareholders accepting
Creeping tender offer	A creeping tender offer is executed by secretly buying shares over a period of time. The acquirer thereby attempts to secure sufficient voting rights to replace the management with one that will approve the takeover
Proxy fight	In a proxy fight, an acquirer will attempt to persuade other shareholders to use their voting rights to install new managers who will approve of the proposed sale of the business

- as a framework for different flows of cash, showing that the net profit of the business results from the difference between sales revenue and various costs;
- as a set of primary and secondary activities that together make up a value chain;
- as a legal form affording certain benefits and obligations under the law, introducing the core idea of legal personality; and
- as an organization structure based on seniority.

The chapter has presented common objectives of the private enterprise, including market standing, innovation, productivity, physical and financial resources, profitability, manager performance and development, worker performance and attitude, and public responsibility. This has been contrasted with a summary of the objectives of owners, introducing the core idea that owners are not always investors but can also be speculators.

This chapter has touched on various topics that are closely associated with famous management thinkers. The book *Competitive Advantage: Creating and Sustaining Superior Performance* by Michael Porter (1985) introduces the concept of the value chain in detail. The book *The Practice of Management* by Peter Drucker (1954) discusses the objectives of businesses in an accessible way. It should be noted that this chapter has adopted terminologies from English Law in its discussion of legal forms. The reader should refer to governmental guidance for more information on the applicability of these terms in other countries.

REVIEW QUESTIONS

1. Which of the following usually fall under the category of private enterprise? **(Question type: Multiple response)**
   - ☐ Non-governmental organizations
   - ☐ Business enterprises
   - ☐ Professional societies
   - ☐ Limited liability companies
   - ☐ Privately held corporations
   - ☐ Companies listed on the stock market
   - ☐ Firms
   - ☐ Non-profit organizations
   - ☐ Governmental departments
   - ☐ Charitable organizations

2. You are analyzing a business manufacturing artisanal potato chips. Over the last accounting period, the company has sold 64,000 kg of chips. To manufacture this amount of product, the company has incurred costs of $0.40 per kg of output for potatoes, vegetable oil, and other ingredients. Total process-related energy costs were $24,500. Additionally, the company has incurred process costs of $1.10 per kg of product to run its machinery and has paid $39,000 to secure the rights to use specific branding. On average, the company has achieved a revenue of $5.00 per kg of product.

(Question type: Calculation)
Calculate the gross profit.

3. Identify operating profit, net profit, added value, gross profit, and sales in the following diagram:
(Question type: Image hotspot)

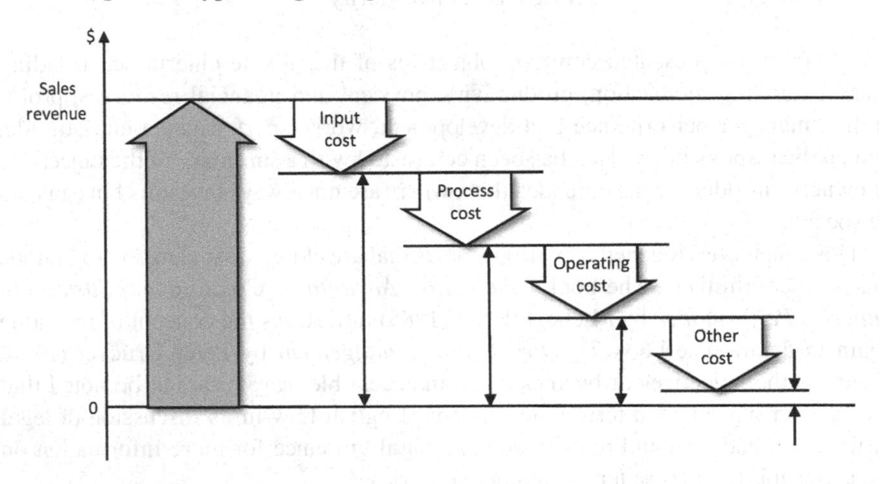

FIGURE 2.7 Identify the correct elements in the figure

4. Which of the following is a secondary activity in Porter's value chain?
(Question type: Multiple choice)
○ Outbound logistics
○ Marketing and sales
○ General strategic opportunities
○ Technology development
○ Customer support and service

5. Which of the following attributes do you expect to vary depending on the level of seniority of a post in management?
(Question type: Multiple response)
☐ Generality of skills
☐ Responsibility
☐ Accountability
☐ Length of time horizon of planning
☐ Adherence to regulations
☐ Management of scarce resources

6. Order the provided corporate ranks according to seniority, with the first being the highest, etc.
(Question type: Ranking)
• Director

- Middle Manager
- Worker
- Managing Director or CEO
- Chairman of the Board

7. Match the provided characteristics of a business to the appropriate legal form:
   (Question type: Matrix)

	Public limited company	Proprietor-ship	Partnership	Private limited company
Owned by one person	O	O	O	O
Multiple owners that are responsible for debt	O	O	O	O
Shares can be bought and sold but not offered to the public	O	O	O	O
Shares can be bought and sold on the stock exchange	O	O	O	O

8. The key advantage of having legal personhood is that owners of such a business cannot fully be held to account for what the company does because they have "limited liability".
   (Question type: True/false)
   Is this statement true or false?
   O True O False

9. Complete the following paragraph providing a general, non-accounting, definition of the term "capital":
   (Question type: Fill in the blank)
   "Capital is any kind of ____________ object, ____________ or ____________, that is owned by a business or an individual and is useful in ____________ or progressing development otherwise".

10. Complete the following paragraph describing the public responsibility of a company:
    (Question type: Fill in the blank)
    "The objective to acting socially responsibly has led to the development of ____________ frameworks. Such frameworks act as mechanisms by which an organization evaluates its ____________ with ethical standards, the spirit of the law and social norms".

REFERENCES AND FURTHER READING

Atrill, P. and McLaney, E., 2018. *Management accounting for decision makers*. 9th ed. Pearson Education. New York.

Boddy, D., 2017. *Management: An introduction*. 7th ed. New York: Pearson Education.

Cambridge Dictionary [Online], 2020. Value. Available at: https://dictionary.cambridge.org/dictionary/english/value. [Accessed 7 December 2020].

Castells, M., 1996. *The information age, Vol. 1: The rise of the network society*. Cambridge, MA and Oxford: Blackwell.

Drucker, P.F., 1954. *The practice of management*. New York: Harper & Row, Publishers, Inc.

Organisation for Economic Co-operation and Development (OECD), 2017. Entrepreneurship at a glance 2017.

Porter, M.E., 1985. *Competitive advantage: Creating and sustaining superior performance*. New York: Macmillan.

3 The Economic Theory of the Firm

OBJECTIVES AND LEARNING OUTCOMES

This chapter introduces how economics provides a general way to understand the activities of businesses and how basic economic principles drive management thinking. After a brief characterization of such principles, this chapter introduces the economic relationships of demand and supply in a standard framework that can be used for a type of model known as static analysis. Drawing the connection to management, the chapter shows how this approach can be applied as the basic theory of the firm, bringing together technical and organizational aspects in the form of cost functions and market characteristics in the form of revenue functions. On this basis, it is shown how businesses set profit-maximizing levels of activities. Specific learning outcomes of this chapter include:

- understanding what economics is and how it relates to the idea of scarcity;
- gaining knowledge of important principles of economics;
- ability to build simple models of supply and demand for use in static analysis;
- understanding of basic forms of markets;
- knowledge of the main relationships in the basic theory of the firm, including cost theory and revenue theory;
- understanding key terms, synonyms, and accepted acronyms; and
- appreciation of important thinkers in economics.

WHAT IS ECONOMICS?

Throughout its existence, economics as a discipline has produced many insights about people, markets, and economies. Economic models and theories often describe the behavior of individuals and businesses. This chapter shows how economic models and theories relate to important management topics. While such models shape business thinking and how managers behave in many situations, the ideas within economics can also be applied in many other contexts.

The word "economics" is derived from the ancient Greek word *oikonomos* (οἰκονόμος) which can be translated as "the one who manages a household" or simply "manager". Like a manager in a business, the manager of a household needs to make decisions on an ongoing basis. Such decisions include how to support the lives of the members of the household by deciding which work or other productive activity to take up. This is similar to the tasks faced by a manager in a business who must

DOI: 10.1201/9781003222903-4

decide who does what and how the available resources are used in pursuit of business goals, as defined in Chapter 1.

The need to organize and manage is stressed in economics because it is assumed that many available resources are *scarce*. The importance of the concept of scarcity for economic thinking cannot be understated. It implies that required resources, be it to the individual, the household, the firm, or the economy, are limited and that therefore not all desirable objects can be obtained and not all plans can be realized. On a fundamental level, the gap between limited resources and human needs, which are often assumed to be limitless in economics, is the reason for the subject of economics to exist. Scarcity can be defined as follows:

Scarcity is the limited availability of objects, which may be required by individuals or by organizations. Scarcity includes the limited availability of financial resources to buy other objects.

In its presentation of economics, this chapter focuses on what is known as *microeconomics*. Microeconomics can be defined as the study and analysis of the ways in which individuals, households, and companies make decisions and how they interact in *markets*. The concept of a market is central to understanding economics and will be defined carefully in this chapter. Microeconomics can be distinguished from *macroeconomics*, which is the study of wider phenomena affecting the whole economy, normally focusing on countries or geographical regions. Microeconomists study, for example, the effect of competition from Asia on the European car industry or how lifetime earnings can benefit from completing a degree at a university. Some technical problems can also be viewed through a microeconomic lens, for example, the question of how customers value an additional variety of a product or how demand is expected to change if the features of a product are altered.

Before proceeding with the characterization of the field of economics, it is necessary to introduce two additional concepts of major importance, the *good* and the property of *efficiency*.

WHAT IS A GOOD?

In economics, and also in business, the term "good" describes a beneficial material, or *tangible*, object, or item. Common synonyms for goods include *articles*, *commodities*, *materials*, *merchandise*, *products*, *supply*, and *wares*. Goods are often created by businesses through activities in farming, construction, manufacturing, or exploitation of natural resources, such as mining or fishing. Goods satisfy human wants or needs by definition. Because of this, they are seen as something that people will always find useful or desirable.

In economics, goods are said to provide *utility* to their owners and it is assumed that owners will want to obtain as much utility as possible. It is important to note that goods can never be *bad* in the sense that the owner loses utility through them. At first, this point appears counterintuitive: after all, what extra benefit could a person possibly have from eating the tenth slice of pizza? Nevertheless, economists normally assume that more of a good is *always* better than less. This assumption, known as the *axiom of non-satiety*, is often criticized. It forms the basis of how mainstream economists and many managers, especially in the investment community, think about rational behavior. For this reason, the idea of non-satiety holds substantial sway in business.

Economics distinguishes between several different types of goods. *Scarce goods* are those that are in limited supply; because most goods are scarce, these goods are also called *economic goods*. Goods that are available in unlimited supply or those that can be acquired without effort or payment, such as air or seawater, are referred to as *free goods*. Interestingly, money itself is also a good. Because its primary purpose is to express the relative amount of value of other goods, it is referred to as a *numeraire good*. Other numeraire goods include precious metals, primarily gold and silver, and cryptocurrencies such as bitcoin.

The non-material, or *intangible*, counterpart to goods is services. In generating a service, a seller provides a valuable activity to the buyer. The benefits received by the buyer through the service are demonstrated by the buyer's willingness to pay for the service. Because services are, by definition, intangible, they cannot be manufactured, transported, or stored for future use. When a service has been completed, it, unlike most goods, irreversibly vanishes. Services include, for example, air travel, taxi rides, haircuts, health care, and accountancy services.

WHAT IS EFFICIENCY?

Just as the ideal state that should be pursued by any legal system is justice, the ideal state pursued by an economic system is efficiency. Efficiency can be defined as a state in which an individual, a business, or society gets the most out of the available resources or goods and services. In this sense, efficiency is the result of an ability to avoid wasting raw materials, energy, effort, money, and time or creating undesirable outcomes such as pollution. Because efficiency is always assessed in relation to an object or an activity, such as the pursuit of a goal, the resources relevant for achieving efficiency depend on the context or the problem at hand.

Economics distinguishes between two specific types of efficiency. The first type is *allocative efficiency*, which describes a state in which every product or service that can be produced or sold is supplied to a buyer who is willing and able to pay for it. Because most resources are considered scarce, allocative efficiency implies that resources are made available to users, for example, companies, in just the right amounts. The second type of efficiency is *productive efficiency*, which is a concept familiar to engineers and technologists. Productive efficiency exists in conditions in which goods or services are generated at the lowest possible average cost or using the minimal amount of resources.

AN OVERVIEW OF THE PRINCIPLES OF ECONOMICS

Economics as a field is varied and contains numerous topics, concepts, and methodologies. To get a first overview of the extent of this field, it is useful to summarize some central and recurring themes. To do this, this section follows the standard textbook approach and summarizes ten such themes that unify the field of economics.[*] It should be noted that, despite being billed as *principles of economics*, these themes arise in numerous other settings and disciplines, for example, in the social sciences and politics. Importantly, they frequently appear in management as well.

Principle 1: Trade-offs Are Unavoidable

The principle of scarcity implies that usually something must be given up to obtain something else. Making sensible decisions, therefore, often requires sacrificing, or *trading off*, one goal against another. That this is extremely important in everyday life can be seen when considering how individuals spend their time. For example, to be able to spend one hour studying, a student cannot spend this hour in the gym or earning money in a part-time job. In the same way, a household with a limited income must trade off buying different goods: money that is spent on clothing cannot be spent on dining out or going on holidays. At the level of society, an important trade-off is between improving efficiency by encouraging intensive competition between businesses and distributing society's wealth more equitably among its members. Addressing this trade-off is an important recurring theme in politics.

Principle 2: The Cost of Opportunity

The existence of scarcity and trade-offs implies that people frequently compare the benefit of one available option against that of another. While this is not always obvious, all individuals make such calculations on an ongoing basis. Economics adds clarity to this process of weighing options by introducing the concept of *opportunity cost*. Opportunity cost is a particular type of cost that expresses the value[†] of the next best option given up, or forgone, by choosing a particular option, usually measured in money terms. It is defined as follows:

IMPORTANT DEFINITION

Opportunity cost is the loss of not enjoying the benefit attached to the best alternative choice or course of action when one alternative is chosen.

Opportunity cost, therefore, reflects the value of the benefits forgone by not choosing alternative options.[‡] Forgoing a very valuable option can create a high opportunity cost if a less valuable option is chosen, which is often a bad idea. Conversely, sacrificing an unattractive, low-value option for a higher value option incurs a negative opportunity cost. Using mathematical notation, the concept of opportunity cost can be expressed through the relationship between the return of the best forgone option *BFO* and the return of the chosen option *CO*:

$$\text{Opportunity cost} = BFO - CO \tag{3.1}$$

[*] Adapted from Mankiw (2020).

[†] As an expression of the benefit or cost to an individual, opportunity cost is an expression of how useful something is and thus reflects *use value*.

[‡] The traditional, and perhaps strictly correct, definition of opportunity cost is that it expresses only the value of the best option forgone, corresponding to *BFO* in equation 3.1. We find the understanding of opportunity cost presented in equation 3.1 far more intuitive and relevant in everyday business situations, however.

The concept of opportunity cost is important for many theories and models used by managers. This book will return numerous times to it, so it is advisable to develop a solid understanding of this idea.

Principle 3: Marginal Benefit and Marginal Cost

Economists often quantitatively capture what is happening in a situation in order to identify the best decision available in a set of circumstances. Such situations can often be addressed by investigating the effects of small incremental changes. For this reason, economists often compare the *marginal benefit*, such as the benefits of one extra unit of activity or product, against the *marginal costs* arising through this extra unit. Such comparisons are frequently used to determine if pursuing this extra unit is worthwhile. As discussed toward the end of this chapter, thinking at the margin is an important tool available to managers in determining the course of action that delivers the maximum benefit or profit to a business.

Principle 4: Incentives Are Important

An *incentive* is something that drives or motivates an individual to perform an action or choose an option. Because incentives are so central to economics, it has occasionally been referred to as the *science of incentives*. Usually, the magnitude of an incentive is defined by the difference between the perceived benefits and costs. Where benefits far outweigh the costs, incentives are assumed to be large. Conversely, where benefits are only slightly larger than the costs, incentives are small. In situations in which the costs outweigh the benefits, there are no incentives toward a particular action or option at all. In economics, it is generally thought that incentives cannot be negative.

Principle 5: Exchanging Things Is Normally Beneficial for Everyone

Adam Smith (1723–1790)

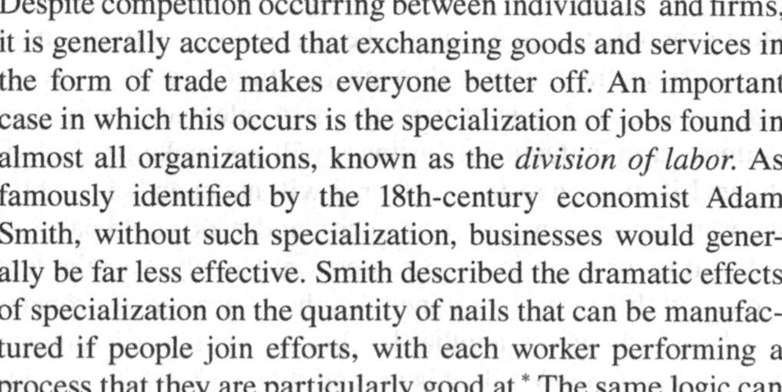

Despite competition occurring between individuals and firms, it is generally accepted that exchanging goods and services in the form of trade makes everyone better off. An important case in which this occurs is the specialization of jobs found in almost all organizations, known as the *division of labor*. As famously identified by the 18th-century economist Adam Smith, without such specialization, businesses would generally be far less effective. Smith described the dramatic effects of specialization on the quantity of nails that can be manufactured if people join efforts, with each worker performing a process that they are particularly good at.* The same logic can be applied to industries in which firms specialize in offering a specific product or

* Despite it being written more than two centuries ago, Smith's (1776) discussion of the division of labor is surprisingly accessible to today's reader and interesting.

service, and importantly also in international trade where countries prioritize particular industries. The idea of the division of labor also forms the basis for Porter's highly influential model of the value chain, as discussed in Chapter 2.

Principle 6: Activity Should Be Organized through Markets

As a major theme in economics, a *market* is any kind of social structure that allows buyers and sellers to interact to exchange goods, services, information, or other objects. It is possible to view a market as a process by which the prices charged for goods and services and the quantities they are sold in are determined. The exchange of such things within a market, with or without money, is referred to as a *transaction*. Because freely operating markets have the tendency to produce a state of allocative efficiency under certain conditions, the common position in economics is that activity should be organized through markets where possible. Markets can be defined as follows:

IMPORTANT DEFINITION

A market is any structure or procedure that allows buyers and sellers to exchange any type of good, service, or information. A market may be specific to a certain good, service, or information.

Markets can vary significantly in form, size, location, and participants. To provide some examples, markets can be organized in the form of an auction, as a stock market, as a private electronic market accessible to subscribers, or as a traditional commodity wholesale market. Even shopping malls or informal discussions between two people can be viewed as markets. Markets that operate illegally are called *black markets*. In some cases, markets are even based on criminal behavior such as violence or extortion. While most markets involve money, it is possible to organize markets entirely without money, for example, by bartering. The terms *industry* or *sector* are sometimes used synonymously with "market".

Principle 7: Government Action Is Sometimes Needed

Most economists agree that markets, economic activity, and society as a whole can benefit from intervention through institutions such as governments. A basic, and easily overlooked, role of the state is to safeguard and enforce property rights. For example, a manufacturing business will not make products if it expects them to be stolen. Likewise, a restaurant owner will not open her establishment for customers if she were not certain that the majority of diners would pay for their meals. In modern political theory, the state possesses a monopoly over the legitimate use of physical force, and this importantly includes the protection of property rights.

There is, however, another level on which governmental action is needed to support the economy. While it is argued that markets are considered beneficial ways of organizing activity, there are certain conditions under which they can fail to reach welfare-maximizing outcomes. In this case, economists speak of *market failure*. A possible reason for such failure is that economic processes often generate unintended by-products which are called *externalities*. Negative externalities impose detrimental effects on others which are not accounted for or compensated by those responsible for their creation. Externalities, also known as *external effects*, can be defined as follows:

An externality is a cost or benefit resulting from an activity that affects a third party who did not choose to incur that cost or benefit. As such, an externality is a by-product of that activity and is often unintended or unvalued by those creating it.

The most pressing negative externality of all economic activity in the present is the global warming as a result of carbon emissions and other forms of environmental pollution. As the by-products of private and public economic activity, these impacts impose a cost to society, future generations, and nature as a whole that is not borne only by the polluters. Chapter 6 introduces a range of approaches designed to measure such external effects. Governments usually attempt to correct market failure of this kind by taxing or regulating the activity creating the negative externality.

A further type of market failure results from *market power.* Market power gives single individuals or small groups the ability to influence market prices at their discretion. An example of such market power would the ability of a business to charge prices that lie above those associated with a competitive market outcome. Market power can be defined as follows:

Market power is the ability of a business to raise the price charged for a good or service over its cost of production without losing all of its customers.

Governments often intervene to limit the market power of businesses. This includes ensuring that businesses do not acquire competitors to obtain market power and breaking up businesses that have significant market power. Such activity is supported through *antitrust* laws.

Principle 8: Overall Wealth Depends on the Ability to Produce

The role of material wellbeing as a prerequisite for leading a happy life should not be underestimated. This is why economic metrics, such as the *per capita gross domestic product (GDP)*, which measures the economic output in an economy per person over a period of time, are important. Countries with a high GDP are, on average, seen to offer better nutrition, superior health care, and longer life expectancy than those with low levels of GDP. Over time, the main determinant of GDP is the productivity of labor in an economy which is a measure of the quantity of goods and services generated, on average, from an hour of a worker's time in that economy.

Principle 9: If the Government Issues Too Much Money, Prices Will Increase

Over the course of the 1970s, the average prices charged for goods and services in the United Kingdom more than tripled. This relative devaluation of money over time is called *inflation*. Inflation occurs when the overall prices of goods increase. A particular branch of government policy, *monetary policy*, aims to influence price levels by either expanding or contracting the amount of money available in the economy, known as the *money supply*. This can be achieved by creating more physical money, i.e. cash, or by issuing debt to companies and individuals. In the recent past, the level of inflation has been around 2% per year in many countries. This indicates that, on average, the price for a good would rise by this amount over the course of the year. While being major themes in economics, inflation and interest rates (discussed in detail in Chapter 15) are also of great importance to managers.

Principle 10: In the Short Term There Is a Trade-off between Output and Inflation

Another important topic in economics is the relationship between the overall level of economic activity, referred to in economics as *output*, and inflation. When governments expand the supply of money, and thereby cause inflation, businesses will find it easier to take out loans to expand their activities. Through this link, mild levels of inflation have a positive, stimulating effect on businesses in the economy, increasing the level of economic activity and decreasing the level of unemployment. This effect is normally only temporary but can last for several years. It gives rise to what is known as the *business cycle*. The term business cycle describes the irregular, yet cyclical, fluctuations in economic activity that may occur over time. The business cycle can be measured in terms of GDP or the level of unemployment.

DEMAND AND SUPPLY IN STATIC ANALYSES

As argued in the previous parts of this chapter, markets are important because they facilitate trade and enable the distribution and allocation of resources among the members of society. Markets occasionally emerge unintentionally and spontaneously but are mostly established purposefully by those interested in buying or selling a good or service. An important and valuable role markets play in this is the determination of the price of a tradable object or service.

When further analyzing how markets work, it is helpful to imagine a market for a specific product, melons, or electric bicycles, for example. It is also helpful to initially assume that many sellers and many buyers are present in this market and that the buyers and sellers are all different in terms of their willingness to buy or sell goods at certain prices.

Demand and Supply

In economics, the standard way of looking at a market is through the two main elements that make it up, *demand* and *supply*. Demand reflects the quantity of a good that a group of buyers is willing and able to buy at a particular price. It can be shown as a mathematical relationship or graphically as a curve. Since the object transacted in the market is a good, in the sense that more of it is always better than

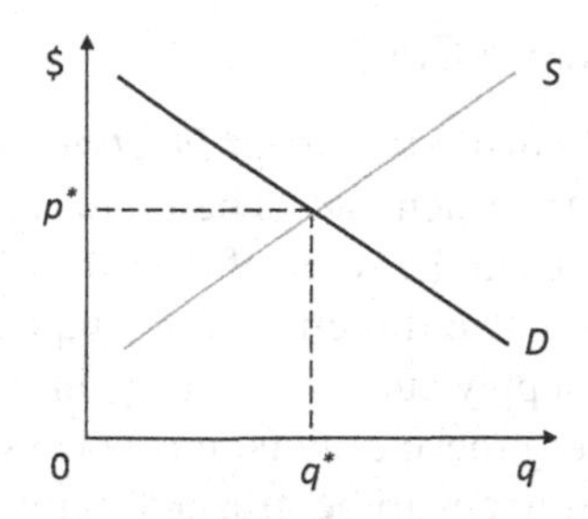

FIGURE 3.1 Demand and supply in a static analysis

less, economic theory dictates that the buyers will want to buy more if the price of the good decreases. This is known as the *economic law of demand*. It is the reason why demand curves *never* slope upward and why the relationship between quantity q demanded and price p is said to be negative.

For supply, which is the quantity a seller is willing and able to offer at a given price, the opposite relationship applies. Standard economic theory dictates that if the price of a product increases, there are greater profits to be made and sellers will want to supply a greater quantity of the good. That is why supply curves are assumed to never slope downward.

While both demand and supply are often referred to as curves, they are normally represented as linear relationships in basic analyses. It should also be noted that such forms of supply and demand analysis ignore the effect of time. That is why models using the traditional supply and demand curves shown in Figure 3.1 are referred to as *static analyses*. The comparison of two static models, for example, before and after a change in an important variable, is referred to as *comparative statics*.

In the framework shown in Figure 3.1, there is an intersection between the supply and demand curves. This intersection defines the point at which the market is in *equilibrium*. This point is of special interest because it defines the quantity q^* and the price p^* at which the buyers' and the sellers' preferences are in balance. If the market is in equilibrium, no additional buyer would be able to buy the product at a price a seller would agree to. The *equilibrium price*, also known as the *market-clearing price*, is the price at which the market mechanism results in allocative efficiency. Note also that determining this price embodies Adam Smith's concept of the "invisible hand", which denotes a process in which a socially desirable outcome is achieved while buyers and sellers purely act in their own self-interest.*

* Critical readers may ask whether it is realistic to assume that the participants act in a fully rational way in the real world. The appropriate response to this is that the depiction of markets in economics is unrealistic since markets are always influenced by institutions, culture, and trust between participants. For a comprehensive, yet difficult to read, treatment of this question, see Fine (2002). What matters for the purpose of this book is how managers think about markets – and this is largely shaped by economics.

A Closer Look at the Demand Curve

Using demand curves, also known as *demand functions*, comparative statics is able to reflect three basic changes to demand. These are movements along the curve, shifts of the curve, and changes in the slope of the curve. Note that all three changes can occur in a positive or negative direction. For simplicity, this chapter will only consider linear demand and supply curves, as shown in Figures 3.2 and 3.3.

However, before discussing these effects, it is necessary to elaborate on a subtle, yet important, point that needs to be grasped to interpret supply and demand curves correctly. Note that by showing quantity q on the horizontal axis and the price p on the vertical axis, the basic framework seems to imply that p is the dependent variable and q is the independent variable. Mathematically, this could be expressed using the following function $p\ (q)$ where x is a negative slope parameter and y is an intercept term:

$$p(q) = xq + y \tag{3.2}$$

This means that the customary graphical way of showing such relationships implies that, instead of being a demand function, D is, in fact, a price function of demand which, technically speaking, is the inverse demand function. This point is valid. Nevertheless, despite showing demand and supply relationships in this way, economics normally views demand and supply as determined by prices. If dealt with

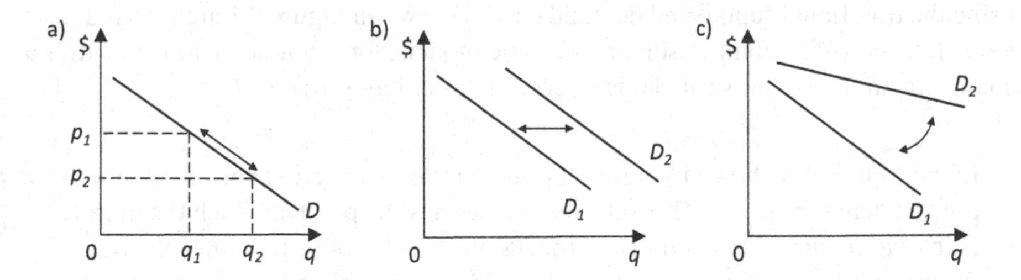

FIGURE 3.2 Movement along the demand curve (a), shift of the curve (b), and change of the slope (c)

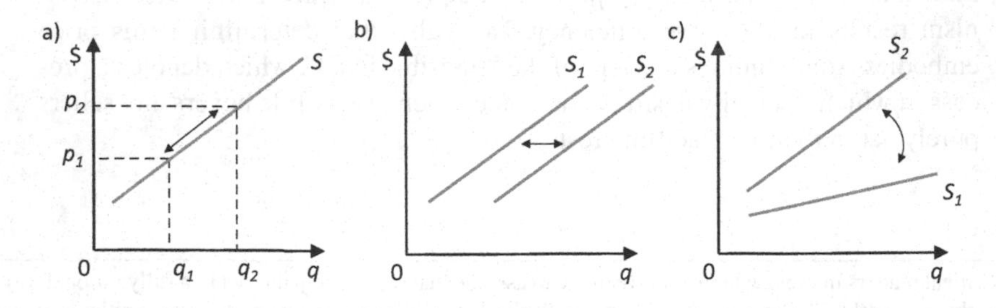

FIGURE 3.3 Movement along the supply curve (a), shift of the curve (b), and change of slope (c)

mathematically, demand functions are generally treated, as done here, in their non-inverse form $q(p)$. Using the negative slope parameter m and the intercept t, demand can be expressed as:

$$q(p) = mp + t \tag{3.3}$$

As can be seen in Figure 3.2a, a movement along the demand curve simply reflects a change in either price or quantity, leading to a negatively related effect on the other. For example, an increase in quantity demanded from q_1 to q_2 leads to a decrease in market price from p_1 to p_2. This is based on the assumption that all participants in the market pay and receive the same price, known as the *law of one price*, implying that the sellers are unable to discriminate between buyers by offering different prices.

Figure 3.2b depicts a shift in the demand curve to the right, indicating that at any given price, the buyers of the good will demand a higher quantity. Such movements, which can occur in both directions, are due to a number of reasons, including changes in:

- buyers' income or level of wealth;
- the price of another good that is related to the good traded in the market;
- buyer preferences, taste, fashion, or expectations; and
- the size of the population.

The third kind of change is shown in Figure 3.2c. Here, the gradient of the demand curve changes. At this point, it is necessary to introduce a further concept to the analysis. The slope of the non-inverse demand function m is related to the concept of *own-price elasticity of demand* (*PED*), which is defined as the percentage change of q over the percentage change of p:

$$PED = \frac{\Delta q / q}{\Delta p / p} \tag{3.4}$$

Since *PED* measures a change relative to a base value, it is important to note that its level is not constant along linear demand curves such as those shown in Figure 3.2. This also means that when estimating *PED*, it is important to clearly identify the base level of price p_0 and quantity q_0 and the new level of price p_1 and quantity q_1. Estimating *PED* is straightforward if p_0, q_0, p_1, and q_1 are given. The formula is simply:

$$PED = \frac{(q_1 - q_0) / q_0}{(p_1 - p_0) / p_0} \tag{3.5}$$

Many different aspects can influence a good's own-price elasticity of demand. This analysis will concentrate on whether a good is perceived as a basic necessity or a luxury good. A basic necessity, such as an unbranded generic type of bread, will exhibit a far lower *PED* than a luxury artisanal bread. This reflects the assumption that an increase in price will have a far smaller effect on the quantity demanded of the basic necessity than on the quantity demanded of the luxury product.

Note here that in the inverse framework shown in Figure 3.2c, low elasticity of demand corresponds to a steep demand curve such as D_1, and vice versa. If *PED* is low (i.e., $|PED| < 1$), economists and managers speak of a price-inelastic demand. If it is high (i.e., $|PED| > 1$), demand is said to be price elastic. If $PED = 0$, demand is referred to as perfectly price inelastic, and if $|PED| = \infty$, demand is called perfectly price elastic.

A Closer Look at the Supply Curve

The logic described for demand curves applies equally to supply curves, also known as *supply functions*, with the important difference that supply functions are upward sloping. Note that the standard way of drawing supply curves in economics also results in inverse supply functions, which are technically price functions of quantity supplied. As before, movement along the curve reflects the relationship between p and q: if p increases, q also increases. This is because sellers will find it more profitable to sell more units in response to price increases, as shown in Figure 3.3a.

Note that supply curves can shift in or out, as shown in Figure 3.3b. The prime reason for shifts in supply curves is change in the available technology, which can make the provision of higher quantities at unchanged prices feasible. Other common reasons for shifts in supply curves are improvements to business processes and changes to the level of uncertainty faced by businesses.

Analogous to the concept of *PED*, it is possible that the slope of a supply curve changes. This is the result of a change in *own-price elasticity of supply* (*PES*), which is defined as the percentage change of the quantity supplied q over the percentage change of p:

$$PES = \frac{\Delta q / q}{\Delta p / p} \tag{3.6}$$

PES can be estimated if the base level of price p_0 and quantity supplied q_0, and the new level of price p_1 and quantity supplied q_1 are known:

$$PES = \frac{(q_1 - q_0) / q_0}{(p_1 - p_0) / p_0} \tag{3.7}$$

Factors that determine the suppliers' ability to change output in response to changes in price determine *PES*. Such factors include, but are not limited to, technology, the quality of management, and the availability of financial resources. Again, remembering that the supply curves in Figure 3.3 show inverse supply functions, low elasticity of supply corresponds to a steep supply curve such as S_2, and vice versa. If *PES* is low (i.e. $PES < 1$), economists speak of price-inelastic supply and if it is high (i.e. $PES > 1$), they refer to price elastic supply. If $PES = 0$, supply is referred to as perfectly price inelastic, and if $PES = \infty$, supply is perfectly price elastic.

A Worked Example of a Market with Linear Demand and Supply Functions

Imagine a hypothetical market for electric bicycles. Analyzing the market as the interaction of a demand function *d* (*p*) and a supply function *s* (*p*), it is possible to numerically calculate the equilibrium price p^* and equilibrium quantity q^*. Assume that an investigation of the market has approximated (non-inverse) linear demand and supply functions as shown in Table 3.1.

TABLE 3.1
Demand and Supply in the Market for Electric Bicycles (Numerical Example)

Demand function	$d(p) = -5p + 10{,}000$
Supply function	$s(p) = 6p - 1{,}000$

These functions can be used to find the equilibrium price p^*, at which $d(p^*) = s(p^*)$:

$$-5p^* + 10{,}000 = 6p^* - 1{,}000 \quad (3.8)$$

$$p^* = 1{,}000 \quad (3.9)$$

At $p^* = 1{,}000$ the equilibrium quantity q^* can be determined:

$$d(p^*) = q^* = -5(1{,}000) + 10{,}000 \quad (3.10)$$

$$q^* = 5{,}000 \quad (3.11)$$

However, at this point, a shortage of raw materials occurs due to a strike. The analysis reveals that the shortage causes 2,200 fewer electric bicycles to be supplied at any price. This equates to a shift of the supply curve to the left. As demand is not affected, the demand function remains the same. Based on the information given, a new supply function s_{new} (*p*) can be stated as:

$$s_{new}(p) = 6p - 1{,}000 - 2{,}200 = 6p - 3{,}200 \quad (3.12)$$

The new supply function changes the equilibrium price and quantity from p^* to p' and from q^* to q' so that the new equilibrium occurs at $d(p') = s_{new}(p')$:

$$-5p' + 10{,}000 = 6p' - 3{,}200 \quad (3.13)$$

$$p' = 1{,}200 \tag{3.14}$$

By inserting p', the new equilibrium quantity q' can be obtained:

$$d(p') = q' = -5(1{,}200) + 10{,}000 \tag{3.15}$$

$$q' = 4{,}000 \tag{3.16}$$

Using Comparative Statics to Analyze Interference with the Market Price

With the basic framework of the static analysis of supply and demand, it is possible to analyze the effects of artificially setting a price, rather than letting the invisible hand of the market determine it through equilibrium price and quantity. For example, the sellers in a market may *collude* and collectively decide to charge a minimum price p_{min} that lies above the equilibrium price p^*. It is important to note that this behavior is illegal under competition law in most markets, though exemptions exist. If such a minimum price is introduced, also known as a *price floor*, the supply in the market will be greater than the demand, as shown in Figure 3.4. In this case, a *surplus* for the good will occur $(q_2 - q_1)$ and economists speak of *excess supply*.

In the opposite case, there may be a maximum price, or *price ceiling*, p_{max}, in the market, for example, if a government decides to set a price limit for political reasons. If $p_{max} < p^*$ then demand at this price level will outstrip the quantity supplied and there will be a *shortage* of the good $(q_2 - q_1)$. Economists refer to this situation as *excess demand*. Both excess demand and excess supply move markets away from a state of allocative efficiency and should, therefore, be avoided in the logic of standard economics.

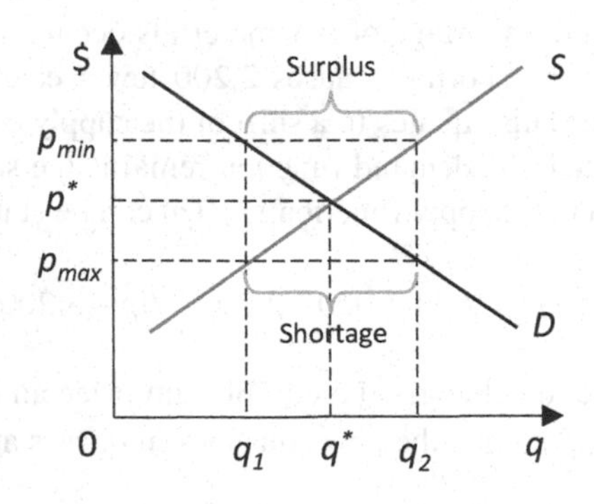

FIGURE 3.4 Excess demand and excess supply

BASIC TYPES OF MARKETS

Economists frequently categorize different kinds of markets by distinguishing between the number of buyers and sellers present in them. According to standard economic theory, the composition of a market has significant implications for its competitiveness and the behavior the participants will adopt. The degree of competitiveness in a market is of the highest importance for managers. As a general rule, if a market is non-competitive, it is likely that the market mechanism will not be able to generate a high enough level of output for allocative efficiency to occur. In other words, it will be unlikely that the market will lead to a socially optimal outcome and market failure is likely to be the consequence. Because such market failure can be very profitable for businesses, it may be the objective of businesses to create certain types of markets. The following section summarizes four basic forms of markets that are of particular interest to economists and managers.

MONOPOLY

Monopolies are markets in which there is only one seller of a product and multiple buyers. Economic theory suggests that in such a market, the monopolist will have the ability to set prices to maximize profits and will thus have market power. In the framework of static analysis, this situation is shown as a downward sloping demand curve faced by the monopolist. This means that by maximizing profit, the monopolist will produce a smaller than efficient level of output and market failure occurs. Monopolies often arise in areas or industries requiring very large upfront investment or expensive infrastructure, such as power generation, the pharmaceutical industry, or consumer-facing internet services.

OLIGOPOLY

In an *oligopoly*, there is a small group of sellers and multiple buyers. Since the market outcome in an oligopoly depends on the strategies employed by the sellers and the buyers, the market can reach many outcomes. Competitive behavior in oligopolies is the subject of its own branch of economics called *industrial economics*. The industry most frequently associated with an oligopolistic market structure is the automotive industry as outlined in Chapter 1.

MONOPOLISTIC COMPETITION

In *monopolistic competition*, there are many buyers and many sellers of a product, but each seller offers a slightly different version of the product. This means that while the products are interchangeable to a limited degree, the sellers will enjoy some market power. The consequence is that the quantity supplied is likely to be below efficient levels. Monopolistic competition occurs, for example, in the consumer goods industry where many sellers differentiate their products through branding.

Perfect Competition

In *perfectly competitive* markets, which is often considered the ideal type of market by economists, there are many sellers and many buyers. Moreover, the products are indistinguishable from each other – meaning that the buyer does not care which seller to buy from. In such markets, the individual seller faces a horizontal demand curve. This means that they can sell either nothing or large quantities, depending on the price they set. This type of market is associated with an efficient outcome in the sense that the quantities bought and sold will be at socially optimal levels. Examples of almost perfect competition can be found in agriculture, where it does not matter which farm a potato comes from, or in markets for standardized bulk commodities.

BASIC THEORY OF THE FIRM

After introducing the basics of the market using static analyses, this section transfers the identified concepts to the business setting. This leads to a set of theories which explain and predict how firms and companies behave under market conditions, known in economics as *the theory of the firm*. Essentially, the theory of the firm aims to explain what profit-maximizing businesses do (or should do!) from the perspective of economics. The view of the firm in mainstream economics is that it is, just like the individual consumer, a rational entity making decisions. According to this view, a firm buys inputs, produces outputs, and sells these with the sole objective of maximizing profit.

This perspective is of the utmost importance for managers because it provides the chief logic to which most managers adhere and against which their conduct is normally evaluated. Moreover, many concepts and techniques flow from the idea that the correct behavior of managers should be rationality-based and directed at maximizing the financial performance of the business or business unit they are responsible for.

Figure 3.5 illustrates how a business can be thought of as an entity that buys inputs to generate and sell outputs in the form of products or services. The inputs bought by the business include what is known in economics as the *three primary factors of production*: land, labor, and capital goods. Additionally, the business may need to buy raw materials. Expanding the framework of financial flows presented in

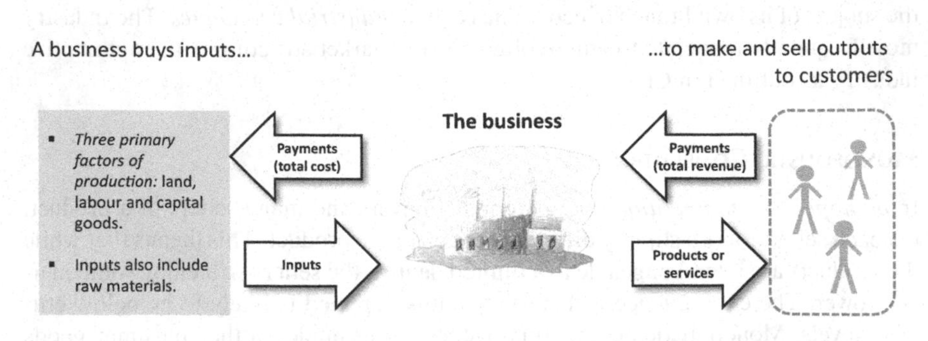

FIGURE 3.5 Illustration of the business as an entity that buys inputs and sells outputs

Chapter 2, the sum of all costs incurred by the firm is referred to as total cost *TC*. The products or services sold to customers result in financial inflows in the form of total revenue *TR*.

According to the theory of the firm, a rationally behaving for-profit business will maximize profit. Profit (π) is defined as the difference of *TR* and total cost *TC*, such that

$$\pi = TR - TC \tag{3.17}$$

Note that the theory of the firm, at least in its basic form, does not distinguish between different kinds of profit as introduced in Chapter 2. This means, for example, that no distinction is made between operating profit and net profit. Moreover, because this kind of model is a form of static analysis, time does not enter. This implies that in such models, the profits available in the distant future are not traded off against profits in the short term.

Cost Theory

To understand the structure of the total cost incurred and total revenue obtained by a business, it is necessary to introduce additional basic concepts, from what is known as *cost and revenue theory*. As seen in Chapter 2, there are many different types of costs, such as operating costs or the cost of goods sold (COGS). More generally, in economics and management, a distinction is made between *fixed costs* and *variable costs*. Fixed costs (*FC*) are those that are not linked to the total quantity of goods and services produced. Normally, fixed costs are not considered fixed indefinitely; they are held fixed over a relevant period of time known as the *short run*. Examples for fixed costs include rent and utility bills, but also costs of tooling and equipment if it is durable. While the term fixed cost is not equivalent to operating cost or overheads, many operating costs and overheads are fixed costs.

Variable costs (*VC*), on the other hand, are costs that change in proportion to the total quantity of goods and services produced. Also referred to as *output*, this quantity is of special interest. Usually measured as a quantity q of discrete units, for example, the number of cars, books, or haircuts produced, the level of q has far-reaching implications. This includes the choice of technology and organizational form by the business. While the term variable cost is not equivalent to input cost, process cost, or the cost of goods sold (as introduced in Chapter 2), many of these costs are variable costs. Examples of variable costs include the costs incurred for raw materials, energy used for production, and components purchased. Figure 3.6 illustrates how fixed costs and variable costs make up total costs in a simple linear framework.

Viewed as a function of quantity q, total costs $TC(q)$ can thus be decomposed into fixed costs *FC*, which are constant at least in the short run, and variable costs as a function of quantity q, $VC(q)$, as follows:

$$TC(q) = FC + VC(q) \tag{3.18}$$

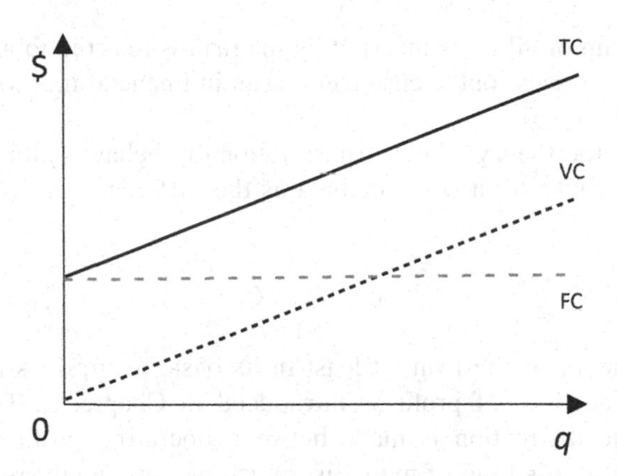

FIGURE 3.6 Total costs (*TC*) decomposed into fixed costs (*FC*) and variable costs (*VC*)

It is worth re-iterating that the categories of variable cost and fixed cost do not necessarily align with the other types of costs, such as those introduced in Chapter 2. While it is safe to say that most input costs are variable and that many operating costs can be assumed to be fixed, such as overheads, the mapping between cost types and their status as variable costs or fixed costs cannot be generalized and depends on the context. For managers, this is a frequent source of ambiguity.

A further cost of great interest in the theory of the firm is *average cost AC*, also called *unit cost*, which is the basic form of cost associated with each unit of output and a quantity of q units. AC is the total cost TC divided by the quantity produced:

$$AC = \frac{TC}{q} \tag{3.19}$$

In many management settings, particularly those related to engineering and product development, methodologies explicitly directed at the minimization of average cost are common. For example, a major approach employed in engineering design is *design for manufacture and assembly (DFMA)*,* which prescribes practical design decisions to reduce AC. Further approaches to systematically minimize AC are *target costing* and *value engineering*.† Since a sufficiently low AC is essential if a business is to survive in the face of competition, innovation efforts frequently target a reduction of AC. This aspect is further explored in Chapter 5.

However, grasping the basic theory of the firm requires one additional type of cost. Like AC, *marginal cost* (MC) is a form of cost associated with each unit of output. As a process-related cost, it describes the cost of the last additional unit of output *at the margin*. Consider, for example, a car manufacturer that has produced a limited series of 500 cars. The MC of this production operation was the extra cost incurred to make the 500th car after making 499 cars. Note that AC is the same for all 500 cars but the MC changes from car to car. In many cases, MC is likely to decrease as

* See, for example, Boothroyd et al. (2002).

† For more details, see Cooper and Slagmulder (1997).

FC is spread over more and more units. This process is called *amortization*[*] of fixed costs. If the total cost function *TC* (*q*) is continuous and differentiable, *MC* is the first derivative of this function with respect to *q*:

$$MC(q) = \frac{dTC}{dq} = TC'(q) \tag{3.20}$$

Drawing *AC* and *MC* functions as curves is a very helpful activity that can generate significant insight. Usually, they are understood to be U-shaped curves, at least in the long run, as shown in Figure 3.7.

The reason for the assumption that the *AC* curve follows a U shape lies in a mix of technical and organizational factors. At low levels of output *q*, *AC* is normally seen to decrease from initially very high levels. This is because of an extremely important phenomenon known as *economies of scale*. Economies of scale reduce *AC* as the level of output *q* grows. In terms of the relationship between *AC* and *MC*, this occurs because *MC* is lower than *AC*, pushing down average costs as *q* increases.

The opposing phenomenon to economies of scale is *diseconomies of scale*, which increase *AC* at high levels of output. Diseconomies of scale occur if *MC* is higher than *AC*, pulling up average costs as output increases. A typical example of a diseconomy of scale is the cost burden of administration and bureaucracy found in large organizations.

Economies of scale and diseconomies of scale can thus be defined as follows:

Economies of scale are cost reductions that are realized when the scale of an operation or process increases. This means that the cost per unit of output decreases as scale increases. Conversely, diseconomies of scale are cost increases that are realized when the scale of an operation or process increases.

IMPORTANT DEFINITION

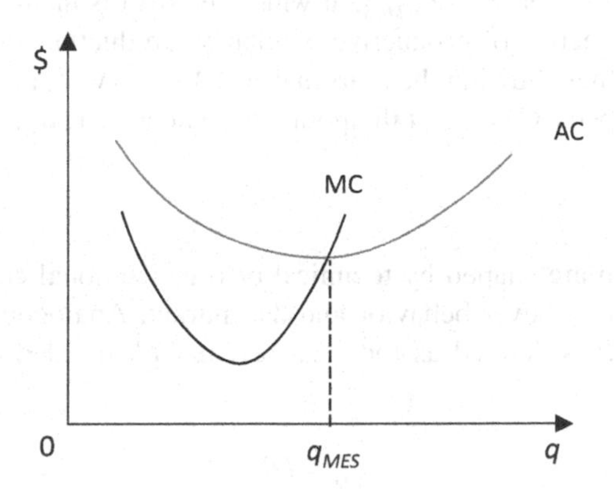

FIGURE 3.7 The average cost (*AC*) curve and the marginal cost (*MC*) curve

[*] The term "amortization" has a second meaning in financial management, as discussed in Chapter 13.

As a fundamental, and perhaps the most important, driver of many decisions in the business world, economies of scale affect nearly all aspects of management. Economies of scale cause the average cost of an activity or process to decrease as the volume of its output rises. A simple example illustrates this: while it may cost $1,000 to print 100 copies of a brochure, it may only cost $2,000 to print 1,000 copies. In this case, *AC* has fallen from $10 to $2 since some important costs of producing the brochure, including the design work and setting up the printing equipment, have no direct relationship with the quantity of brochures printed and can be amortized over a greater quantity.

CORE IDEA

Economies of scale are often directly based on technological characteristics such as the throughput of production machinery and cost effects of production volume due to the indivisibility of important equipment. Economies of scale may also be grounded on non-technological aspects, such as learning among the workforce, organizational factors, or the degree of market power commanded by a business. The pursuit of economies of scale is deeply ingrained in the way managers and investors think, since realizing economies of scale is often the key to running a profitable business or defeating competitors. This is the reason businesspeople often speak of *scaling up* a business or of the *scalability* of a particular technology or business model. This idea is discussed in detail in Chapter 7. Note here that this aspect has gained even more importance through the emergence of information technology. This is because of the special cost structure associated with information technology, as discussed in detail in Chapter 4.

The U-shaped average cost curve shown in Figure 3.7 also implies that there is a point of minimum efficient scale q_{MES}, at which the firm is incurring the minimum *AC*. From the perspective of productive efficiency, production should be set up to make q_{MES} units. When drawing the marginal cost *MC* curve, it is typically shown to intersect the U-shaped *AC* curve at the point of efficient scale q_{MES}.

Revenue Theory

Unlike costs, which are shaped by technical or organizational considerations, revenue is determined by buyer behavior and the market. Analogous to average cost, average revenue *AR* is defined as the total revenue *TR* divided by the quantity *q* produced:

$$AR = \frac{TR}{q} \tag{3.21}$$

Normally, the average revenue function *AR* (*q*) is shown as a downward sloping graph, indicating that as output quantity *q* increases, the customers' willingness to pay for each unit decreases. The *AR* curve thus equates to the inverse demand curve faced by the firm. Note that this does not automatically equate to the demand curve in a market as there may be other competing suppliers in the market. Only in the special cases of a monopoly or perfect competition, does the *AR* curve equate to the demand curve in the market. This implies that in the standard analysis in which a downward sloping *AR* curve is shown, the firm is assumed to have some influence on the market price and, therefore, must have a degree of market power.

Analogous to marginal cost, marginal revenue *MR* is the additional, extra, revenue obtained by increasing the output quantity q by one unit. In the example, the *MR* associated with the 500th car is the extra income generated through the sale of the 500th car after selling 499 cars. If the total revenue function $TR(q)$ is continuous and differentiable, *MR* is the first derivative of this function with respect to q:

$$MR(q) = \frac{dTR}{dq} = TR'(q) \tag{3.22}$$

If shown as linear relationships, the *MR* function is often expressed graphically as a downward sloping line with a steeper gradient than the *AR* line as shown in Figure 3.8. Note that since *MR* is shown as a downward sloping line, the associated total revenue function $TR(q)$ will be nonlinear.

Determination of Quantity, Price, and Profit

With this basic framework at hand, it is possible to use static analysis to show how a rational, profit-maximizing firm will determine quantity q to maximize its profit π. Note that in the traditional theory of the firm, it is assumed that the business will always maximize profit. In the framework developed in this chapter, this is done by applying what is known as the *first-order criterion of profit maximization*. This condition is applied by setting a quantity q such that:

$$MR = MC \tag{3.23}$$

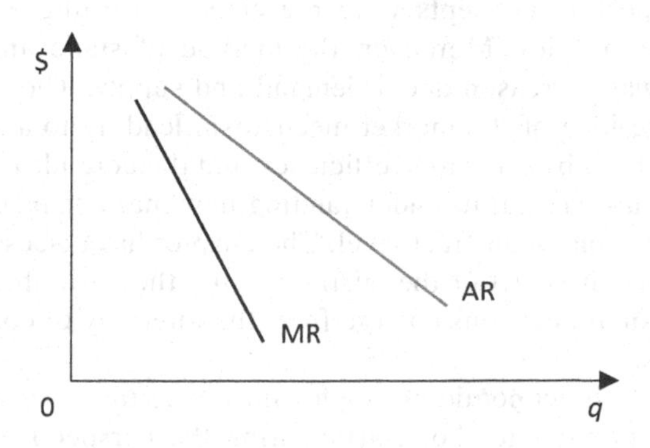

FIGURE 3.8 The AR and MR curves

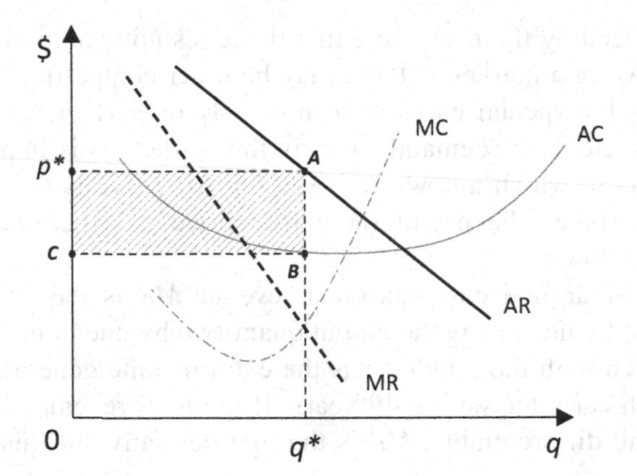

FIGURE 3.9 Determining quantity, price, and profit through static analysis

The graphical solution presented in Figure 3.9 combines the *AC*, *MC*, *AR*, and *MR* functions. The profit-maximizing level of output q^* is, subject to (3.23), where the *MC* and *MR* functions intersect. Thus, in this situation, the producer will maximize π by making q^* products, incurring a total cost of $q^* \times c$ and receiving a total revenue of $q^* \times p^*$. Obviously, the firm will make a loss if $AR\ (q) < AC\ (q)$. The profit received by the manufacturer corresponds to the shaded rectangle $ABcp^*$. If the *AC*, *MC*, *AR*, and *MR* functions can be established as well-behaved functions of q, it is possible to numerically obtain q^* and π.

A matter of significance of this analysis is that, subject to a downward sloping *AR* function, the rational manager will not choose the socially optimal level of output $q_{MES,}$ which would minimize *AC*. This result is important since it demonstrates that profit-driven management decisions may not lead to behavior that is desirable for customers or society in general.

CONCLUSION

This chapter has introduced the perspective of mainstream economics on business. Several important concepts were presented, including opportunity cost, markets, and externalities. Moreover, the method of static analysis has been introduced as a way to reason about demand and supply. The chapter has also presented the core idea of the market mechanism leading to a state of equilibrium which is marked by allocative efficiency and the core idea of economies of scale. Both ideas are central to understanding how many managers think about the pursuit of profit on an abstract level. The chapter has discussed this form of decision-making at the level of the business using the basic theory of the firm, showing how rational decisions emerge from the interplay of cost functions and revenue functions.

Most textbooks on economic principles introduce the discussed concepts in greater detail. A popular textbook articulating the perspective of mainstream economics is *Principles of Economics* by N. Gregory Mankiw (2020). An older yet very accessible treatment of microeconomics, including the theory of the firm, is provided by *Principles of Microeconomics* by Peter Else and Peter Curwen (1990). For additional historical context, the book *An Inquiry into the Nature and Causes of the Wealth of Nations* by Adam Smith (1776) is highly interesting. More recently, mainstream economics has come under attack for being unrealistic in its approach. A comprehensive but difficult to read treatment of this kind is provided by *The World of Consumption: The Material and Cultural Revisited* by Ben Fine (2002). While such criticism is valid, the point of this chapter is to explain how managers *think* about rational decision-making – which is largely determined by economics. For technologically minded readers, introductions to methods for the minimization of average cost, such as *Product Design for Manufacture and Assembly* by Geoffrey Boothroyd and colleagues (2002) or *Target Costing and Value Engineering* by Robin Cooper and Regine Slagmulder (1997) may be of interest.

REVIEW QUESTIONS

1. Economists assume that more of a good is always better than less.
   (Question type: True/false)
   Is this statement true or false?
   ○ True ○ False

2. Complete the below statement on the market mechanism.
   (Question type: Fill in the blanks)
   "The intersection between demand and supply defines the point at which the __________ is in __________. This point is of special interest in economics because it defines the quantity and the price at which the buyers' and the sellers' __________ are in __________. At this point, no additional __________ would be able to buy the product at a price a __________ would agree to".

3. Which of the following events can be associated with shifts in a supply curve?
   (Question type: Multiple response)
   ☐ A change in the price of a related good
   ☐ Improvements in business processes
   ☐ A change in quantity
   ☐ Changes in available manufacturing technology
   ☐ An increase in the price elasticity of demand

4. You have been tasked with the analysis of two different types of bread products. One is an own-brand value product and the other is a high-priced luxury item.

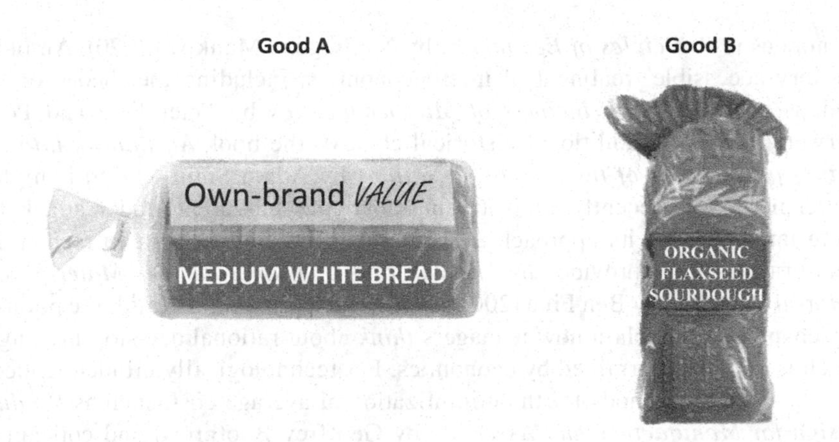

FIGURE 3.10 Illustration of two bread products

Which of the following statements about these two goods would you expect to be true or false according to standard economic theory?

(Question type: Dichotomous)

True	False	
O	O	From the data provided, it is not possible to make statements on absolute levels of price elasticity
O	O	Relative to good A, good B will be perceived as a basic necessity
O	O	Price elasticity of demand for good B will be zero
O	O	Relative to good B, good A will be perceived as a luxury product
O	O	The price of good A will affect the quantity demanded of good B
O	O	Price elasticity of demand for both products will be greater than zero because they are normal goods
O	O	Demand for good A will be perfectly elastic
O	O	Demand for good A will be more elastic than demand for good B

5. A total quantity of 8 million cans of mushroom soup will be demanded if it is priced at $0.65 per can. The quantity demanded will not be reduced if the price is increased to $0.69 per can. Using these data, identify which of the following statements are correct:

   (Question type: Multiple response)

   ☐ The own-price elasticity of demand is 0
   ☐ The own-price elasticity of demand is –1
   ☐ Demand for this good is perfectly elastic
   ☐ It is not possible to state anything about own-price elasticity
   ☐ Demand for this good is perfectly inelastic
   ☐ The own-price elasticity of demand is –100,000

6. You are investigating the market for a new type of premium cheese. You are given an estimated linear demand function:
   - $D(p) = -450p + 11{,}500$

   On the basis of the existing conventional manufacturing process, the following supply curve is estimated:
   - $S(p) = 400p - 1{,}000$

   You are tasked with the evaluation of a technological improvement in the cheese-making process. This change will allow the manufacturers to increase the supply of cheese by a factor of three for any increase in the price level. Note that the intercept of the supply curve is unaffected by this change.

   (Question type: Calculation)

   Calculate the equilibrium price level after the technological change.

7. You are considering a new type of satellite navigation device. You have analyzed the market and identified the following information:
   - The supply and demand functions in the market are linear
   - If price were to be zero, you would expect to shift 5,500 units
   - For each additional $ in price, you would expect to sell 4 units less
   - You would need to supply at least 9,000 units before you would find it worthwhile to enter the market at any price level
   - For each additional $ in price, you would be willing to supply 15 more units

   (Question type: Calculation)

   Use this information to calculate the equilibrium quantity in the market.

8. If a total revenue function is linear and runs through the origin, the marginal revenue is constant.

   (Question type: True/false)

   Is this statement true or false?

   O True O False

9. If a total cost function is linear and there is a non-zero fixed cost, the average cost is constant.

   (Question type: True/false)

   Is this statement true or false?

   O True O False

10. The following figure shows the average cost (AC) curve, the marginal cost (MC) curve, the average revenue (AR) curve, and the marginal revenue (MR) curve. Draw the area that corresponds to the total revenue made under profit maximization.

    (Question type: Area)

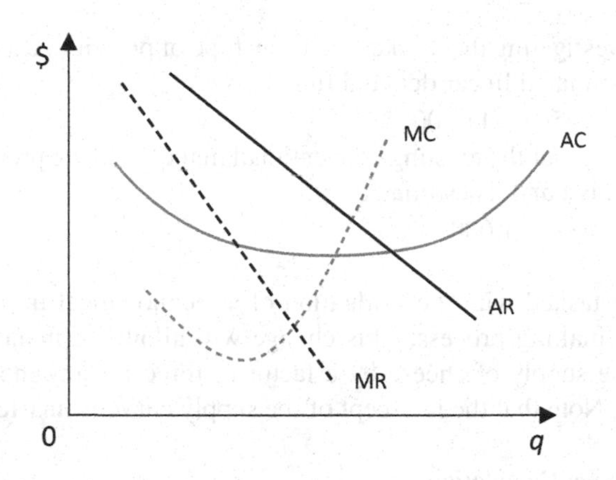

FIGURE 3.11 Draw the appropriate area in the figure

REFERENCES AND FURTHER READING

Boothroyd, G., Dewhurst, P. and Knight, W., 2002. *Product design for manufacture and assembly*. 2nd ed. New York: Marcel Dekker.

Cooper, R. and Slagmulder, R., 1997. *Target costing and value engineering*. Portland: Productivity Press.

Curwen, P. and Else, P., 1990. *Principles of microeconomics*. London: Unwin Hyman.

Fine, B., 2002. *The world of consumption: The material and cultural revisited*. 2nd ed. London: Routledge.

Mankiw, N.G., 2020. *Principles of economics*. Boston: Cengage Learning.

Smith, A., 1776. *An inquiry into the nature and causes of the wealth of nations: Volume One*. London: Strahan and Cadell.

4 The Economics of Digitalization and Automation

OBJECTIVES AND LEARNING OUTCOMES

The objective of this chapter is to show the deep impact of the most important technological development of our time, digitalization, on the business context and management. After defining information technology and showing the dramatic decline of the cost of computing over time, which is reflected in what is known as "Moore's law", this chapter introduces the economics of information technology and outlines important models and applications. It further briefly outlines the societal challenges posed by these developments. Specific learning outcomes include:

- understanding what digitalization is and how it relates to information technology and computers;
- understanding the sweeping importance of Moore's law for business and wider society;
- ability to characterize the impact of digitalization using the framework of economics, resulting in a theory known as the economics of information goods;
- knowledge of the relevance of flexible manufacturing systems (FMS) and fundamental aspects of artificial intelligence (AI);
- grasp of the societal challenge posed by automation, especially in terms of its impact on labor markets;
- understanding relevant terms, synonyms, and accepted acronyms; and
- appreciation of important thinkers and innovators in this field.

BACKGROUND AND DEFINITIONS

Joseph Schumpeter (1883–1950)

FAMOUS THINKER

According to the famous early 20th-century economist Joseph Schumpeter, technological innovation, which is often closely associated with engineering activities, is the main cause for change in economies over time. Schumpeter's grand insight was that technological change is inseparable from economic processes and is, therefore, an important part of any kind of economic development. More specifically, Schumpeter recognized that technological innovation causes economies to change from within, resulting in an enduring pattern in which existing structures, such as firms relying on established technologies, are periodically overturned by new

DOI: 10.1201/9781003222903-5

structures based on emerging technologies. Schumpeter famously labeled this ongoing process *creative destruction.*

As characterized in Chapter 1, the industrial revolution beginning in the late 18th century was an era marked by significant creative destruction, originating from multiple developments in parallel in engineering, chemistry, and metallurgy. In combination, these developments produced a sudden and sharp change in the trajectory of human progress reshaping the face of commercial activity. An evolution of management thinking ultimately followed from this.

In the industrial revolution, the invention of the steam engine as a source of energy was a decisive moment. It allowed the construction of machinery and equipment to overcome the limitations of human and animal power. In turn, steam power enabled new forms of transportation and then mass-production. Closely linked to this new abundance of energy was the concept of *automation*, which can be defined generally as follows:

Automation is any kind of technology by which a process, procedure, task, or activity is performed with minimal human assistance or intervention.

Since processes and procedures are not limited to doing work but can also be used to control work, an important part of automation is *automatic control.* For the purposes of this book, automatic control denotes the use of control systems for the operation of equipment with minimal or reduced human intervention. Up to the mid-20th century, the industrial uses of automation and automated control concentrated on machinery, factory equipment, boilers, pressure vessels, heat treatment processes as well as telephone networks, naval control systems, and transportation systems.

This chapter is about the next, and still ongoing, stage in the spread of automation. This phase is marked by the emergence of *information technology*, often referred to collectively as *computers*, which will be characterized in detail in this chapter. As this chapter will also show, the twin processes of *digitization* and *digitalization* are fundamental to a range of broad and sweeping effects in the economy and in wider society. Digitization and digitalization can be defined as follows:

Digitization is the process of converting relevant information and procedures into digital form so they can be acted upon by computers, including their transmission through computer networks. Digitalization is the process of using and leveraging digitization for various purposes and activities.

To characterize the consequences of digitalization for the world of business and commerce, which are very complex and far-reaching, it is necessary to first introduce what digital information technology actually is. The most straightforward way of building such an understanding is to present a brief account of the most important discoveries underpinning modern computers.

WHAT EXACTLY IS INFORMATION TECHNOLOGY?

Information technology generally refers to the use of digital computers to store, retrieve, transmit, and manipulate information in electronic form by a limited group of users. The term *information technology system (IT system)* denotes any computer-based information handling or communications system designed to perform these tasks. This label includes all hardware, software, and peripheral equipment, including devices that bring information into the system (*input devices*), devices that make information available to users (*output devices*), and information storage devices.

John von Neumann (1903–1957)

Prior to the arrival of information technology and computers in the modern sense, there has been a long history of technologies and practices aimed at storing and manipulating information and data by non-electronic means. For example, the Antikythera mechanism, which is believed to be a mechanical computer designed for astronomical calculations, was constructed in ancient Greece in the period 205–87 BC. A much more recent example for a programmable digital, but not yet electronic, computer is the Zuse Z3, which was constructed in Germany in 1941.

The Zuse Z3, and other designs of its time, conform to what is known as the *von Neumann architecture*, proposed in 1945 by mathematician, physicist, computer scientist, engineer, and polymath John von Neumann. According to the von Neumann architecture, a computer contains a number of distinct devices: a central processing unit (CPU), memory, input devices, and output devices. The key difference between this architecture and competing computer architectures proposed at the time is that the von Neumann architecture makes use of the same memory for both data and program instructions. This revolutionary idea allowed the function of a computer to be changed as quickly as the data in its memory. To this day, the von Neumann architecture is accepted as the standard way in which computers are structured. Figure 4.1 summarizes the von Neumann architecture.

Up to the 1950s, computers did not yet use *transistors* to perform logical operations. This made early computers complex, large, difficult to operate, and very expensive. As the basic building blocks of today's computer hardware, transistors are small solid-state semiconductor devices employing quantum effects to act as logical switches. Their invention constituted a true leap forward in technology. Since their operating principle cannot be understood without a grasp of quantum theory, all semiconductor-based computers are in fact quantum devices.

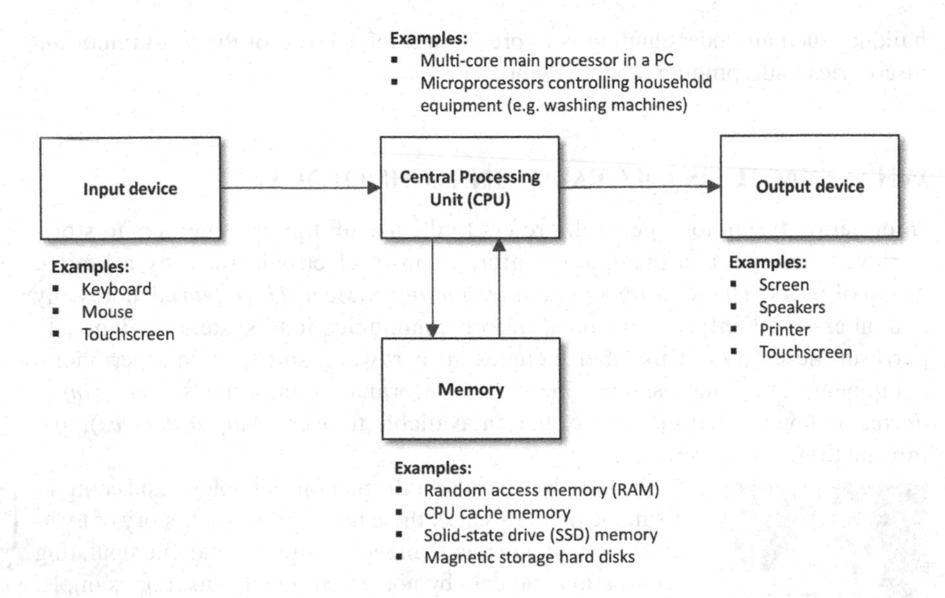

FIGURE 4.1 Simplified von Neumann architecture

Building on prior work in transistor technology by William Shockley, John Bardeen, and Walter Brattain in the early 1950s, the breakthrough invention in electronics occurred in 1959 at Bell Labs in the United States when the *metal–oxide–semiconductor field-effect transistor* (*MOSFET*) was invented by the engineers Mohamed Atalla and Dawon Kahng. Since their invention, MOSFETs have been manufactured in huge numbers. They are the most widespread electronic device ever manufactured and are considered to be the single most significant invention in electronics. Transistor technology forms the precursor to many other inventions in the late 20th and early 21st centuries. Virtually every electronic device available today is based on MOSFETs, including the ubiquitous smartphone.

William Shockley (1910–1989)

An additional invention that was required to make the sweeping success of information technology possible was a way to manufacture complex arrangements of transistors and circuitry cheaply and reliably. This was achieved through the *integrated circuit (IC)*. An IC is a circuit in which all or some of the circuit elements are inseparably associated and electrically interconnected. The defining characteristic of ICs is that they are considered to be indivisible for the purposes of construction or sale. Invented at the company Fairchild Semiconductor in the United States in early 1959, ICs were a remarkable achievement in

John Bardeen (1908–1991)

manufacturing technology. The key idea enabling IC manufacture is to deposit circuitry (including MOSFETs) on a flat chip of silicon using an optical process known as *photolithography.* This was unlike all prior manufacturing technologies that had always relied on the addition of components in discrete, manual, or automatic operations. Using an optical process, manufacturers of ICs can add extra logical elements to their integrated circuits while incurring only a tiny, perhaps negligible, additional effort.

Walter Brattain (1902–1987)

The significance of ICs is perhaps best understood in the terms of economics. Recall the concept of marginal cost from Chapter 3. The unprecedented effect of the invention of ICs was that it was now possible to make transistors at almost zero marginal cost. Based on this extremely low marginal cost, the invention of ICs unlocked the manufacture of extremely large quantities of cheap, reliable, fast, and small electronic devices. This led to affordable yet powerful computer hardware becoming a reality in a wide range of applications, contributing to the spread of digital technology into every aspect of our everyday lives. This process is still far from complete.

Mohamed Atalla (1924–2009)

The innovations introduced so far (von Neumann architecture, MOSFETs, and ICs) concern only the hardware side of things. A crucial missing ingredient is an applicable theory of the software to run on this hardware, without which the new technologies would be meaningless. The intellectual basis for computer software was established by the mathematician Claude Shannon around 1948. Shannon recognized that, at the most basic level, digital computers are systems capable of processing simple logical instructions based on the manipulation of zeroes and ones as symbols, represented by the absence or presence of a voltage. As shown by Shannon, adopting a symbolic representation of information as *binary digits*, or *bits*, allows the error-free processing of symbols in imperfect systems, up to a threshold known as the *Shannon limit.* This enables, in principle, the construction of reliable computers from not-always-reliable parts, such as the transistor. This idea also gave rise to the field of information theory and information processing in the 1950s and 1960s. Over time, combining simple logical instructions allowed circuits to solve logical equations that form the basis for more complex operations. Gradually, Shannon's ideas led to the software programs and applications we know today.

Dawon Kahng (1931–1992)

Table 4.1 summarizes the four key inventions and discoveries forming the basis for information technology in its current state as discussed in this chapter. Computers incorporate many other inventions and technologies, such as digital storage and display technologies, so this treatment is not exhaustive. The table also provides some additional references to useful introductory literature discussing the implications of the outlined innovations.

Claude Shannon (1916–2001)

MOORE'S LAW

So why is all of this important for management? After all, physical, non-computational goods such as food, housing, and clothing continue to be essential for the lives of all people, and most would like to have such material goods in abundance. The reason for the importance is that computers are becoming essential too, with more and more tasks being carried out digitally, changing and often enhancing the physical world. One of the first people to truly grasp the magnitude of the coming changes was a computer scientist working at Fairchild Semiconductor named Gordon Moore. In 1965, Moore wrote that "integrated circuits will lead to such wonders as home computers – or at least terminals connected to a central computer – automatic controls for

TABLE 4.1
Four Key Inventions Underpinning Information Technology

Invention	Summary	Inventor(s)	Useful references
Von Neumann architecture	The dominant computer architecture describing the main components of computers and how they work together	John von Neumann	Gershenfeld (2008)
MOSFET	Metal–oxide–semiconductor field-effect transistor, the main electronic component technology used in computers, based on quantum mechanics	William Shockley, John Bardeen, Walter Brattain, Mohamed Atalla, Dawon Khang	Gilder (1990), Clegg (2014)
Integrated circuit (IC)	A flat ("planar") electronic circuit in which the elements are inseparably associated and interconnected, allowing miniaturization and low cost	A group of inventors based at Fairchild Semiconductor	Gilder (1990)
Information theory	Computers as tools to manipulate information as binary digits, or "bits", allowing error-free processing in imperfect systems	Claude Shannon	Gershenfeld (2008)

Gordon Moore (born 1929)

automobiles, and personal portable communications equipment".[*] These statements are remarkably prescient. However, in the same article, Moore made another, even more influential, forecast concerning the possibilities of adding more and more logical elements, such as transistors, to ICs.

"The complexity for minimum component costs has increased at a rate of roughly a factor of two per year [...]. Certainly over the short term this rate can be expected to continue, if not to increase. Over the longer term, the rate of increase is a bit more uncertain, although there is no reason to believe it will not remain nearly constant for at least ten years."[†]

In this statement of what has since then become known as Moore's law, "complexity for minimum component costs" refers to the amount of computing power that can be bought for a constant amount of money. Moore's insight was that over the brief period that ICs had existed up to 1965, the power of computers relative to their cost had approximately doubled each year. More importantly, Moore predicted that this exponential growth would continue in the future. Astonishingly, Moore's law still remained approximately intact in the 2010s. What is more, up to the early 2010s, Moore's law has held not only for ICs but for other aspects of information technologies as well, such as processing speed, transmission rates of digital data networks, and the capacity of data storage hardware.

Figure 4.2 illustrates the exponential growth indicated by Moore's law for a range of technologies; the reader should note the logarithmic *y*-axis. To correctly understand the significance of this exponential growth in cost performance, it is necessary to understand that a general shift has taken place in the early 2000s. Many devices and sensors, such as microphones, cameras, and accelerometers, have been converted from analog or mechanical devices into digital devices, essentially becoming parts of ICs themselves. Through this process of miniaturization, they too have become subject to the exponential improvement pattern reflected in Moore's law. This means that many devices have become drastically faster, cheaper, smaller, and lighter, often permitting entirely new functionalities. This phenomenon has increased the power and usefulness of individual computers while also increasing the power and reach of digital information. In turn, this has led to mobile devices such as smartphones,

[*] Moore (1965, p.114).
[†] Moore (1965, p.115).

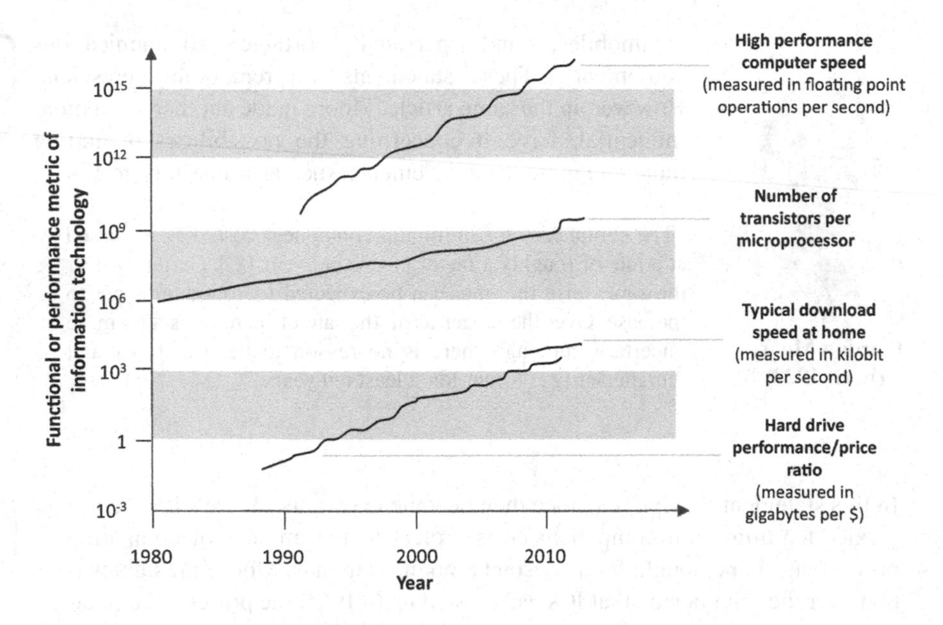

FIGURE 4.2 Exponential growth in computing performance, adapted from Brynjolfsson and McAfee (2014)

which have greatly promoted the development of groundbreaking innovations such as social media and mobile apps.

THE ECONOMICS OF INFORMATION GOODS

In Chapter 1, it was argued that developments in management thinking often lag behind technological change. The managerial changes resulting from digital technologies are no exception to this. Fundamental insights into the economics and business effects of computers emerged in the 1990s, decades after the arrival of the electronic computer. These new theories centered on *information goods* and the economics of digital networks used to supply and transmit information goods. Information goods can be defined as follows:

Information goods are those made of binary digits, referred to as "bits", rather than of physical materials. This means that anything that can be encoded in a sequence of bits can be an information good.

Following this definition, cartoons, books, databases, films, newspapers and magazines, music, stock prices, sports scores, and websites can all be considered information goods. As with any other good, the usefulness of information goods can be varied and depends on the user. Some information goods will be used for entertainment while others will be used in business. Nevertheless, as a generalization, people are willing to pay for information goods. An immediate difference between physical goods and information goods, however, lies in the cost structure of their production. While being very cheap, almost free, to copy, the initial production of information goods can be very expensive. More specifically, in the terms of economic cost theory, the production of an information good can involve a high fixed cost. Expressed as a marginal cost, the first unit of an information good is likely to incur a high marginal cost whereas all subsequent units will incur a low, perhaps even negligible, marginal cost. This is why, for any existing information good, marginal cost is said to be low. Considering Moore's law, it is important to note that the marginal costs of information goods are still decreasing. In fact, the reproduction of many information goods is now so cheap that they are routinely made available free of charge. Consider, for example, free online newspapers or video clips available without payment on video-sharing websites, such as YouTube or Vimeo.

A further important feature of information goods is that they can be reproduced perfectly. This means that once a digital original has been created, there is no loss in quality resulting from the act of copying whatsoever. This characteristic does not apply to most other kinds of goods that can be copied only with a loss in quality. Consider, for example, that creating replicas of statues or paintings will almost inevitably lead to a loss in quality, depending on the skill of the artisan creating the copy.

Since information goods are perfectly and almost freely reproducible, it would not make business sense to price them according to average cost, for example, by using a mark-up pricing strategy in which the price is set by adding an increment to the cost of generating the information good. Instead, information goods should be priced according to the customer's valuation of the good, which will depend on its usefulness to the individual customer.* This implies that when selling information goods, the seller should adopt a strategy that aims to offer the goods to different customers at different prices. This approach is known as *differential pricing* or *price discrimination*.

The effects of differential pricing on the total value obtained by the producers and the customers can be investigated through static analysis. Recall from Chapter 3 that processes in a market can be modeled by drawing (inverse) functions of demand and supply. Demand functions are negative, downward-sloping relationships expressing the negative relationship between a price and the quantity customers are willing and able to buy of a given product at that price. Supply functions describe a positive relationship between a price and the quantity producers are willing and able to sell at that price. Figure 4.3 expresses this relationship by showing a linear demand curve D and a linear supply curve S, intersecting at equilibrium price p^* and quantity q^*.

In competitive markets with well-informed buyers, it is not normally possible to charge different prices for the same product or service. This situation is known as the *law of one price*, as discussed in Chapter 3. If the law of one price holds, the price paid by every consumer in market equilibrium is p^* and the producers collectively obtain a revenue of $p^* \times q^*$. This means that a number of customers are able to buy the product or service below the price they are willing and able to pay, which is known as their *reservation price*. This leads to extra value being obtained by some

* McAfee and Brynjolfsson (2017).

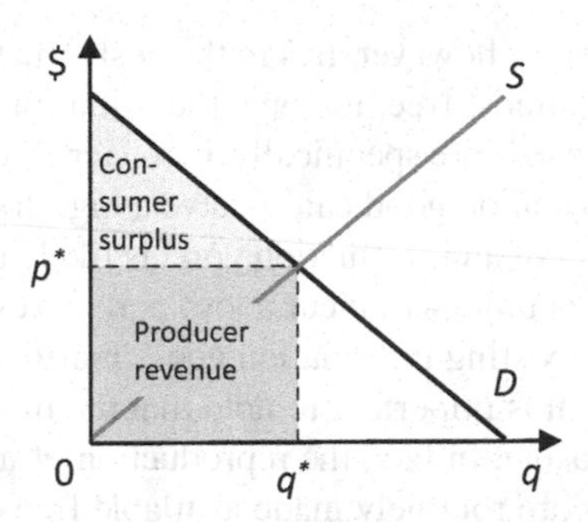

FIGURE 4.3 Consumer surplus and producer revenue in static analysis

customers as defined by the difference between their reservation price and the price they have paid, p^*. The sum of this extra value obtained by all customers is known as the *consumer surplus*. Figure 4.3 shows the consumer surplus received by the buyers as the light gray triangle under the demand curve.

According to the economics of information goods, producers of information goods will want to charge prices based on the customers' individual valuation of the product. Obviously, this goes against the law of one price. Figure 4.4 depicts two possible outcomes of differential pricing, one which is completely successful with the producer managing to price each unit at each customer's reservation price (Figure 4.4a) and one which is partially successful, with the producer establishing a two-tiered pricing scheme (Figure 4.4b).

In Figure 4.4a, each unit is priced at each customer's reservation price so the customers will still find it worthwhile to buy the product at this price. However, the producer will capture all the benefits available from using the good in the form of producer revenue, given by the area $p^* \times q^* + 0.5 \times \left(\dot{p} - p^*\right) \times q^*$. Since the producer captures the entire value in the market, the consumer surplus is zero. Figure 4.4b depicts the situation in which the producer has introduced a two-tiered pricing scheme with the prices p^* and $p^* + m$ and successfully allocated the products according to each customer's reservation price such that no customer who is willing and able to pay misses out. In this case, the producer revenue is given by $p^* \times q^* + m \times q'$, where q' is the quantity sold at the higher price level, and consumer surplus is given

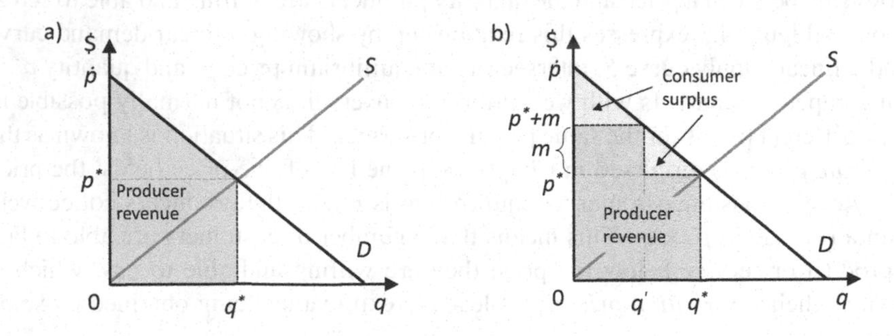

FIGURE 4.4 Fully successful differential pricing (a) and two-tiered pricing (b)

by the remaining area $0.5\times\left(\dot{p}-p^{*}\right)\times q^{*}-m\times q'$. As can be seen from this simple graphical analysis, the producers benefit from pricing strategies aiming to charge different prices to different buyers.

A further idea emphasized in the economics of information goods is how different goods are related to each other. In economics, there are three basic relationships between goods: *complementarity, substitutability,* and *unrelatedness*. Among these, complementary goods play a decisive role. They can be defined as follows:

A complementary good is one whose appeal increases if the popularity of another related good also increases. In other words, the demand for a complement increases if the price of another good decreases.

Complementarity forms an extension of the concept of own-price elasticity of demand presented in Chapter 3. If good *A* is a complement to good *B*, an increase in the price of *A* will result in a movement along the demand curve of *A* but it will also shift the demand curve of good *B* inwards. This means that less of both goods is demanded. Often-cited examples for pairs of complements are ground beef and hamburger buns, or left and right shoes. Complementarity is a very common relationship between pairs of goods, especially if one of the goods is an information good.

In economics, this relationship is formally captured using the concept of *cross price elasticity of demand* (*XED*). This type of elasticity measures the responsiveness of demand for a good *A* following a change in the price of good *B*. Recall from Chapter 3 that the gradient of the demand function is related to the level of own-price elasticity of demand. *XED* establishes this relationship across two goods. It is defined as the percentage change of the quantity q_A demanded of good *A* over the percentage change of price p_B of good *B*:

$$XED=\frac{\Delta q_A\,/\,q_A}{\Delta p_B\,/\,p_B} \tag{4.1}$$

To estimate *XED*, it is necessary to identify a base level of quantity q_{A0} and price p_{B0} and an altered level of quantity q_{A1} and price p_{B1}. This means *XED* is not constant along linear demand curves and it measures a percentage change from one state to another. Estimating *XED* is straightforward if q_{A0}, p_{B0}, q_{A1}, and p_{B1} are given. The formula is:

$$XED=\frac{(q_{A1}-q_{A0})\,/\,q_{A0}}{(p_{B1}-p_{B0})\,/\,p_{B0}} \tag{4.2}$$

For complementary goods, such as hamburgers and buns, an increase in the price of one good (hamburgers) will lead to a decrease in demand for the complement by shifting its demand curve (buns) inwards. This means the *XED* for complements will be negative. The stronger this relationship, the more negative the value of *XED*.

For *substitute goods*, or *substitutes*, the situation is vice versa. A substitute can be defined as follows:

IMPORTANT DEFINITION

A substitute is a good whose appeal increases if the popularity of another related good decreases. In other words, the demand for a substitute increases if the price of another good increases.

If good *A* is a substitute to good *B*, an increase in the price of *A* will result in a movement along the demand curve of *A* but it will also shift the demand curve of good *B* outwards. In other words, the less of good *A* is demanded, the more of good *B* is demanded. Accepted examples of substitutes are potatoes from different farms, margarine and butter, or red and blue pencils. For substitutes, *XED* will be positive, and the stronger this relationship, the higher the value of *XED*. The third relationship cited, unrelatedness, simply indicates the absence of either complementarity or substitutability.

Because the cost of reproducing information goods with digital technology is extremely low, with a marginal cost approaching zero, it is quite common for businesses to provide information goods free of charge, at least in their most basic version. This creates very interesting possibilities if goods are complements. Consider, for example, a mobile software app and a smartphone, which are sold in two separate markets with the two demand functions D_{app} and $D_{smartphone}$ for the app and the smartphone, respectively. Using static analysis, Figure 4.5 illustrates the situation arising when the software producer makes the app available for free ($p_{free} = 0$) instead of charging the equilibrium market price p^*_{app}.

Since the app and the smartphone are complements, the demand curve in the market for smartphone shifts outward from $D_{smartphone}$ to $D_{smartphone\ new}$. This forces the market for the smartphone from its equilibrium with p^* and q^* into a new equilibrium with p^*_{new} and q^*_{new}, thereby leading to the sale of a greater quantity of smartphones. It might be worthwhile for the producer of the smartphone and the software producer to find an agreement in which the app is offered free of charge to the end user, perhaps by paying the software producer royalties or license fees. Because the software app is an information good, unlike the smartphone, the software producer can

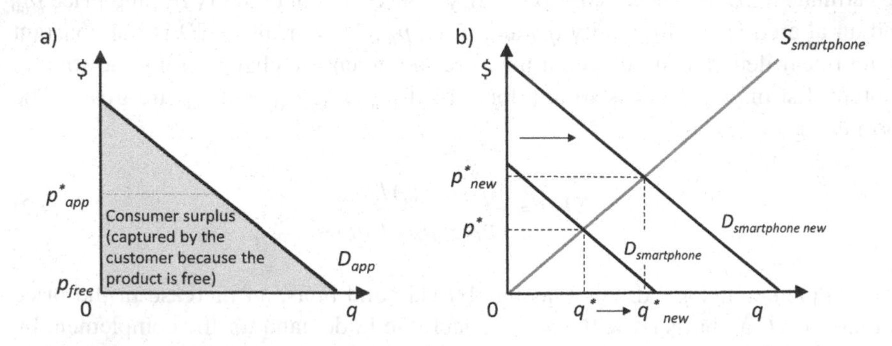

FIGURE 4.5 Complementarity relationship between a mobile app (a) and a smartphone (b)

expand the quantity supplied at extremely low additional costs, meaning that only very small additional costs are incurred as the app is supplied with the increased number of smartphones. Such an ability to exploit the very low costs of reproduction of information goods is commonly framed in terms of economies of scale, as introduced in Chapter 3. For this reason, businesses providing information goods are often said to *scale* extremely well, which is a very attractive characteristic to managers and investors.

The significance of this example is to show that the economics of information goods, with the ability of reproduction at almost zero marginal cost, can have a strong influence on the markets of some physical goods. This general pattern is now very common and has led to significant changes in how many industries operate. Most importantly, this phenomenon has contributed to the rise of platform business models, as discussed in detail in Chapter 7.

INFORMATION ECONOMICS AT WORK: FLEXIBLE MANUFACTURING SYSTEMS AND ARTIFICIAL INTELLIGENCE

Following the discussion of the general nature of information technology and the introduction of the economics of information goods, it is instructive to briefly survey two specific and highly influential technologies that are based on information technology. The first technology is *flexible manufacturing systems (FMS)*, which emerged as the first implementation of information technology in manufacturing in the second half of the 20th century. The second technology, which is currently receiving enormous attention in industry and the investment community, is *artificial intelligence*. This section briefly summarizes the main characteristics of both technologies, as seen from the perspective of the economics associated with information technology.

Flexible Manufacturing Systems

As an important manufacturing technology to emerge from the early stages of computerization in the 1960s, flexible manufacturing systems have some degree of flexibility enabled by digital control. This allows flexible manufacturing systems, such as computer numerically controlled (CNC) machining tools, to react or adapt when planned or unplanned changes occur. In line with the goals of automation, production via flexible manufacturing systems is largely automated which carries the benefit of reducing overall labor costs. As discussed in Chapter 9, managers are interested in flexible manufacturing systems because they offer several additional benefits that enable specific operations systems. Traditionally, flexible manufacturing systems are seen to incorporate three main components:

1. machines or machine tools using digital instructions to control and execute an automated manufacturing process;
2. material handling systems that are capable of organizing a changing and optimized flow of workpieces, connecting and serving the machines or machine tools as required; and

3. a computer system that processes the required information and controls the movement of materials and the processes carried out by the machines.

The main advantage of flexible manufacturing systems over other manufacturing approaches, such as those found predominantly in mass-production, as discussed in Chapter 1, is their ability to respond to fluctuations in the availability of resources such as machine capacity, and to adapt to changes in the type and quantity of the products being demanded. In the light of the economics of information goods, this implies that some machines, such as CNC machining systems, are themselves complements to associated information goods, such as digital designs and machine instructions generated by design software packages. In consequence, in many applications flexible manufacturing has improved manufacturing efficiency and consequently lowered production costs. It can also form a key component of make-to-order strategies that allow customers to customize products, which will be discussed in Chapter 9.

In the field of industrial economics, flexible manufacturing systems have found their expression in an extension to the idea of economies of scale. Recall from Chapter 3 that economies of scale reduce the average cost of a process as the level of output increases. A special kind of economics of scale, *economies of scope*, occurs if the technological ability to handle a variety of products allows the production of multiple products or product variants alongside each other at a lower cost than if produced separately. Economies of scope can thus be defined as follows:

IMPORTANT DEFINITION

Economies of scope are cost reductions that arise when the cost of joint production of multiple products or product variants is lower than the cost of producing each output separately. Economies of scope form a special kind of economies of scale.

As outlined in this section, the ability to handle variety in the activities of a business, as a necessary technological prerequisite for economies of scope, is generally seen to benefit greatly from information technology.

Artificial Intelligence

Computers and information technology are extremely well suited for the execution of precise, step-by-step instructions for accomplishing tasks, also known as *algorithms*. Since the emergence of information technology, however, the visionaries in the field, such as von Neumann, have dreamed of improving computers to go beyond following pre-defined orders. Consequently, it has been a major goal of computer science to create information systems, consisting of hardware and software, which can accomplish true acts of reasoning and intelligence that are similar to or surpass those performed by humans. Considering this ambition, the technology of *artificial intelligence* (*AI*) can be defined as follows:

IMPORTANT DEFINITION

Artificial Intelligence is the science and engineering of making intelligent machines, capable of accomplishing true acts of reasoning and intelligence.

While defining artificial intelligence in this way is common, it is also very problematic. This is due to two reasons. First, the definition is *recursive* because it uses the term "intelligence" to define artificial intelligence, thereby avoiding a statement of what intelligence actually is. Second, the definition implies the possibility that intelligence can somehow be made. While artificial intelligence systems are often associated with rational patterns of behavior that humans are (occasionally!) capable of, it is in reality still unclear if anything resembling real intelligence can be designed or artificially made.

It is therefore fundamentally inappropriate to characterize machines as doing humanlike things such as discerning, learning, perceiving, understanding, and so on. Computers, at any currently conceivable level of complexity, remain lifeless objects. It is certain that they do not learn, decide, recognize, or prefer anything, at least not in the way humans do. The ascription of human characteristics to inanimate objects, and also animals and plants, is referred to as *anthropomorphization*. While anthropomorphization may be helpful in explaining the purpose of artificial intelligence systems, it is important to remember that it is not a description of how such systems operate. This is captured in the joking admonition that computers should not be anthropomorphized – they hate it.*

The field of artificial intelligence can be divided into two broad branches. Briefly discussing the difference between these two approaches is instructive. The first branch follows a rule-based, *symbolic* understanding of artificial intelligence. In this view, intelligence, even of the artificial kind, is expressed in words, numbers, and other symbols that an intelligent entity can understand and manipulate. Despite significant initial promise, it has so far not been possible to construct capable symbolic artificial intelligence systems. The most advanced symbolic systems reach levels far below human performance and reliability in tasks such as speech recognition and image classification. It has been theorized that the symbolic approach has (so far) failed for two broad reasons. The first reason is that the world is complex and that a huge number of rules would need to be known by the system to produce useful results. The second reason lies in the fact that the processes of human reasoning itself are not understood. As humans, we understand things tacitly as subjects in the world but cannot fully explain why and how. This objection provides some credence to the belief that it will never be possible to build digital systems that perform true feats of intelligent reasoning, in turn also dooming the presented definition of artificial intelligence.

The second branch of artificial intelligence employs statistical methods to recognize patterns, using experience in the form of digital data, repetition, and feedback loops. Despite not involving learning in the non-metaphorical sense, this approach has come to be known as *machine learning*. Using techniques such as neural networks, it is possible to construct systems that improve over time through ever more data and more powerful computers. Coupled with other developments such as *cloud computing*, which allows additional computing power to be accessed flexibly through the digital networks, and the recent explosion of the amount of digital data available

* See McAfee and Brynjolfsson (2017).

known as the *big data* phenomenon, the statistical approach has begun yielding impressive results in the 2010s. It is expected that these systems will become applied in many different areas of business and will have a very significant effect on commerce and culture in the immediate future.

AUTOMATION AS A SOCIETAL CHALLENGE

Throughout human history, it has been feared that technological progress in the form of automation will disadvantage some groups of people. During the industrial revolution, as discussed in Chapter 1, many people agreed that certain developments associated with the technological changes they witnessed were unacceptable, such as the use of child labor and harmful working conditions. In response to these concerns, the 19th century saw the emergence of a secret organization of English textile workers known as the *Luddites*. This group was named after the character Ned Ludd who was an apparently fictional man from the Nottingham area in England.

During this period, factory owners in the textiles industry recognized the newly available machinery as an opportunity to introduce cheap but harmful labor practices. In response, the Luddites vandalized the machinery as a form of activism. The resulting movement began in Nottingham and led to a local rebellion lasting from 1811 to 1816. Factory owners and the government responded to the Luddite movement with force, resulting in the shooting of many activists. The term "Luddites" is still used today to describe individuals opposed to industrialization, automation, digitalization, or new technologies in general.

In the 20th century, the debate around the social impact of technology gradually centered on its impact on the equality of wealth and income. According to some economic theories, income inequality automatically decreases in advanced phases of economic development, irrespective of political choices and other aspects. This process is thought to occur through a leveling up of all wages and has often been summarized metaphorically by comparing economic growth with a rising tide that lifts all boats. While this seems to have been the case in the decades following World War II, more recent developments indicate that income inequality is again increasing throughout the world and the median real income is falling in developed countries. This does not bode well for the future.*

The discourse on the negative social impacts of the advance in technology focuses on the fact that automation, now in the guise of digitalization, reduces the demand for some forms of labor. However, new technologies undoubtedly also create new jobs so this burden is spread unequally across different groups, professions, and also countries. For this reason, the question of which jobs will be automated away is of high social importance and extensive research in this direction is currently ongoing. Underlining the issue, in 2019, the Office for National Statistics (ONS) in the United Kingdom estimated that approximately 1.5 million jobs in England are at a high risk of being partially or fully automated in the future.†

* For an extensive and influential treatment of this issue, see Piketty (2014).

† Office for National Statistics (2019).

The study indicates that women, young people, and part-time workers are the most vulnerable to the effects of automation, mainly because they tend to work in lower-skilled roles. While the complete automation of all jobs will remain a fiction, routine and repetitive tasks that can be executed rapidly and cheaply by digital systems are particularly at risk. The estimated probability of automation across all occupations assessed by the ONS study is shown in Figure 4.6. It indicates that the three jobs with the highest risk of automation are waiters and waitresses, shelf fillers in shops, and sales assistants, all of which are classed as *elementary occupations*, meaning that they are jobs requiring routine tasks. The three jobs at the lowest risk of automation are all considered high skilled; they include doctors and medical practitioners, senior professionals in education, and university teaching staff. This obviously comes as a great relief to the authors of this book.

The main insight from this study, and similar investigations, is that the successful creation of ideas and meaningful innovative activities are tasks that will require some form of human input in the future. While automated and programmed approaches to reasoning, as part of technologies such as machine learning, are likely to address a wide variety of tasks in the future, it is safe to assume that human input will remain needed. Nevertheless, some authors believe that finding a partnership with digital systems – complementarity in the language of economics – will be a decisive characteristic for personal and business success in the future.*

CONCLUSION

This chapter has given a compact overview of the economics of digital technology and some of its major impacts. It was argued that the economic power of information technology stems from a particular cost structure. The idea of strongly diminishing marginal costs of computing encapsulated in Moore's law was highlighted as a core idea. The chapter has further shown how the economic approach of comparative statics can be used to explain business decisions involving information technologies, including pricing decisions. In its discussion of flexible manufacturing systems and artificial intelligence, the chapter has explored the notion of economies of scope as an extension of economies of scale. The chapter also articulated a brief and concerned outlook on the social implications of the continuing digital revolution.

The chapter has drawn heavily on two books by the researcher duo Erik Brynjolfsson and Andrew McAfee, *The Second Machine Age* (2014) and *Machine, Platform, Crowd* (2017). Both books are very accessible and highly recommended. A further very accessible and influential introduction to the economics of digital technology is *Information Rules* by Carl Shapiro and Hal Varian (1999). The history of computing provided by Neil Gershenfeld in *Fab* (2005) is also very instructive, as is the detailed history of computer hardware provided by George Gilder in *Microcosm* (1989). Despite being a more general and extensive treatment of current economic developments, *Capital in the Twenty-First Century* by Thomas Piketty (2014) is relevant for readers interested in the broader economic implications of rising inequality.

---

* See Brynjolfsson and McAfee (2014).

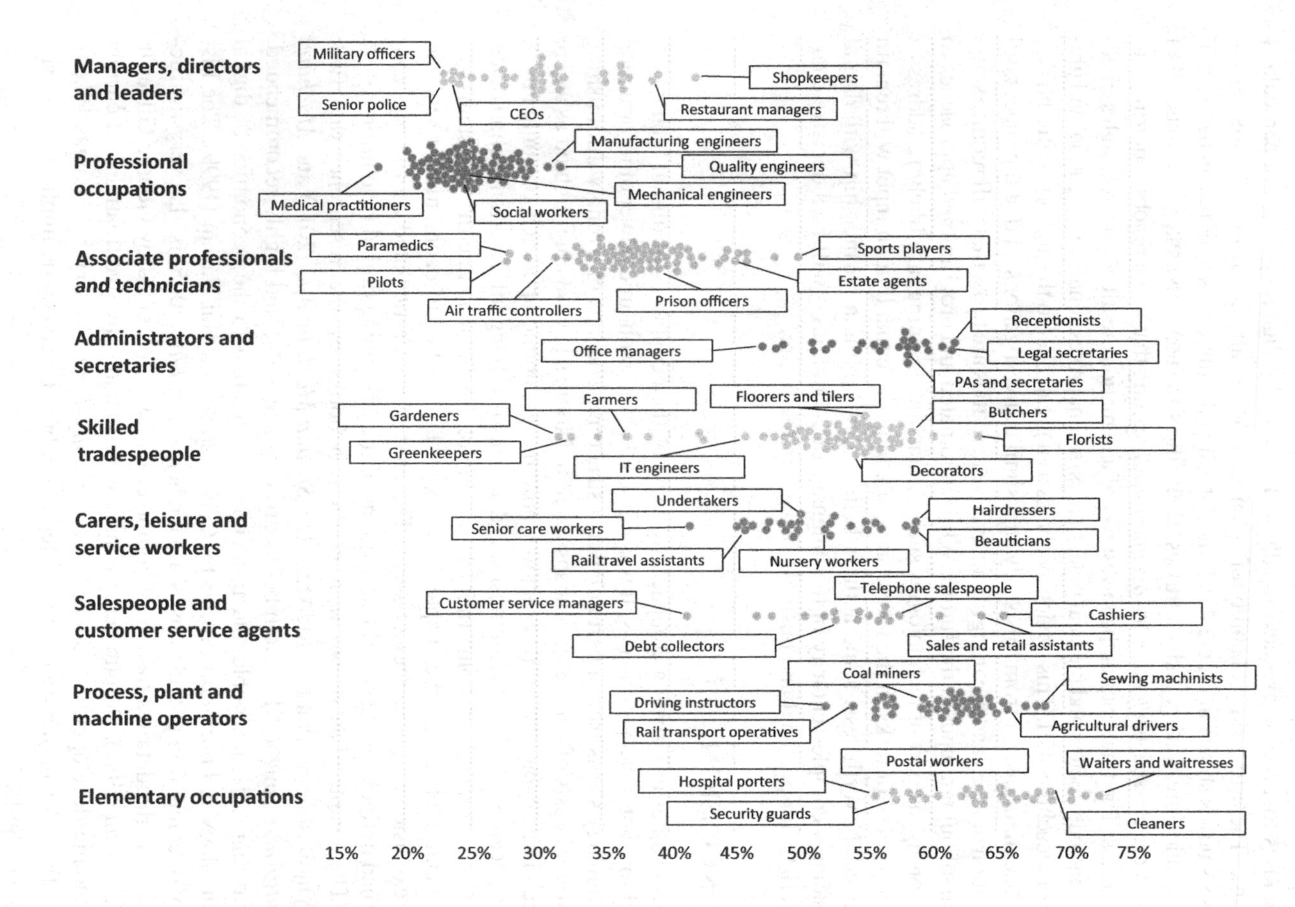

FIGURE 4.6 Probability of automation in England 2011–2019, adapted from ONS (2019)

REVIEW QUESTIONS

1. Complete the following paragraph describing the importance of transistors by picking the appropriate text to fill in the gaps.
   (Question type: Fill in the blanks)
   "Transistors form ___________ of computers today and represent a ___________ in technology because they are ___________ devices. They employ ___________ effects in order to act as ___________".

2. Match the provided descriptors to the appropriate characteristics of information technology.
   (Question type: Matrix)

	MOSFET	IC	Information as bits	Delayed differentiation	Von Neumann architecture
Symbolic representation allowing error-free processing	O	O	O	O	O
Solid-state transistor technology, based on quantum mechanics	O	O	O	O	O
Electronic elements deposited on a flat substrate	O	O	O	O	O
Basic structure of most computers	O	O	O	O	O
This is not a characteristic of information technology	O	O	O	O	O

3. Complete the following figure of a simplified von Neumann architecture by inserting the correct elements and by finding examples.
   (Question type: Labeling)

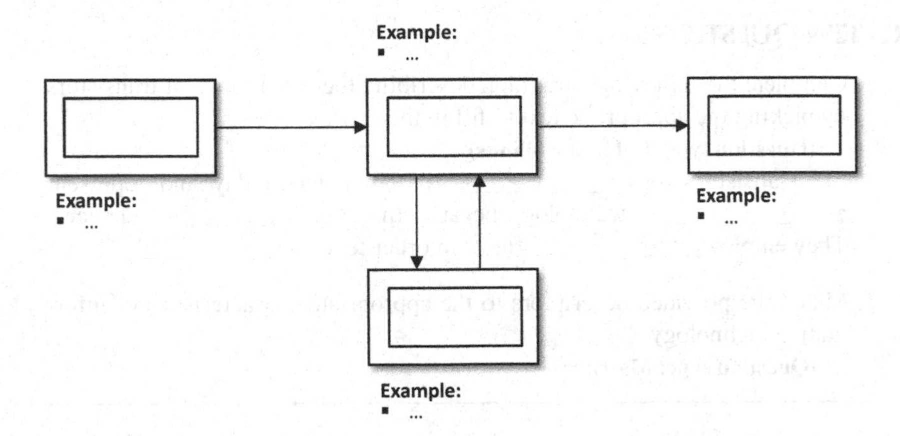

FIGURE 4.7 Insert the correct labels and three examples each

4. Which of the following are characteristics of information goods?
   (Question type: Multiple response)
   - ☐ Extremely high variable costs
   - ☐ Information goods are often priced according to costs
   - ☐ Anything that can be expressed interpretively can be an information good
   - ☐ Anything that can be encoded as digital information can be an information good
   - ☐ Comparatively high fixed costs
   - ☐ Low variable costs
   - ☐ Zero marginal costs
   - ☐ Information goods change the world
   - ☐ Low marginal costs

5. Which of the following statements about Moore's law is incorrect?
   (Question type: Multiple choice)
   - ○ It is about the possibility of adding more logical elements to ICs
   - ○ It has remained valid for a remarkably long time
   - ○ It was first formulated in 1965
   - ○ It is about the expansion of Random Access Memory (RAM)
   - ○ It refers to the amount of computing power that can be bought for a constant amount of money

6. To generate some extra income, you have written an ebook about your hobby. Through careful analysis, you have established that demand for your ebook is as follows:
   - The demand curve is a straight line
   - At a price of zero, the ebook would be downloaded 9,000 times
   - For each 1c increase in price, the number of downloads decreases by 14
   - The estimated equilibrium price is $4.20

- It is estimated that one-half of your customers will buy the ebook through a subscription and that this group has a higher willingness to pay

You have decided to use differential pricing and charge your subscription customers $0.08 more than the equilibrium price.

(Question type: Calculation)

Calculate the total consumer surplus.

7. A study of flexible manufacturing systems has analyzed the markets for CNC control software and small CNC machining systems in two time periods. This analysis has produced the following estimates for the CNC control software.

	Average price	Total market revenue
Period 1	$6,800	$850,000
Period 2	$2,600	$1,300,000

For the CNC machining systems, the following estimates were obtained:

	Average price	Total market revenue
Period 1	$11,000	$2,250,000
Period 2	$11,000	$4,850,000

(Question type: Calculation)

Calculate the cross price elasticity of demand for the CNC machining systems with respect to the price of the CNC control software.

8. Which of the following elements are normally found in flexible manufacturing systems?

(Question type: Multiple response)

- ☐ Manual labor inspection stations
- ☐ Material handling systems
- ☐ Computer control systems
- ☐ Work machines such as CNC systems
- ☐ Trolleys instead of forklifts
- ☐ Injection molding machines
- ☐ Word processing software

9. Which of the following statements about machine learning is correct?

(Question type: Multiple choice)

- ○ It involves a learning entity
- ○ It makes use of symbolic artificial intelligence
- ○ It is based on advanced statistics

○ It is inferior to symbolic artificial intelligence
○ All automation is a form of artificial intelligence

10. There is overwhelming evidence that increasing automation benefits everyone in society.
    (Question type: True/false)
    Is this statement true or false?
    ○ True ○ False

REFERENCES AND FURTHER READING

Brynjolfsson, E. and McAfee, A., 2014. *The second machine age: Work, progress, and prosperity in a time of brilliant technologies*. New York: WW Norton & Company.
Clegg, B., 2014. *The quantum age: How the physics of the very small has transformed our lives*. Cambridge: Icon Books Ltd.
Gershenfeld, N.A., 2005. *Fab: The coming revolution on your desktop--from personal computers to personal fabrication*. New York: Basic Books.
Gilder, G., 1990. *Microcosm: The quantum revolution in economics and technology*. New York: Simon and Schuster.
McAfee, A. and Brynjolfsson, E., 2017. *Machine, platform, crowd: Harnessing our digital future*. New York: WW Norton & Company.
Moore, G.E., 1965. Cramming more components onto Integrated circuits. *Electronics*, 38(8), pp.114–117.
Office for National Statistics, 2019. *Which occupations are at highest risk of being automated?* Report. Available at: https://www.ons.gov.uk/employmentandlabourmarket/peopleinwork/employmentandemployeetypes/articles/whichoccupationsareathighestriskofbeingautomated/2019-03-25 [Accessed 01 June 2020].
Piketty, T., 2014. *Capital in the twenty-first century*. Cambridge: Harvard University Press.
Shapiro, C., Carl, S., and Varian, H.R., 1998. *Information rules: A strategic guide to the network economy*. Cambridge: Harvard Business Press.

5 Technology, Innovation, and Disruption

OBJECTIVES AND LEARNING OUTCOMES

This chapter provides an overview of contemporary perspectives and methods relevant to the management of technology and innovation. The chapter opens with a general account of how technology can be defined, making clear that a general characteristic of technology is that it serves specific, goal-oriented purposes. Further context is provided through the presentation of two well-known frameworks for the classification of technologies. This is followed by an introduction to the concept of innovation, highlighting that the outcomes of innovation are always subject to a degree of uncertainty. As a relatively recent addition to theories of innovation in management, the chapter briefly introduces the topic of disruptive innovation, showing that forestalling such disruption forms an important objective for managers. The chapter also gives a summary of the most important sources of innovation and presents a range of methods and approaches to analyze the financial effects of innovation. Further, it outlines two modern approaches to innovation developed to reduce uncertainty, agile development and customer development. The chapter closes with a brief summary of different forms of intellectual property as a legal framework to reward innovation. Specific learning outcomes include:

- understanding of how technology can be thought of in a general way;
- knowledge of the pattern in which new technologies typically spread, the presence of uncertainty, and the economic consequences of this process;
- appreciation of the importance of disruption and the resulting innovator's dilemma and how it may be possible to overcome this issue;
- understanding technology push and market pull innovation;
- understanding common cash flow and average cost perspectives on innovation;
- knowledge of agile development and customer development as modern commercial innovation techniques;
- overview of basic forms of intellectual property;
- understanding key terms, synonyms, and accepted acronyms; and
- appreciation of important thinkers in innovation and technology.

WHAT IS TECHNOLOGY?

As argued for information technologies in Chapter 4, technological progress is a main driver of change in the economy and society over time. The concept of creative

DOI: 10.1201/9781003222903-6

destruction explains change through technological innovation and that commercial success leads to market power held by innovators. However, while technology is very important for industry and the world of commerce, it is surprisingly hard to characterize in a general way.

An influential view on the nature of technology was promoted by communication theorist and sociologist Everett Rogers in the second half of the 20th century. Interested in how technologies become accepted over time and in different settings, Rogers defined technology in terms of the knowledge it embodies and its effect on improving human understanding of processes occurring in the world. Rogers' definition is as follows:

Everett Rogers (1931–2004)

"A technology is a design for instrumental action that reduces the uncertainty in the cause-effect relationships involved in achieving a desired outcome."*

As can be seen from this definition, technology is seen as something intentionally designed or devised for a specific, goal-oriented purpose. Rogers referred to this as *instrumental action*. Because technology allows users to reduce uncertainty in practical situations to achieve their goals, it increases knowledge. The significance of this definition is that it makes clear that the remit of technology is not limited to material or technical objects, such as a wrench or a computer, but extends to other non-material objects, such as business strategies and business models, organizational approaches, financial services, and even cultural products such as films or paintings if they are made to achieve specific outcomes.

Another – very different – view on technology is becoming increasingly prominent as it is emerging that current patterns of industrial activity and human consumption are not environmentally sustainable (discussed in greater detail in Chapter 6). This view centers on the idea that there is more to technology than being a tool to achieve a specific outcome, as reflected in Rogers' definition. This issue was expressed by philosopher Martin Heidegger after World War II in a highly influential philosophical essay.

Martin Heidegger (1889–1976)

Heidegger argued that it is an error to think of technology as something that exists to support human needs. Rather, technology should be seen as a certain way of thinking and revealing hidden insights. In this, a central feature of technology is that it orders and regiments the things and people interacting with it. But because technology follows its own logic, the resulting ordering is not automatically beneficial to humans. Instead, Heidegger

* Rogers (2003, p. 16).

argued, human interests and the natural world become subordinate to the technological way of thinking and thus to the technology itself. Heidegger calls this process *enframing*. In other words, rather than adding to human capability and freedom in the form of new tools, technology also constrains and limits its adopters and users by forcing them (and anything else that is required by the technology) to be at its disposal, irrespective of whether this is beneficial to those involved.

The implication of Heidegger's theory is that the technological way of thinking tends to devalue anything and anyone who is standing ready for use by the technology. Likewise, any object or person that cannot be employed by the technology as an input is at risk of being discarded as valueless. Heidegger stressed that this problem lies in the nature of technology itself and cannot be avoided. All that can be done is to become aware of this danger.*

Heidegger's groundbreaking philosophical assessment of technology makes the important and general point that, while delivering benefits and advantages, technology unfolds in its own way and tends to change those who interact with it. This can have sweeping negative consequences for human culture and happiness, as illustrated in the discussion of the social consequences of technological progress in Chapter 4. This also forms an explanation of why humanity is struggling with the pressing issue of climate change due to carbon emissions.

IDENTIFYING DIFFERENT KINDS OF TECHNOLOGY

FAMOUS THINKER

Joan Woodward (1916–1971)

Considering the organizational lag model presented in Chapter 1, it is perhaps no surprise that prior to the 1950s, it was entirely unclear to most scholars and managers whether a single ideal organizational form exists that all businesses should adopt. This changed with the work of social scientist Joan Woodward in the 1950s and 1960s. Woodward showed that the appropriate way of organizing a business depends, above other factors, on the production methods and technologies used by the business. Figure 5.1 encapsulates this idea by associating different kinds of organizational forms with appropriate technologies. Some of the terms used in Figure 5.1 are defined in Chapter 9; the concepts of mass-production and manufacturing lines have already been encountered in Chapter 1.

Woodward's work established the idea among organizational theorists and managers that it is worthwhile to have a structured view of technologies and that the type of technology employed matters for managers when deciding on how to organize the activities of a business. As will be seen in Chapter 9, this line of thought forms a major topic for operations managers.

* Heidegger (1977). This paragraph forms a cursory attempt to capture the meaning of Heidegger's essay for the purpose of this book. Heidegger's work is very difficult to read, so patience is advised when consulting the original text.

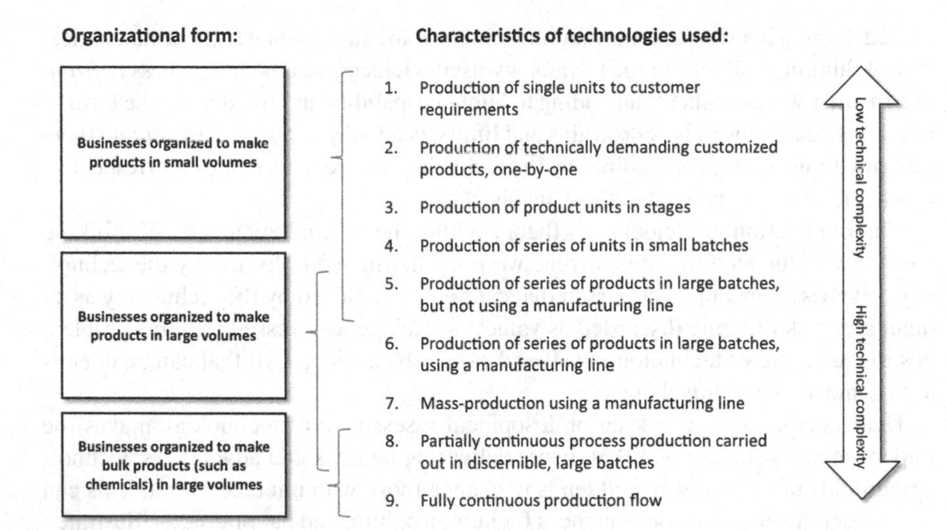

FIGURE 5.1 Classification of production systems, adapted from Woodward (1965)

Following Woodward's seminal work, additional classification systems for technologies were proposed. A framework that is of great relevance in the present, as will become clear in the discussion of the platform business model in Chapter 7, is the model proposed by James Thompson in 1967. This model is particularly useful because it not only reflects manufacturing technologies, as Woodward's original model does, but is broad enough to account for technologies used in information processing and in the services industries. Thompson's framework distinguishes between the following three kinds of technologies:

1. *Long-linked technologies*

   Long-linked technologies are those that feature step-by-step processes of hand-off between distinct elements. Dominant in manufacturing, especially in mass-production, long-linked technologies exhibit an *activity flow* that can range from raw material generation to the assembly and packaging of the finished product. They are often characterized by highly standardized processes dedicated to a particular application with uniform inputs and outputs. The archetypal long-linked technology is the manufacturing line, as discussed in Chapter 1. Note also the similarity between long-linked technologies and the primary activities in Porter's value chain discussed in Chapter 2, suggesting that businesses themselves resemble long-linked technologies. In this sense, long-linked technologies also bear close similarity to so-called pipeline business models, as discussed in Chapter 7.

2. *Mediating technologies*

   Mediating technologies are those that facilitate interactions between individuals or businesses. A prime example of current mediating technology is internet search engines, which allow users requiring internet search services to connect to the providers of the digital content. In turn, the search engine may also allow paying advertising customers to engage with the users. As this example suggests, the emergence of the internet has greatly enhanced the power and reach of mediating technologies. Mediating technologies are marked by standardized, yet adaptable, transformation processes and dissimilar inputs and outputs. This development is discussed in detail in the introduction to platform business models in Chapter 7.

3. *Intensive technologies*

   The third type of technology in Thompson's framework is intensive technologies. Intensive technologies produce changes and transformations through intensive interaction with an object in a specific situation. This requires intensive technologies to employ non-standardized, adaptable transformation processes and the ability to handle dissimilar and specific inputs and outputs. Typical examples for intensive technologies can be found in medicine, where adaptable systems are used to diagnose and treat patients, and in experimental research, where experiments are undertaken to deliver novel insight under significant uncertainty.

Figure 5.2 summarizes Thompson's classification of technologies based on the dimensions of standardization in transformation processes and uniformity in inputs and outputs. More detail on the view of business operations as transformation processes is provided in Chapter 9.

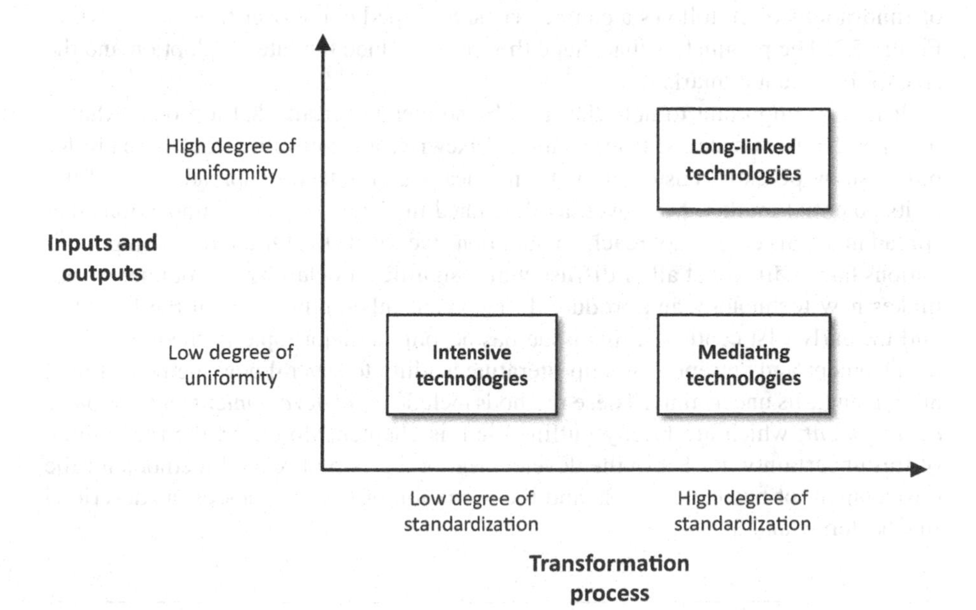

FIGURE 5.2 Thompson's classification of technologies

THE GENERAL PROCESS OF INNOVATION

A useful model of the way in which technological change proceeds over time is the so-called Schumpeterian trilogy,* named after economist Joseph Schumpeter (introduced in Chapter 4). This model views technological change as a sequence consisting of three stages. In the first stage, *invention*, new ideas are generated, for example, through the recombination of existing ideas or the discovery of genuinely new phenomena. This is followed by a second stage in which the newly invented technologies are transformed into products or services that can be used by prospective users in a process known as *innovation*. Innovation can be defined as follows:

Innovation is the process of transforming an idea that is new or perceived as new into products or services that can be sold in a market.

The final stage in this three-step model is the spread of the new products or services within the population of available adopters, also known as a *potential market*. This is referred to as the *diffusion* process. Importantly, the impact of the new technology occurs during this stage. Diffusion can be defined as follows:

Diffusion is the process by which products or services spread within a potential market. Only products and services that have undergone an innovation process can enter into diffusion but there is no guarantee that diffusion will occur.

Because the diffusion stage is so important, it has been studied for many different kinds of products. The main result from this research is the insight that the diffusion of innovations often follows a characteristic S-shaped curve over time, as shown in Figure 5.3. The parameters that shape this curve include the rate of adoption and the size of the potential market.

It is very important to note that it is by no means certain that a product that is invented leads to successful innovation. Likewise, an innovation that is ready for purchase by potential customers is by no means guaranteed to spread successfully in its potential market. Moreover, as illustrated in Figure 5.3, even innovations that spread in a market may not reach an adoption level of 100%. Often, promising innovations fail to diffuse at all or diffuse with a significant delay. This root uncertainty makes new technology and product development inherently risky. In the late 20th and the early 21st centuries, this issue has become a major topic in the technology development and entrepreneurship literature leading to several new methods aimed at reducing this uncertainty. These methods include *agile development* and *customer development*, which are briefly outlined in this chapter. Moreover, the recognition of this uncertainty has led to the development of a group of related methods for the development of business models and the formation of new businesses, as described in Chapters 7 and 8.

* Stoneman (1995).

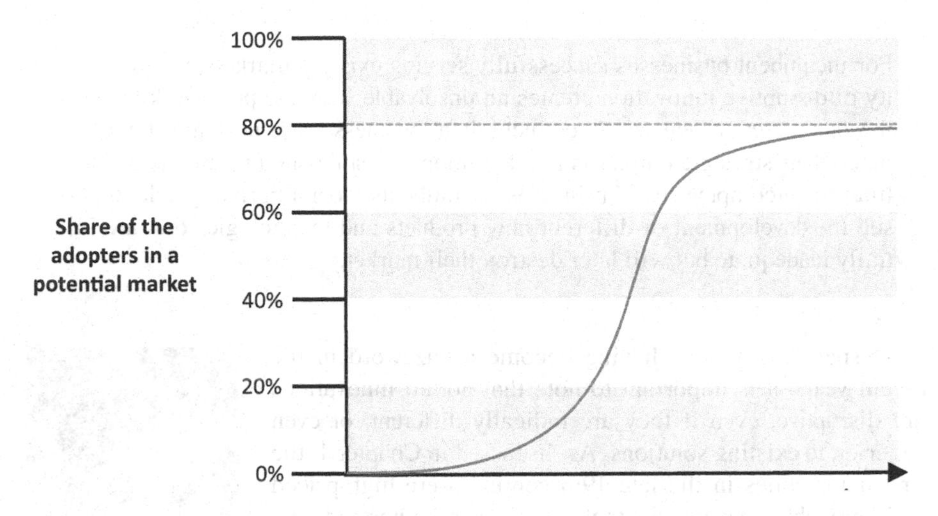

FIGURE 5.3 Typical S-curve of technology diffusion

DISRUPTIVE INNOVATION

Clayton Christensen (1952–2020)

In the 1980s and 1990s, many large and established businesses came under increasing pressure from successful new businesses making greater use of information technology. Unsatisfied with the three-part explanation of how new products and services spread, management scholar Clayton Christensen responded to these developments by introducing the notion of *disruptive innovation*. Unlike other types of innovation, disruptive innovation creates new markets by providing new and unexpected value to buyers, for example, in the form of a radically different new product. Christensen observed that new businesses, referred to as *entrants*, do not always gradually spread into existing markets, in which established businesses, referred to as *incumbents*, operate. Sometimes, the new value offered by an entrant is so great that it destroys an existing market and the incumbents. In such cases, the incumbents are unable to use their advantages and resources, including their market position, knowledge, and capital to defend their position. Disruptive innovation can thus be defined as follows:

Disruptive innovation, or disruption, is the process of creating radically different new products or services that destroy an existing market and the advantages held by the businesses in this market.

CORE IDEA

For incumbent businesses successfully serving existing markets, the possibility of disruptive innovation creates an unsolvable strategic problem known as the *innovator's dilemma*. Since incumbent businesses are normally forced to place their strategic emphasis and attention on established profitable product lines to outcompete rival businesses, incumbents are not normally able to pursue the development of different new products and technologies that are initially inadequate but will later destroy their market.*

FAMOUS THINKER

Richard D'Aveni (born 1953)

Despite "disruption" having become a buzzword in the recent years, it is important to note that not all innovations are disruptive, even if they are radically different, or even superior, to existing solutions. As discussed in Chapter 1, the first automobiles in the late 19th century were high-priced and unreliable luxury items that co-existed with horse-drawn vehicles for several years. The overall market of individual transportation remained largely unchanged until Ford developed the low-priced Model T. Only the mass-produced, and hence affordable, car was a disruptive innovation because it led to the rapid disappearance of the market for horse-drawn transportation.

Further exploring the strategic implications of new technologies, the notion of disruptive innovation has recently been extended by strategy scholar Richard D'Aveni. In an investigation of digital manufacturing technologies, D'Aveni stressed that the outcome of an innovation process depends on the features of the innovations involved and the frequency of innovations over time.† This dual distinction is illustrated in Table 5.1.

In this 2-by-2 matrix, traditional patterns of innovation are captured in the upper row, where non-disruptive innovations occur with low or high frequency. This produces patterns of no change or regular minor changes to the structure of the industry. Importantly, this turbulence is not strong enough to dislodge the incumbents from their dominant market positions. Chiefly, this is because the occurring innovations are aligned to the advantages of the incumbents and strengthen their position. Such advantages are likely to relate to traditional sources of competitive advantage such as those identified by Porter (1985), as discussed in Chapter 2.

In contrast, the lower row in Table 5.1 shows patterns of disruptive innovation in which innovations destroy the advantages or competencies held by the incumbents. The bottom-left cell depicts a situation in which disruptive innovations occur at a low frequency that allows incumbents to retain their dominant position over a prolonged period of time before being replaced. The bottom-right cell shows a pattern marked by frequent disruptive innovation. In this situation, a succession of new entrants

* An insightful and very accessible discussion of disruption processes is provided by Wu (2010).
† D'Aveni (2018).

TABLE 5.1
Frequency of Innovation versus Type of Innovation

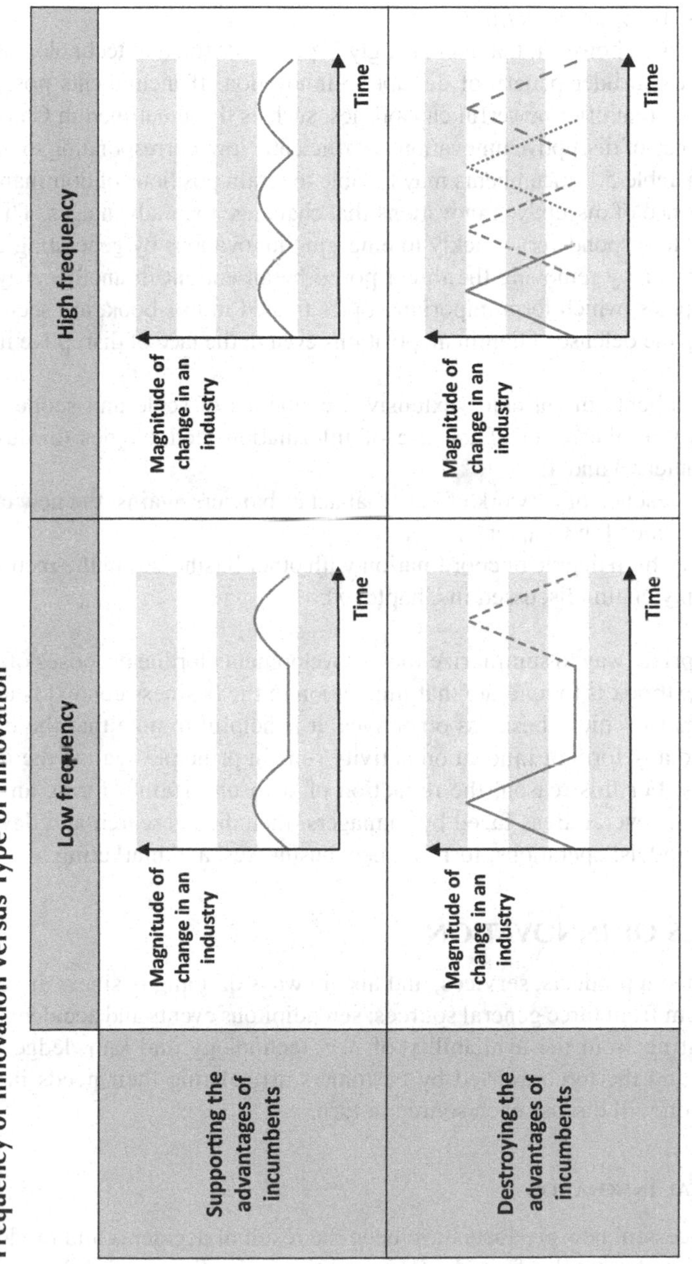

Source: Adapted from D'Aveni (2018).

captures the industry and is then displaced by the following entrant. This ongoing cycle of disruption has become a pattern found in some industries marked by disruptive innovation. Since this pattern is characterized by extreme competitiveness, it is referred to as *hypercompetition*.*

It is thought, however, that increasingly capable information technologies are leading to a new, milder pattern of disruptive innovation. If incumbents possess digital technologies that offer powerful capabilities, such as those outlined in Chapter 4, and the frequency of disruptive innovation is sufficiently low, corresponding to the bottom-left cell in Table 5.1, incumbents may be able to retain positions of dominance despite the emergence of disruptive innovations that challenge their advantages. This requires the ability to respond very quickly to emerging innovations by generating competing innovations or by removing the threat posed by an entrant in another way. The following aspects, which form important topics treated in this book, are seen as factors supporting the defense of dominant positions even in the face of disruptive innovation:

- incumbents might enjoy extensive economies of scale and scope, particularly as the result of the use of information technologies (discussed in Chapters 3 and 4);
- the presence of network effects that act as barriers against the new entrants (introduced in Chapter 7); and
- a very high degree of coordination with other businesses in the incumbent's supply chain (discussed in Chapter 7).

An appropriate way to summarize these developments for the purposes of this introductory textbook is to reiterate that innovation in the business context is undertaken as a means to achieve business objectives. It is helpful to note that the uncertainty inherent to any form of innovation activity runs in principle against the interests of businesses. For this reason, the reduction of such uncertainty forms an important objective in several areas faced by managers, including research and development, business models, operations, forming new businesses, and marketing.

SOURCES OF INNOVATION

Innovations in products, services, and also in ways of doing business are commonly seen to stem from three general sources: serendipitous events and accidents, the pressure resulting from the availability of new technology and knowledge inside the business, and the force exerted by customers articulating their needs in a market. This section will discuss each source in turn.

Accidental Innovation

Many successful new products have been the result of accidents and random events. An often-cited example of accidental innovation is the discovery of the first naturally

* D'Aveni (2018).

occurring antibiotic drug penicillin by the pharmacist Alexander Fleming in 1928. Fleming was a researcher experimenting with the influenza virus, who happened to be known for carelessness in his laboratory work. Following his return from a vacation, Fleming discovered that a mold had grown on a culture plate containing bacteria. As an observant scientist, Fleming noticed that the bacteria in the proximity of the mold had died. This led to the innovation of penicillin as a drug in 1942. As such events may or may not occur and may or may not be recognized correctly by those involved, accidental innovation cannot be planned and therefore cannot form the basis of a systematic innovation methodology.

Technology and Knowledge Push Innovation

Many businesses relying on innovation activities employ processes designed to translate ideas into marketable products. While the creation of the original idea, which may be related to the scientific discovery of a phenomenon, is largely random and unplannable, businesses and other organizations frequently attempt to implement structured processes to translate such ideas into marketable products and services. This process is known as *technology push* or *knowledge push* innovation and can be defined as follows:

Technology push innovation, also known as knowledge push innovation, is an innovation process that deliberately selects an underlying idea for processing into a marketable product or service on the basis of the perceived quality of that idea.

Due to its sequential nature, technology push innovation is closely associated with the traditional planned method of innovation, which sees such processes as the execution of a fixed sequence of steps. The traditional planned method is also referred to as the *over-the-wall* model, the *waterfall* method, or the *sequential mode of innovation* because a product under development passes through a series of distinct stages, often undertaken by distinct teams or different organizations, as summarized in Figure 5.4. While the sequence of steps is fairly consistent, the duration of each step and its complexity can vary significantly.

In manufacturing and engineering, the planned innovation process routinely involves organizations of different kinds, including businesses, universities, publicly

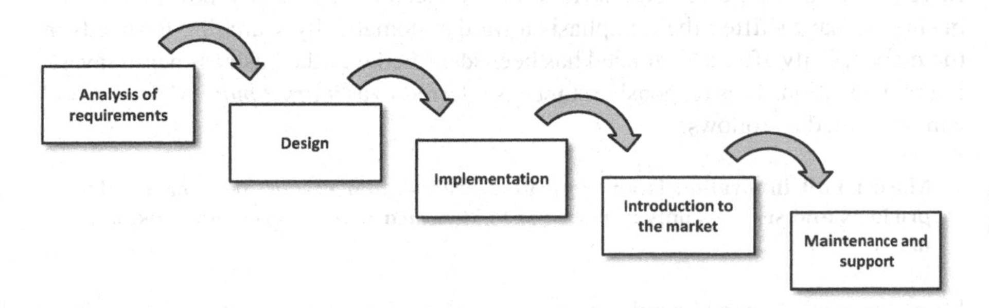

FIGURE 5.4 The sequential mode of innovation, adapted from Blank and Dorf (2012)

funded research institutes, and *research and technology organizations (RTOs)*, which are specialized organizations dedicated to developing scientific research outputs. In this context, a typical technology push journey from an idea to a marketable product is completed in five levels of research and industrial development: pure research, basic research, applied research, product development, and product design. Common motivations, the organizations commonly involved, and the main funding sources associated with these stages are summarized in Table 5.2.

While a significant amount of industrial research and development is funded by businesses, there is, throughout the process, a role for governments to provide and allocate funds. Each stage of the planned innovation process will have separate objectives and will involve different departments or organizations and draw on researchers and developers with different skills. Once industry becomes involved in the innovation process, normally in the later stages, the technical and marketing departments within businesses work together very closely to develop marketable products. This is done to avoid further developing products and services for which there is no market interest. Similarly, such coordination avoids the marketing of products or services that are not technically or economically viable.

Over the recent decades, pure technology push innovation has come to be considered a problematic and outmoded method for innovation. This perception is largely driven by the realization that the percentage of products that are innovated but subsequently fail in the market is very large. Frequently cited and unverifiable estimates suggest that between 70% and 95% of all new products fail in this way. Obviously, this is an enormous waste.

The high rate of failure is due to many problems such as a lack of quality of the underlying idea, design flaws, a lack of distinctiveness compared to other products, an inability of the customers to understand the offering, and problems with supplying the product in a sufficient quantity. The insight that technology push modes of innovation are risky and ultimately wasteful has stimulated many new approaches based on identifying customer requirements and has shifted the focus on a different source of innovation, as described in the following section.

Market Pull Innovation

In response to the perceived shortcomings of technology push innovation, many businesses have shifted their emphasis toward systematically searching for needs in the market. Only after a clear need has been identified will the business begin investing in innovation. This responsive stance is referred to as *market pull innovation* and can be defined as follows:

Market pull innovation is an innovation process that creates new marketable products and services only in response to identified market forces and customer needs.

Large businesses spend heavily on analyzing the needs of customers and how these needs change over time, for example, by running research or innovation centers

TABLE 5.2
Five Stages of Technology Push Innovation

Stage	Common motivations	Organizations typically involved	Main funding source
Pure research	• Scientific interest • Individual curiosity	• Universities	• Publicly and charitably funded through research grants
Basic research	• Development of a scientific theme or program • Response to a technical challenge or scientific anomaly	• Universities • Research institutes	• Publicly and charitably funded through research grants
Applied research	• Bringing innovation to market to address a generally perceived need • Demonstrating an innovation • Identifying additional research and development requirements	• Research institutes • Research and technology organizations • Businesses	• Publicly and charitably funded through research grants • Private funding • Often combining public and private funding
Product development	• Private profit • Meeting general commercial objectives	• Businesses	• Privately funded (possibly subsidized)
Product design	• Private profit • Meeting specific commercial objectives	• Businesses	• Privately funded (possibly subsidized)

devoted exclusively to this task. Generally, such initiatives try to understand the customers' experiences as they use products or services. Businesses wishing to address their customers' expectations, therefore, need to develop an awareness of their customers' knowledge and abilities and, expressed in the terms introduced in Chapter 4, develop complementarity to the traits of their customers. Collecting information from customers to inform innovation has also taken a significant leap forward with the emergence of the internet, which allows businesses to collect detailed information on customer behavior and traits.

CAPTURING THE FINANCIAL IMPACTS OF INNOVATION

On a practical level, it is essential for managers to understand the financial costs and benefits of innovations. Such analyses must be performed on two distinct levels. The first level is the analysis of the direct impact of innovation on the financial situation of the innovating business, ideally making clear how the activity results or will result in increased commercial success and profitability. The second level assesses the performance of innovations themselves in the form of new processes or products. Innovation leads to better processes employed by the business, for example, in the form of improvements to manufacturing technology. Alternatively, it can lead to better products or services created by the business that will be more valuable to customers.

Costs and Benefits of Innovation to the Business

In most industries, competitive pressure dictates that businesses need to engage in some form of innovation to survive. As explained in this chapter, however, innovation activities are inherently risky no matter what approach is taken by managers. Moreover, innovation programs tend to be expensive, especially in the close-to-market stages of design and product development when resource-consuming detailed work is undertaken to generate a marketable product or service. However, the likelihood of technological success in the innovation process also increases in these final stages.

Once a business has decided on how to implement an innovation, it will usually plan to incur initial expenses and later recoup these expenses through the sale of the new product or service. A successful innovation project will likely result in the characteristic cash flow pattern shown in Figure 5.5. The initial costs of innovation can be small or extremely high, depending on the industry and the nature of the innovation. Similarly, the time required to complete the innovation and to generate financial returns to recoup the initial expense can vary dramatically. Consider, for example, that the development of a new civil aircraft may take several years. In such cases, the recovery of the initial expense may take place over decades following the introduction of the new product to the market.

To gauge the commercial performance of the innovative activities in a business, several general metrics can be constructed from basic financial measures, such as those introduced in Chapter 2. These metrics mostly reflect the level of expenditure on innovation and how well the business converts innovation expenditure into revenue from the sale of the new products.

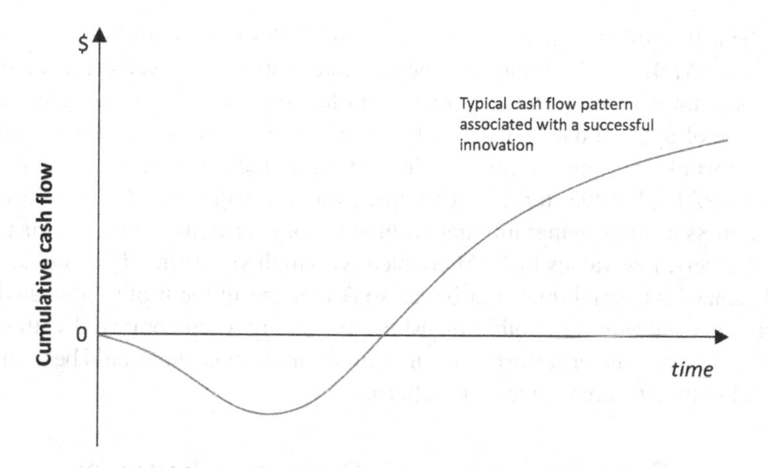

FIGURE 5.5 Cash flow profile of successful innovation, adapted from Martin (1994)

An underlying simple metric is the *new product sales ratio*, which is the share of sales revenue originating from new products (*sales revenue*$_{new\ products}$) in overall revenue *sales revenue*. This is typically expressed as a percentage:

$$\text{New product sales ratio} = \frac{\text{Sales revenue}_{\text{new products}}}{\text{Sales revenue}} \times 100 \tag{5.1}$$

A high new product sales ratio indicates that the results of innovation activity make up a significant part of the income received by the business. This may also indicate that the established product lines in the business are obsolete or are losing traction in the market.

A more advanced metric is the *research-and-development-to-product conversion* (*RDP*). This ratio provides information on how well the innovation expenditure translates into the sale of new products. *RDP* is formed by dividing the *new product sales ratio* over the share of innovation cost in sales revenue and can be expressed as a percentage term:

$$\text{RDP} = \frac{\text{New product sales ratio}}{\left(\dfrac{\text{Innovation cost}}{\text{Sales revenue}}\right)} \times 100 \tag{5.2}$$

A further useful ratio is the *new-products-to-margin conversion* (*NPM*), which indicates how much the sales of new products or services contribute to the *gross profit to sales ratio* of the business, which is introduced in Chapter 14. Also known as *gross margin*, this is a basic financial ratio expressing the relationship between gross profit and sales revenue obtained by the business. *NPM* is formed by dividing the *gross profit sales ratio* over the *new product sales ratio*:

$$\text{NPM} = \frac{\text{Gross profit to sales ratio}}{\text{New product sales ratio}} \tag{5.3}$$

While the nature of an analyzed business and the industry it operates in will have an effect on the *NPM* and *RDP* metrics, the presented ratios can nevertheless be used to gauge the innovation performance among similar companies. Available data suggest that a value of approximately 1 would be expected for *NPM* and a value of approximately 5.5 would be expected for *RDP* for typical businesses with average innovation performance.* High values for *NPM* combined with low values for *RDP* would suggest that a business is adept at maximizing profit but poor at bringing new products to market. Conversely, low values in *NPM* coupled with high values in *RDP* would suggest that a business is a serial innovator but is weak in terms of the profits generated from this activity. Low values in both dimensions, especially when compared with similar businesses, suggest underperformance in terms of innovation and should be a cause for concern. Figure 5.6 summarizes this pattern.

Assessing the Cost Performance of the Outcomes of Innovation

The previous section has concentrated on the effect of innovation on the cost performance and profitability of the business itself. However, it is often necessary to investigate improvements in terms of the cost performance of specific processes, products, or services. The technique of *cost-volume-profit analysis*, also known as *breakeven analysis*, is routinely employed for analyses of this kind. Cost-volume-profit analyses can be characterized generally as a group of techniques used to compare functional

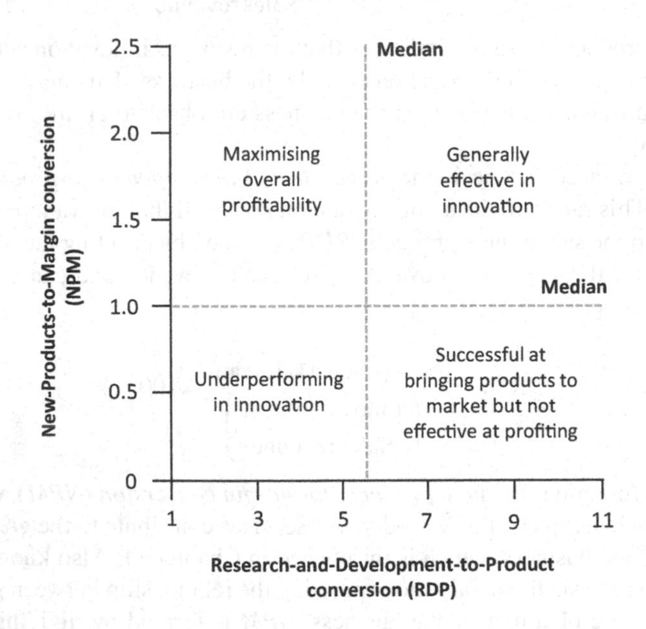

FIGURE 5.6 Median levels of NPM and RDP and interpretation, adapted from Aase et al. (2018)

* Aase et al. (2018). Note that RDP is not presented as a percentage term in this case.

relationships sharing the same independent variables to identify the points at which these functions intersect, also known as points of *breakeven*. This form of analysis is introduced in greater detail in Chapter 14.

Such analyses can be used for the investigation of the cost performance of new technologies versus old technologies along various measures of the scale or magnitude of activities. For a new kind of industrial machinery, for example, such an analysis would typically reflect the cost per unit processed; for a new kind of electric vehicle, the analysis could estimate the cost per mile.

Recall from Chapter 3 that the average cost AC denotes the cost incurred for each unit of activity or output q. Specified as a function of q, different $AC(q)$ functions can be compared to evaluate the relative cost performance of different processes or products. This allows the identification of threshold quantities at which specific alternatives begin to exhibit better cost performance. For example, assume that there are two technologies: an old technology exhibiting the average cost function $AC_{old}(q)$ and a newly innovated technology exhibiting the average cost function $AC_{new}(q)$. Both cost functions can be plotted in the same figure in order to compare relative cost levels and identify one or more points of *breakeven* q_{BE}, as shown in Figure 5.7.

This example suggests that below breakeven quantity q_{BE}, the old technology exhibits a lower average cost. Beyond this level of activity, however, the new technology exhibits a better cost performance. If the goal of the innovation is to deliver a new process that exhibits a lower cost at high levels of output, greater than q_{BE} in this example, then the innovation project undertaken can be considered successful. From the engineering and design points of view, however, such comparisons are naïve. The reason for this is that there will normally be interdependency between the type of technology employed and the characteristics of its output, such as appropriate specifications and available materials. Hence, a comparison simply based on cost without exploring knock-on effects on other factors may be misleading.

Additionally, for the evaluation of new technologies, it will be of interest to explore how the average cost responds to changes at particular levels of activity or scale. This can be explored by calculating the marginal average cost function $MAC(q)$ associated with

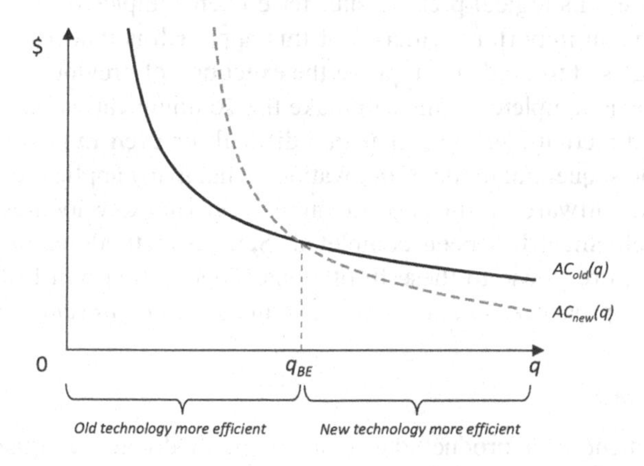

FIGURE 5.7 Breakeven analysis for two competing technologies

different technologies. If the average cost function $AC(q)$ is continuous and differentiable, $MAC(q)$ is the first derivative of this function with respect to q:

$$MAC(q) = \frac{dAC}{dq} = AC'(q) \quad (5.4)$$

Assuming a linear total cost function using the parameters a and b in the form $TC(q) = a + bq$, the following average cost function can be stated:

$$AC(q) = \frac{a + bq}{q} = aq^{-1} + b \quad (5.5)$$

The first derivative of this function can be formed either through the quotient rule or by simple derivation:

$$AC'(q) = -\frac{a}{q^2} = -aq^{-2} \quad (5.6)$$

This function can then be used to assess the rate of change of AC associated with a particular technology at a particular level of output q, shedding light on the impact on cost performance of changing quantities when a technology is operated at a specific scale. Moreover, in the case of U-shaped AC functions, the determination of points at which $MAC(q) = 0$ yields insight into the most efficient scale for the operation of new technology.

ITERATIVE APPROACHES TO INNOVATION

In many industries, such as software and motorsports, the sequential mode of traditional innovation projects, closely associated with the technology push mode of innovation discussed in this chapter, has come to be seen as unsuitable. In this approach, innovation is completed as a step-by-step process such that any particular step cannot proceed unless its logical prerequisites have been completed. Apart from being inherently risky, an important criticism of this approach is that insights from later steps cannot be used to guide or improve the execution of previous steps since they have already been completed. This can make the communication between different developmental functions, groups, or teams difficult or even impossible. A further limitation of the sequential mode of innovation is that many applications, such as the aforementioned software products or racing cars, permit development cycles after the initial development has been completed. Several methods of innovation have been proposed in response to these limitations. This section will briefly introduce two important approaches, *agile development* and *customer development*.

AGILE DEVELOPMENT

Agile development is a product development methodology designed to shorten the development time of products and to allow the continuous collection of new information to guide the developmental process. As an approach to innovation, agile development advocates the continuous adaptation of plans, incremental development, early exposure of users to new features, and a welcoming attitude toward ongoing change. A major difference between the traditional sequential mode of innovation and agile development is that the latter forms an iterative process in which development activities are repeatedly executed and assessed in a cycle over time.

The main organizing principle of agile development is to separate product development tasks into small increments of work to limit the requirement for detailed formal planning. Normally, multiple iterations are required to develop a new product or service sufficiently for release. While the sequential mode of innovation is usually carried out by specialist functional groups, each completing a step in the innovation process, agile development progresses through a collaboration of cross-functional teams, including planners, analysts, designers, engineers, and testing specialists, possibly also involving customers.

Agile development is now a standard methodology in the software industry. It is increasingly spreading to product and service development in other industries and is used in the development of marketing services in particular. A standard implementation of agile development will include the following sequence of seven basic activities:

1. *Assemble teams to deliver the work*
   In agile development, the delivery of the objectives is executed by cross-functional teams under the guidance of a product manager. The size of an agile team should be as small as possible but as large as necessary.
2. *Form the product backlog*
   The *product backlog* forms an important decision-making tool in agile development. It is a prioritized list of *deliverables** for implementation, including desired product features. The product backlog allows the team to refine the schedule of the work so that the most critical tasks can be completed first. As a living document, the product backlog should be updated frequently.
3. *Create iterations*
   Using the information contained in the backlog, the overall work is broken down into *iterations*, also known as *sprints*. These small increments of work are scheduled in advance and typically last from one to four weeks. It is possible that the iterations are continued until the developmental requirement is met and this can be demonstrated by the team.
4. *Perform risk mitigation measures*
   Unexpected and unpredictable events are anticipated in agile development. This creates the need to manage risks actively through mitigation in every iteration. Risk mitigation techniques are introduced in this book in Chapter 12.
5. *Hold scrums*
   Recognizing that agile development techniques place special demands on communication between team members, meetings are normally conducted in the *scrum* format. A scrum is a short daily meeting led by a *scrum*

* The term "deliverables" is commonly used in project management and is defined in Chapter 16.

master in which team members present their progress. Scrums are normally held every day. In some businesses, these meetings are referred to as *stand-ups.*

6. *Perform testing*

   Agile development requires that testing be executed in parallel with other product development activities. Specific team members test product increments generated in each iteration, allowing the rapid identification and resolution of problems.

7. *Collect customer input and feedback*

   As opposed to the sequential mode of innovation, agile development stresses the need to expose the user or customer as early as possible to the innovation. This is done to minimize the risk of not meeting the true requirements of the customer, which is a significant risk in many development projects.

Customer Development

As a further response to the perceived shortcomings of the sequential model of innovation, the approach of customer development has been popularized by Blank and Dorf (2012) and Ries (2011). The insight underlying both books is that, at least in small and new companies, the success of the product or service development processes is intertwined with the overall fate of the business. This places the question of how a product or service relates to real customer needs at the center of the business. It also implies that product development is a process of discovery in which hypotheses about the business are formulated and tested, not unlike academic research.

Like agile development, customer development aims to expose customers as quickly and cheaply as possible to a product or service under development. This involves the creation of a series of stripped-down initial versions of products or services. Known in customer development as *minimum-viable products* (*MVPs*), these are objects that can function as a product with the smallest possible number of features. They embody the hypotheses held by the managers about the customer needs and how they can be satisfied. Once a *MVP* has been released, managers attempt to obtain as much actionable information as possible in order to refine their hypotheses and the design of the product or service.

Similar to agile development, customer development is an approach in which hypotheses held by managers about a product are tested iteratively. This is structured in two stages in which iterative cycles occur. The first stage is *customer discovery.* In this stage, the developers' original vision is translated into hypotheses about the business. The second stage is *customer validation.* This stage involves testing whether the business and the product or service are viable. At the end of each iteration, a minimum-viable product is generated and tested. Based on the outcome of this testing, managers decide either to *pivot,* which requires a change to the product or service and the associated hypotheses, or to *persevere,* which means moving forward to the next stage or to begin executing the business by introducing the product to the market. This process is summarized in Figure 5.8.

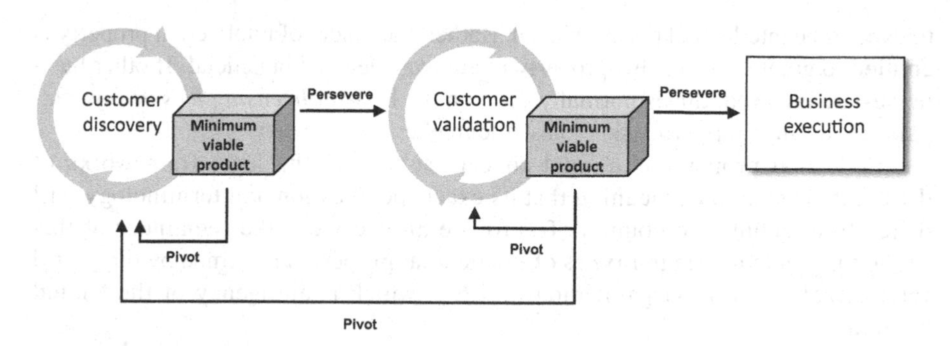

FIGURE 5.8 Shortened customer development process, adapted from Blank and Dorf (2012)

THE PROTECTION OF INNOVATIONS THROUGH INTELLECTUAL PROPERTY

The final topic covered by this chapter concerns the legal protection of the outcomes of innovation. If innovation activities are successful and yield a promising product or service offering, the resulting profits may well be significant. In this case, new entrants and imitators will soon be attracted. Since innovative activity normally carries significant positive externalities* that benefit everyone, not only the innovator, it is generally acknowledged that innovators deserve extra legal protection to recoup their investments.

This protection is provided by *intellectual property law*, which gives people and businesses property rights over the information and ideas they create, usually lasting for a limited period of time. By doing so, intellectual property law creates temporary monopolies in which the creators of the new product or service are granted the opportunity to generate significant returns from their innovation. The resulting property rights are known as *intellectual property (IP)* and can be owned, granted access to, and sold by their owners. Intellectual property can generally be defined as follows:

Intellectual property is a form of property that results from original creative thought. As a legal object, the characteristics of intellectual property are defined by the laws of a country.

Unsurprisingly, any business that has produced a valuable innovation will wish to protect its intellectual property because it represents knowledge, effort, investment, and the promise of future profits. It is important to note that a business does not have

* See Chapter 3.

to exploit the intellectual property itself. Rather, the holder of intellectual property is entitled to grant usage of this property in any way deemed beneficial. If other businesses are involved, this is normally done on the basis of *licensing fees* or *royalties* paid in exchange for permission to use the intellectual property.

Intellectual property is a legal concept subject to the legal frameworks of different jurisdictions, meaning that its exact specification and terminology will differ from country to country. To provide an overview, the remainder of this section presents four major types of intellectual property as defined by the World Intellectual Property Organization (WIPO), which is an agency of the United Nations.*

Copyright

Recognized in most countries, *copyright* grants authors rights that only they can exercise, such as the right to adapt, publish, or broadcast their work, or to prevent the circulation of distorted copies. Copyright relates to artistic and literary works, including music, paintings, books, sculptures, sound recordings, films, broadcasts, computer programs, and electronic databases. A special feature of copyright is that it comes into existence automatically without registration.

Patents and Utility Models

Patents are the most widespread and arguably most important form of intellectual property. Patents are granted by a state or a patent office acting on behalf of multiple states. Once a patent is granted, the owner has the legal right to prevent any other business or person from commercially using the invention covered by the patent for a limited duration, which is 20 years in most countries. The patent application must disclose the invention publicly to be granted protection. Moreover, patents must be applied for in each country separately and their rights can be enforced only in a country in which it was granted. It is important to note that not all inventions are patentable. Patent laws generally state that the following conditions must be satisfied:

1. *The invention must be patentable*

   The criterion of *patentability* requires that the invention must fall within patentable subject matter as defined by national law. Normally this implies that the invention must exhibit an adequate level of ingenuity and inventiveness. This does not imply, however, that the invention must be complex. There are many exclusions from patentability. Normally things such as scientific theories, mathematical methods, plants or animals, discoveries in nature, medical methods, and any invention running against public order, morality, or public health are excluded.

* World Intellectual Property Organization (2016).

2. *Industrial applicability*

   The object of the patent application must have a practical use or be capable of some form of industrial application.
3. *Novelty*

   The invention must exhibit some new attribute that is not contained in the available body of existing specialist knowledge, referred to as *prior art*.
4. *It must contain an inventive step*

   The patent application must contain a non-obvious inventive step that cannot be made by a person with normal knowledge in the relevant technical field.
5. *The invention must be disclosed sufficiently*

   The invention must be made public in a way that is sufficiently clear and complete so that it could be used by an individual who is competent in the relevant technological skills.

Unsurprisingly, patents are wrapped up in complex and specialist law. This means that patent applications are normally prepared on behalf of inventors and businesses by a specialist patent lawyer, whose services tend to be costly. Moreover, it is not certain that a patient application will be granted by the patent office. If granted, a patent may need to be defended in court, resulting in significant additional costs. In combination, these factors make patenting a very expensive and risky process.

Utility models offer similar protections to patents but will normally cover technically less complex inventions. The process of applying for utility models is usually simpler and far less costly than for patents. While utility models vary strongly from country to country, they typically require the applicant to satisfy the criterion of novelty. However, the requirement for an inventive step is normally reduced and utility models have a shorter term of protection (with a typical duration of seven to ten years).

Industrial Designs

A product that is not inventive but has a unique appearance can be registered as an industrial design. Industrial designs refer to the appearance of designed objects, including arrangements of lines, shapes, or colors that give a distinctive aesthetic quality to an object. The object must normally be reproducible with industrial or commercial methods.

Design registration protects the creative nonfunctional elements in the appearance of an object that potentially determines its success in the marketplace. The duration of this protection is limited in time, ranging normally from 10 to 25 years. To be eligible for registration, a design must normally:

- be new at the date of registration; it must not have been previously published or disclosed in any way;
- be materially different and distinctive in shape, pattern, or ornament when compared with existing designs; and
- belong to an identifiable type of article, such as a toy car, kettle, or vacuum cleaner.

Trademarks

A trademark is a sign that can be associated with or attached to a product or service to distinguish it from the products or services offered by other businesses. Trademarks traditionally contain a combination of words, letters, numbers, images, shapes, and colors. However, an increasing number of countries allow for the registration of other features such as three-dimensional signs, sounds, or smells. The owner of a trademark has the exclusive right to use it and to prevent others from doing so. The period of protection varies but can normally be renewed indefinitely. Trademarks are generally used by managers to distinguish a product or service offering, signal a particular quality, and adherence to standards to buyers. In this way, trademarks are often used to support branding activities, as outlined in Chapter 11.

CONCLUSION

This chapter has presented a diverse collection of theories and models relating to technology and innovation. It has covered different definitions of technologies, common patterns in the diffusion of technology and innovations, the sources of innovation, various financial perspectives on innovation, modern iterative approaches to innovation, and forms of intellectual property. Two core ideas were highlighted: first, Christensen's theory of disruptive innovation, which characterizes the processes currently occurring in many industries. Second, intellectual property, which is a legal provision granting innovators a way to extract benefits by giving them temporary monopoly rights. Without these protections, extensive commercial innovation would be unthinkable. As a general overarching theme in innovation, the chapter has stressed the root uncertainty inherent to new technology development and innovation.

The book *Diffusion of Innovations* by Everett Rogers (2003) provides a well-known treatment of the processes through which innovations spread. The essay *The Question Concerning Technology* by Martin Heidegger (1977) is considered one of the most profound texts ever written about technology. The introduction by William Lovitt contained in the cited English translation of Heidegger's text is helpful when attempting to understand this demanding essay. Traditional classifications of technology are concisely summarized by David Jaffee in his book *Organization Theory: Tension and Change (2001)*, including additional, and pertinent, frameworks to capture the characteristics of digital technologies. The book *Managing Innovation and Entrepreneurship in Technology-Based Firms* by Michael Martin (1994) forms an accessible treatment of the sequential mode of innovation, including introductions to the concepts of technology push and market pull. For modern iterative methods, the *Manifesto for Agile Software Development* by Kent Beck and colleagues (2001) is an influential document. The books *The Startup Owner's Manual* by Steve Blank and Bob Dorf (2012) and *The Lean Startup* by Eric Ries (2011) are accessible standard pieces on customer development and are highly recommended.

REVIEW QUESTIONS

1. Which of these is not a motive for a company to innovate?

   (Question type: Multiple choice)

   O To continue the expansion of the company
   O To ensure that the products are up to date
   O To maintain an efficient company structure
   O To maintain technical competitiveness
   O To ensure the future viability of the company

2. All innovations become fully accepted in the marketplace if it can be demonstrated to the customer that they achieve their design objectives.

   (Question type: True/false)

   Is this statement true or false?

   O True O False

3. Complete the below definition of disruption.

   (Question type: Fill in the blanks)

   "Disruption is the process of creating ___________ products or services that can be sold in a market. In this process the new products or services destroy the ___________ and the advantages held by ___________."

4. Which of the following are characteristics of the problem known as the innovator's dilemma?

   (Question type: Multiple response)

   ☐ It is created by the possibility of disruptive innovation events in a market
   ☐ It is the result of established businesses prioritizing existing products
   ☐ It will inevitably destroy a market
   ☐ It is a puzzle caused by excessive spending on research and development
   ☐ It threatens the existence of incumbent businesses
   ☐ It threatens the existence of entrant businesses

5. A manager is concerned with the overall performance of innovation programs in a business. In particular, she wants to know how well the money spent on innovation results in the sale of new products. She has collected the following information for the last accounting period:

   - The sales revenue is $1.2m
   - The sales revenue originating from new products is $0.4m.
   - The total spend on R&D is $0.2m

   (Question type: Calculation)

   Calculate the appropriate metric for further analysis.

6. You are investigating a new manufacturing process used to produce molded components. To benchmark the technology against alternatives, you assess how average cost changes when output is expanded beyond 180 units. You have estimated the following total cost function:

   - $TC(q) = 14400 + 13q$

(Question type: Calculation)
Calculate the rate of change of average cost at this level.

7. Which of the following are not basic activities in the agile development approach?
   (Question type: Multiple response)
   ☐ Form the product backlog
   ☐ Hold scrums
   ☐ Create iterations
   ☐ Put in place Kanban systems
   ☐ Prioritize documentation
   ☐ Collect customer input and feedback
   ☐ Perform testing
   ☐ Target cost engineering
   ☐ Assemble teams to execute the work
   ☐ Perform risk mitigation measures

8. The following diagram summarizing the method of customer development contains errors. Identify them.
   (Question type: Image hotspot)

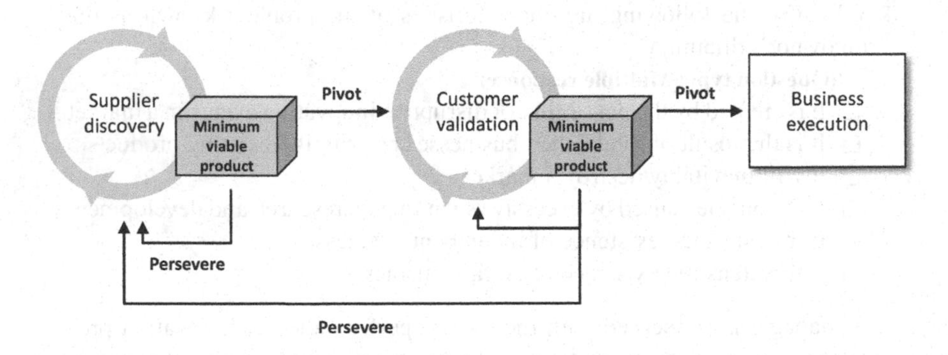

FIGURE 5.9 Identify the errors in this image

9. Pursuing patent protection is an inherently expensive and risky strategy.
   (Question type: True/false)
   Is this statement true or false?
   ○ True ○ False

10. Match the below types of intellectual property to the provided products and product characteristics.
    (Question type: Matrix)

	Copyright	Patent	Trademark	Industrial design
A painting	○	○	○	○
A new technology for power generation	○	○	○	○
A distinctively designed vacuum cleaner	○	○	○	○
The logo printed on business stationery	○	○	○	○

REFERENCES AND FURTHER READING

Aase, G., Swaminathan, S., and Roth, E., 2018. Taking the measure of innovation. *The McKinsey Quarterly.*

Beck, K., Beedle, M., Van Bennekum, A., Cockburn, A., Cunningham, W., Fowler, M., Grenning, J., Highsmith, J., Hunt, A., Jeffries, R., and Kern, J., 2001. Manifesto for agile software development.

Blank, S. and Dorf, B., 2012. *The startup owner's manual: The step-by-step guide for building a great company.* Pescadero: K&S Ranch.

D'Aveni, R., 2018. *The pan-industrial revolution: how new manufacturing titans will transform the world.* Boston: Houghton Mifflin.

Heidegger, M., 1977. *The question concerning technology and other essays.* New York: Harper & Row.

Jaffee, D., 2001. *Organization theory: Tension and change.* Boston: McGraw-Hill Humanitics, Social Sciences & World Languages.

Martin, M.J., 1994. *Managing innovation and entrepreneurship in technology-based firms.* Hoboken: John Wiley & Sons.

Porter, M.E., 1985. *Competitive advantage: Creating and sustaining superior performance.* New York: Macmillan.

Ries, E., 2011. *The lean startup: How today's entrepreneurs use continuous innovation to create radically successful businesses.* New York: Currency.

Rogers, E.M., 2003. *Diffusion of innovations.* 5th ed. New York: Free Press.

Stoneman, P., 1995. Introduction. In: P. Stoneman (ed.), *Handbook of the economics of innovation and technological change.* Hoboken: Blackwell.

Thompson, J.D., 1967. *Organizations in action.* New York: McGraw-Hill.

Woodward, J., 1965. *Industrial organization: Theory and practice.* Oxford: Oxford University Press.

World Intellectual Property Organization, 2016. *Understanding industrial property.* 2nd ed. Geneva.

Wu, T., 2010. *The master switch: The rise and fall of information empires.* New York: Vintage.

6 Life Cycle Thinking

OBJECTIVES AND LEARNING OUTCOMES

Decision-making in management is complicated by the fact that consequences of actions often occur at different points in the future. Moreover, these consequences routinely go beyond financial impacts and are likely to affect others or the environment of the business. The objective of this chapter is to familiarize the reader with the practice of life cycle thinking to evaluate financial and other impacts occurring in different time periods. As shown, methods of life cycle thinking are applicable in many different situations, for example, when deciding which project to choose from a list of potential options, when anticipating cash flows produced by a business or project over time, or whether to buy or lease a significant piece of equipment. Going beyond financial impacts, life cycle thinking is indispensable in the evaluation of how parties outside of the business are affected and for the assessment of the impacts on the natural environment. The last point is becoming increasingly pressing for managers due to the rapidly unfurling climate crisis. Specific learning outcomes of this chapter include:

- understanding how life cycle thinking can be defined;
- understanding profile evaluation as a simple evaluation tool to assess courses of action with disparate effects;
- knowledge of the typical cash flow shape associated with commercial projects and of its parameters;
- understanding whole-life costing and the ability to construct basic whole-life cost analyses;
- understanding the basic features of cost-benefit analysis and the idea of the stakeholder;
- understanding fundamental aspects of life cycle assessment techniques;
- understanding key terms, synonyms, and accepted acronyms; and
- appreciation of important thinkers and innovators in the field of life cycle thinking.

WHAT IS LIFE CYCLE THINKING?

It is often incorrectly assumed that the most important focus of decision-making in a business lies on immediate financial returns. This view is dangerously shortsighted because it will likely lead to decisions that are damaging in the long term. This realization gives rise to the question of how to think about impacts occurring over time. Over the recent decades, the practice of *life cycle thinking* has emerged to address such questions systematically. It is based on the observation that most things

DOI: 10.1201/9781003222903-7

created through human action exist in time and go through a more or less predictable sequence of life cycle stages, just as humans do (including managers!). The life cycle of an object, be it a business, product, or service, can be defined as follows:

A life cycle is a series of distinct stages through which an object, such as a business, a product, or a service offering, passes during its existence.

As this definition emphasizes, the underpinning idea of life cycle thinking is that products, services, projects, and businesses exist in time. Life cycle thinking is based on the premise that an optimal decision maximizes the positive impacts occurring throughout the life cycle while minimizing the negative impacts. This implies that it is necessary to adopt a perspective enabling decision-makers to weigh positive and negative effects occurring across the series of stages that make up the life cycle. Naturally, this form of decision-making depends on how important the individual impacts, good or bad, are relative to each other.

This chapter assembles a range of techniques that are used in the management context to assess the impacts of decisions over time, both on individuals and on the wider environment. In these methodologies, time does not always enter directly as a variable. For example, in the method of profile evaluation, which is presented first, timing is implicit in the various outcomes assessed. Other methods, such as the cash flow shape model, assess changes as time progresses. Moreover, while some methods are focused on financial impacts, such as whole life costing, other methods, such as life cycle assessment, take into account a much broader set of impacts, including damage to the natural environment.

PROFILE EVALUATION

Managers often face choices that will lead to distinct outcomes. These decisions could be certain courses of action, such as developing a certain product or service, or making a specific investment. In many cases, there is also the option to do nothing. However, it is normally not immediately obvious how beneficial or desirable the options are relative to each other. A family of simple methods that can be used in such situations is known as *profile evaluation*.

Profile evaluation begins by grouping a set of impacts or outcomes resulting from a proposed course of action under major headings. For a product development project proposed in a manufacturing business, which will be presented as an illustrative example in this section, this typically includes activities involving finance, production and operations, technology, and marketing and selling.

The second step begins by assigning a *time horizon* to each outcome, defining when it takes effect. A common distinction would be between immediate, short-term, and long-term outcomes. On this basis, each outcome is rated according to its quality or desirability, for example, by applying subjective categories such as "good", "neutral", or "poor" which can be coded numerically as "1", "0", and "–1". Each outcome is awarded a weighting factor reflecting its importance, such that, for example, "1" corresponds to low importance, "2" corresponds to intermediate importance, and "3" corresponds to high importance. By multiplying the rating of quality with the weighting factor, each characteristic is awarded a score, known in the evaluation context as a *figure of merit.*

Table 6.1 illustrates the application of profile evaluation to the example project in commercial manufacturing. The results can also be presented graphically as a column chart resulting in a distinctive profile, as done in Figure 6.1.

In this way, the profile evaluation approach results in a snapshot of the current and subjective understanding of a project option or course of action on the basis of the available information. Constructed for multiple options, profile evaluation can be used to decide between multiple alternatives, for example, to identify candidates for further, more detailed assessment. Staged investigations of alternatives in this way are known as *downselection.*

Profile evaluation can be made more elaborate, for instance, by expanding the investigated outcome categories. However, profile evaluation is in principle a simplistic method based on subjective information, both in terms of quality ratings and weighting factors, and should therefore not be overcomplicated. In particular, profile evaluation is vulnerable to missing the true consequences of the failure of important elements. For example, the consequences of a very poor rating in an essential outcome could be masked by other more highly rated characteristics, incorrectly suggesting an overall positive assessment.

THE CASH FLOW SHAPE TECHNIQUE

Most decisions in businesses will be structured with an eye on generating a financial return. As discussed in Chapter 2, many forms of business activities can be evaluated by observing the patterns of the inflows and outflows of money over time. This has led to the identification of a typical pattern of cash flow, known as a *cash flow shape,* associated with successful commercial initiatives, including product development projects, investments, and even new businesses. This pattern of cash flow is based on the assumption that the costs incurred in the short term are offset against the revenues generated by the same activity or project in the future. Activities or projects that behave in this way are known as *private ventures.**

The observation that the rate of sales occurring over time for products or services tends to also follow this general pattern has led to the idea of the life cycle of a market for a product, which is known more widely as the *product life cycle.* The product life cycle is usually characterized as consisting of four stages:

- invention and innovation;
- diffusion and growth;
- stabilization and maturity; and
- decline.

* Lanigan (1992).

TABLE 6.1
Illustrative Example of a Profile Evaluation

Heading	Characteristic	Time horizon	Quality rating	Numerical rating	Weighting factor	Score
Financial aspects	Availability of funding	Immediate	Good	1	2	2
	Timespan until revenue starts flowing	Short term	Poor	–1	3	–3
	Level of investment required	Immediate	Very good	2	2	4
Production aspects	Impact on production costs of other products	Immediate	Good	1	1	1
	Spare production capacity available	Immediate	Neutral	0	2	0
	Requirement to retrain the workforce	Immediate	Very good	2	2	4
	Ability to use the existing suppliers	Immediate	Very good	2	2	4
	Ability to support with existing production capability	Immediate	Good	1	3	3
	Environmental impact	Long term	Very good	2	3	6

(Continued)

TABLE 6.1 (CONTINUED)
Illustrative Example of a Profile Evaluation

Heading	Characteristic	Time horizon	Quality rating	Numerical rating	Weighting factor	Score
Technological aspects	Suitability of available technology	Immediate	Good	1	2	2
	Perception of required technology	Short term	Neutral	0	1	0
	Impact of intellectual property held by competitors	Short term	Neutral	0	1	0
	Potential to generate intellectual property	Short term	Good	1	2	2
Marketing and sales aspects	Public acceptance	Long term	Very good	2	3	6
	Potential for licensing deals with other businesses	Long term	Good	1	3	3
	Ability to price the product competitively	Short term	Neutral	0	3	0
	Complementarity with the existing product line	Short term	Good	1	3	3
	Effect on competitors	Long term	Neutral	0	2	0
	Public acceptance of the product	Long term	Very good	2	2	4

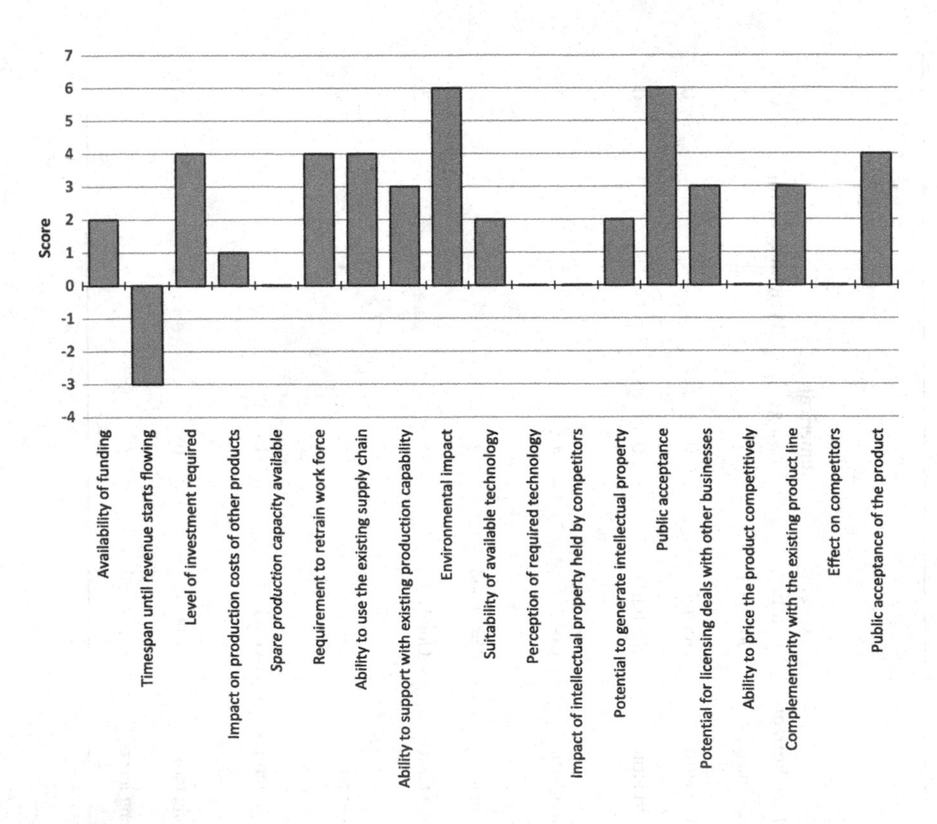

FIGURE 6.1 Profile evaluation example as a column chart

In turn, the idea of a generalizable product life cycle has led to its own distinct field in management known as *product life cycle management (PLM)*, focusing on strategies supporting a product or service as it progresses through its life cycle, from development to obsolescence*. It can be defined as follows:

Product life cycle management is the activity of managing the full life cycle of a product or service offering from its inception to its termination or replacement.

The underlying rationale of product life cycle management is that the conditions in which a product or a service offering exists are subject to change over time and must be managed appropriately. As evident from the following, product life cycle management shares significant conceptual similarity with cash flow shape analysis.

Figure 6.2a illustrates the idealized pattern of the sales rate over time of a product or service throughout its life cycle. The sales rate is measured in the quantity q sold at a given time. After a product or service has been invented and its innovation has been completed, it may enter its growth stage during which its diffusion begins and the sales rate accelerates, as described in Chapter 5. This is followed by a phase of maturity and stabilization

* The label "product life cycle" is confusing since it can easily be misunderstood to refer to the life cycle of an individual product unit. What is meant here is the life cycle of a product offering or product type in a business.

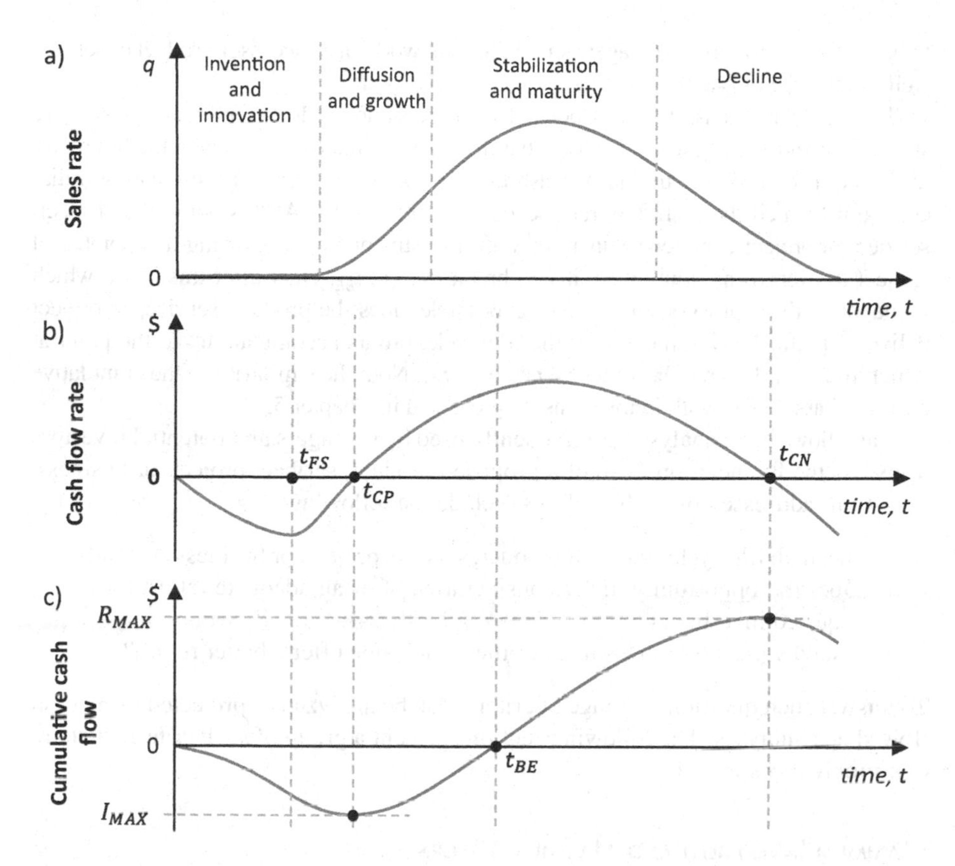

FIGURE 6.2 Typical cash flows associated with a successful project or service, adapted from Lanigan (1992)

during which the sales rate slows down and eventually begins to decrease. Finally, the product enters its decline during which the quantity sold gradually decreases to zero.

Figure 6.2b shows the cash flow rate at a given time. As can be seen, the cash flow rate in the project is initially zero and becomes increasingly negative in the initial stages of the project during which money is invested into development. The rate of spending typically increases in the beginning due to the increasingly expensive activities required to bring the product to market or to set up a project or business. The cash flow rate typically improves once the first unit is sold or the first returns are generated at time t_{FS}. If everything goes to plan, the cash flow rate soon becomes positive after this, at time t_{CP}. This is likely to occur during the diffusion and growth stages of a product or service. Following this, it is expected that the product, service, or project will remain cash flow positive for an extended period of time during which the initial expenses can be recouped and profits can be made. At some point, the sales rate will start to decline, bringing the cash flow rate back to zero at point t_{CN}, after which the product, service or project will become cash flow negative once more. This will typically happen during the decline stage of the life cycle. Naturally, managers will want the time span during which the cash flow is positive to be as long as possible. Managers should initiate work on a successor product or service well in advance of t_{CN}.

The cumulative cash flows associated with the same product, service, or project are shown in Figure 6.2c. Note, however, that the scales of the three y-axes in the figure are different. The graph of cumulative cash flow shows the pattern of accumulating initial costs until which the cash flow rate becomes positive at t_{CP}. At this point, the product, service, or project experiences its maximum investment I_{MAX}. Following this point, and subject to everything going well, it will break even at t_{BE}. Only after this point, which managers will want to occur as soon as possible, does the product, service, or project deliver a profit. During the rest of the life cycle, profits accumulate up to the point at which the cash flow rate becomes negative, at t_{CN}. Note the similarity to the cumulative cash flow associated with innovations, as presented in Chapter 5.

Cash flow shape analyses are frequently used by managers and potential investors to assess the financial promise of a proposed product, service, project, or business. Questions addressed by such analyses include the following:

- In which life cycle stage is a product, service, project, or business currently?
- Does the opportunity under consideration offer an adequate return for the required investment?
- Would some other investment of the same value offer a better return?

To answer such questions, a range of criteria can be analyzed or projected using cash flow shape analyses. The following sections present a group of such criteria that are commonly investigated.*

MAXIMUM INVESTMENT AND MAXIMUM RETURN

The point of maximum investment I_{MAX}, which occurs at time t_{CP}, is the point where the cumulative negative cash flow is at its highest level in the life cycle. This is obviously a very risky stage and ideally leads to breakeven at t_{BE} as revenue is generated. The point of maximum return R_{MAX}, which occurs at time t_{CN}, denotes when the product, service, or project has delivered its maximum level of profit and managers should end the project. In reality, ceasing activity at the optimal point is highly strategic and challenging – the failure to achieve this in many high-profile companies is at least partly responsible for the observed cycle of making staff redundant and re-hiring staff at a later point. Combining I_{MAX} and R_{MAX} allows the calculation of the *end ratio*, which measures how many times the maximum cumulative cash outlay is returned over the lifetime:

$$\text{End ratio} = \frac{R_{MAX}}{I_{MAX}} \tag{6.1}$$

Naturally, managers will want the end ratio to be as high as possible. It is important to note, however, that the end ratio does not take into account the duration of the life cycle. Generally, managers will also prefer the duration of the cash flow positive sales lifetime $t_{CN} - t_{CP}$ to be as long as possible.

* Lanigan (1992).

Average Rates of Investment, Recovery, and Return

To be able to provide the required funds, managers will need to know how much money is needed as the project progresses. A useful metric to gauge the requirement of cash in the early stages is the *average rate of investment* up to the point at which the cash flow rate becomes positive, t_{CP}. This can be determined as follows:

$$\text{Average rate of investment} = \frac{I_{MAX}}{t_{CP}} \tag{6.2}$$

At the point of breakeven, the cumulative positive cash flow reaches a level that balances the cumulative negative cash flow. Here, the initial investment has been recovered and the business is entering profitability. The speed with which this is achieved – the faster the better – is measured by the *average rate of recovery*, which is estimated as follows:

$$\text{Average rate of recovery} = \frac{I_{MAX}}{t_{BE} - t_{CP}} \tag{6.3}$$

The average rate with which cash will be generated by the product, service, or project after the business has become cash flow positive is also of great interest to managers. Managers will normally try to maximize the *average rate of return*, which is defined as follows*:

$$\text{Average rate of return} = \frac{R_{MAX}}{t_{CN} - t_{CP}} \tag{6.4}$$

It is important to note that the cash flow model, as introduced in this chapter, does not take into account interest rates or changes to the relative value of money over time (both aspects are introduced in Chapter 15) and inflation (see Chapter 3). These omissions complicate the situation in so far as that any investment must offer a return greater than the level of inflation to produce a real return and should offer a return greater than the zero-risk interest rate to be attractive.

WHOLE-LIFE COST

Whole-life cost is a financial concept that helps managers identify and compare the various costs of a product, service, project, or system. Also known as the *total cost of ownership (TCO)*, the whole-life cost reflects the total cost throughout multiple stages of the life cycle. Using the example of a piece of industrial equipment, such a life cycle

* It should be noted that outside of cash flow shape analysis, another metric is also known as the average rate of return. This metric assesses the profitability of an investment over its expected life. It is defined generally as a percentage:

$$\text{Average rate of return (alternative)} = \frac{\left(\dfrac{\text{Cumulative profit}}{\text{Duration of the investment}}\right)}{\text{Cost of the investment}} \times 100$$

typically includes design, manufacturing, purchase, installation, operation, maintenance, and disposal. A general definition of whole-life cost is as follows:

IMPORTANT DEFINITION

Whole-life cost is the total cost of owning an object over its life cycle as determined by financial analysis. It includes costs from purchase to disposal.

Whole-life cost may take into account a diverse range of costs such as shipping and logistics, opportunity costs, taxes, tax incentives, and customer-oriented costs such as those arising through supplier visits. Because it aims to be comprehensive, often including the costs of replacement and upgrades at the end of a life cycle, whole-life cost is particularly useful in the evaluation of capital investments such as expensive items of machinery. In this section, the concept of the whole-life cost will be introduced using the illustrative example of the purchase of an automobile with different finance options.

The initial decision in the construction of a whole-life cost analysis is selecting an appropriate figure of merit. In the evaluation of the whole-life cost of a personal vehicle, an often-used figure of merit is the total cost per mile traveled. While this is a particularly common evaluation metric, alternative analyses may be interested in total cumulative cost, cost per unit of time, or a cost which closely matches the way in which the vehicle generates value for the business. For example, commercial passenger transportation systems, such as trains or aircraft, may be more appropriately evaluated in terms of the cost per mile per passenger, also known as the cost per passenger mile.* Similarly, a container vessel may be usefully evaluated in terms of the cost per container mile.

The illustrative example presented in this section is based on the estimation of the cost per mile traveled in a personal car. To construct this whole-life cost model, the following costs are assumed:

- the purchase price of the car is $20,000;
- the car will be kept and operated for a period of time with a maximum of five years;
- annual fuel, maintenance, and other operating costs are $1,500;
- the car is subject to loss of value† over time; and
- the car will be driven 10,000 miles per year.

The example will compare the total expense of two scenarios related to the ownership of the car. In the first scenario, the car is *bought outright* from a dealership with available cash. The second scenario is a *closed-end lease agreement*. In such an arrangement, the customer may or may not pay a deposit and then pays the dealer a leasing fee over a period of time. Once a particular sum has been paid or a period has elapsed, the customer may have the right to purchase the car at some agreed price or return it to the dealership.

When assessing financial flows over time, it is often useful to view such flows as occurring at discrete points in time. The interval at which these events occur should

---

* A common metric of this kind in the aviation industry is *cost per available seat mile (CASM)*.

† Loss of value of this type is known as *depreciation*. This concept will be introduced in detail in Chapter 13.

be chosen to suit a particular need, with the most common duration being 12 months. In the illustrative example, all flows of money and value are assumed to occur exclusively at the *end of year* (*EOY*). On the basis of the EOY figures, this example will construct the metric of *cost per mile*$_i$ for each duration of ownership i for the two scenarios.*

Outright Purchase

In the most straightforward case, the customer pays a purchase cost of $20,000 upfront to acquire the car and then operates the vehicle for a maximum of five years. This way of acquiring an object is known as *outright purchase*. As the car is fully owned, it could be disposed of after each year at *residual value*, which is the remaining value of the car after its use. During each year of its ownership, running the car, therefore, incurs a loss of value as well as operating costs and is used to drive a number of miles, as shown in Table 6.2.

Using these data, the whole-life cost per mile for each ownership period with a duration of i years, ranging from one to five, can be calculated using the following formula, where TLV_i is the total cumulative cost of the loss of value, TOC_i is the total cumulative operating cost and TM_i is the total cumulative mileage traveled:

$$\text{Cost per mile}_i = \frac{TLV_i + TOC_i}{TM_i} \tag{6.5}$$

Based on the assumptions listed, Table 6.3 shows the values of TLV_i, TOC_i, and TM_i used in this calculation of *cost per mile*$_i$ for each ownership duration. Note that the values shown in Table 6.3 do not show annual values but the total values at a duration of i years.

TABLE 6.2
Costs and Miles Traveled at the End of Each Year

EOY	Purchase cost ($)	Residual value at EOY ($)	Loss of value ($)	Operating costs ($)	Miles traveled
1	20,000	12,000	8,000	1,500	10,000
2	–	6,000	6,000	1,500	10,000
3	–	3,000	3,000	1,500	10,000
4	–	2,000	1,000	1,500	10,000
5	–	1,500	500	1,500	10,000

* For simplicity, the example discussed in this chapter does not discount the value of future flows of money or the benefits received from the use of the car to net present value. This important aspect is introduced in detail in Chapter 15.

TABLE 6.3
Whole-Life Cost of Outright Purchase

	Ownership duration, *i*				
	1	2	3	4	5
Total loss of value after *i* years, TLV_i ($)	8,000	14,000	17,000	18,000	18,500
Total operating costs after *i* years, TOC_i ($)	1,500	3,000	4,500	6,000	7,500
Total mileage after *i* years, TM_i	10,000	20,000	30,000	40,000	50,000
Cost per mile$_i$ *($)*	0.95	0.85	0.72	0.60	0.52

Closed-End Lease

In the closed-end lease scenario, the customer agrees to pay $7,000 to the dealership as a leasing fee during each of the first three years. In the fourth year, the customer has the option to buy the car for an additional $1,000 to continue using it or to return it to the dealership. This is known as a *purchase-option price.* If the car is returned to the dealership, the arrangement ends at the end of the third year. Note that at the optional purchase of the vehicle in year 4 the car has a residual value ($2,000 at EOY 4) which is greater than its purchase-option price. In this situation, it is said that the customer *has equity* in the car; this is typically done to incentivize customers, also known as *lessees* in this case, to buy the vehicle after the lease period has elapsed. In the framework developed in this example, the fact that the residual value of the vehicle is greater than the purchase-option price results in a negative loss, or gain, of value for the customer at EOY 4. This situation is summarized in Table 6.4.

It is important to note that up to this optional sale at EOY 4, the dealership retains ownership of the car. This implies that in the first three years, the cost of the decrease

TABLE 6.4
Costs and Miles Traveled at the End of Each Year

EOY	Leasing fee ($)	Purchase cost ($)	Residual value at EOY ($)	Loss of value ($)	Operating costs ($)	Miles traveled
1	7,000	–	–	–	1,500	10,000
2	7,000	–	–	–	1,500	10,000
3	7,000	–	–	–	1,500	10,000
4	–	1,000	2,000	–1,000	1,500	10,000
5	–	–	1,500	500	1,500	10,000

in value arises to the dealership, not to the customer. Thus, the whole-life cost arising to the customer at an ownership duration of i years can be calculated by including the total leasing fees TLF_i as follows:

$$\text{Cost per mile}_i = \frac{TLF_i + TLV_i + TOC_i}{TM_i} \tag{6.6}$$

Table 6.5 shows the values used in the calculation for each ownership duration. Note that the cost per mile remains constant in the first three years. This is because the customer pays a fixed rental fee rather than incurring the value losses of the car, which are absorbed by the dealer.

However, due to the significant loss of value occurring in the early years of the life of the car, the dealership is unlikely to allow the customer to terminate the closed-end lease agreement before the agreed period has elapsed. In this case, it may be agreed that if the customer terminates the agreement at EOY 1 or EOY 2, a one-off penalty of $2,000 must be paid. This is in fact quite lenient since in many cases, the customer will be contractually obliged to keep paying the fees until the lease period has ended irrespective of whether the car is still needed.

Table 6.6 shows the modified calculation of the cost per mile for each successive year of ownership, using the total leasing fees with penalty $TLFP_i$. Note that, due to the one-off $2,000 penalty charged if the lease period is terminated prematurely, the cost per mile is no longer constant in the first two years.

COMPARISON

Now that the whole-life cost metric for outright purchase and closed-end lease has been estimated, it is possible to compare the cost per mile for each ownership duration. As illustrated in Figure 6.3, the mode of acquisition has an effect on the total cost of owning the car. It is evident that if the car is kept for the full duration of five years, outright ownership is preferable. However, if there is no penalty for the

TABLE 6.5
Whole-Life Cost of Close-End Lease with Optional Purchase after Three Years

	Ownership duration, i				
	1	2	3	4	5
Total leasing fees after i years, TLF_i ($)	7,000	14,000	21,000	21,000	21,000
Total loss of value after i years, TLV_i ($)	–	–	–	-1,000	-500
Total operating costs after i years, TOC_i ($)	1,500	3,000	4,500	6,000	7,500
Total mileage after i years, TM_i	10,000	20,000	30,000	40,000	50,000
Cost per mile$_i$ ($)	0.85	0.85	0.85	0.65	0.56

TABLE 6.6
Whole-Life Cost of Closed-End Lease with Penalty

	Ownership duration, *i*				
	1	2	3	4	5
Total leasing fees after *i* years with penalty, $TLFP_i$ (\$)	9,000	16,000	21,000	21,000	21,000
Total loss of value after *i* years, TLV_i (\$)	–	–	–	-1,000	-500
Total operating costs after *i* years, TOC_i (\$)	1,500	3,000	4,500	6,000	7,500
Total mileage after *i* years, TM_i	10,000	20,000	30,000	40,000	50,000
Cost per mile$_i$ (\$)	1.05	0.95	0.85	0.65	0.56

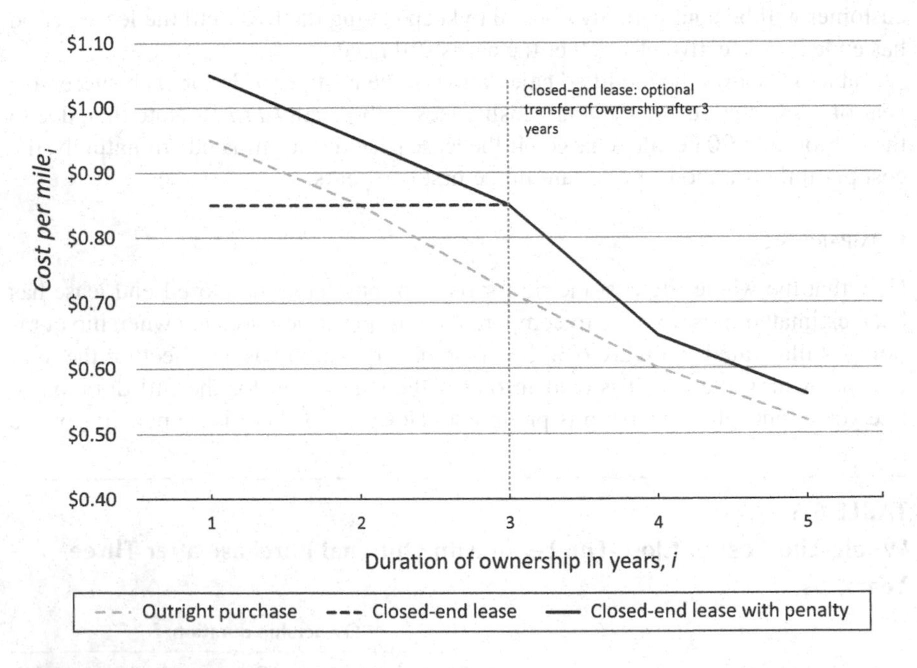

FIGURE 6.3 Whole-life cost per mile compared

premature termination of the closed-end lease agreement, the example suggests that at least in the first year, when the loss of value is high compared to the leasing fee, the leasing option is preferable.

While the example presented in this section focuses on different payment options for the same car, other analyses may construct whole-life cost models for different types of objects. For example, a manufacturing business may be interested in whether it is preferrable to buy an expensive and highly durable piece of equipment that lasts for a long period or if it is more cost-efficient to buy a succession of cheaper units with shorter life spans. As indicated in the illustrative example developed in this section, an important additional question that can be answered with whole-life costing models is whether to buy or rent equipment and machinery.

As in the presentation of profile evaluation, it is important to note that the provided illustrative example has not taken into account interest rates, changes to the relative value of money over time, and inflation. Especially where the duration of ownership is long, as is the case with ships and aircraft, these additional aspects can have a significant effect. Moreover, the presented example has not considered the accounting and tax implications of outright purchase versus leasing agreements. In outright purchases, a significant share of the overall cost arises in the form of depreciation, whereas in leasing arrangements, a large share of the cost is incurred as leasing fees, which are expenses. As introduced in Chapter 13, these costs enter financial accounts in different ways and may attract a different level of tax.

COST-BENEFIT ANALYSIS AND STAKEHOLDERS

The methods presented in this chapter up to this point are aimed at establishing the *private*, or *internal*, advantages and disadvantages arising to an owner from a product, service, object, project, or business over time. Since the impacts of business activities are seldom restricted to those directly involved, the scope of such analyses must often be expanded to include additional, outside, individuals or parties. A widely applied systematic approach to doing so is *cost-benefit analysis (CBA)*, which can be defined generally as follows:

Cost-benefit analysis is the systematic assessment of the advantages and disadvantages of alternative courses of action to identify the option that achieves the greatest benefits while protecting value for all affected parties.

FAMOUS THINKER

Jules Dupuit (1804–1866)

Setting cost-benefit analysis apart from the methods of life cycle thinking presented so far, its goal is to determine choices which increase overall welfare rather than maximizing private, internal benefits arising to specific individuals. Despite gaining acceptance from the mid-20th century, the concepts underlying cost-benefit analysis were introduced much earlier. The mid-19th century engineer and economist Jules Dupuit is credited with using such methods to determine the appropriate amount of toll for a bridge. In the present, cost-benefit analyses form a widely adopted method used by businesses and other organizations to assess the general attractiveness of potential courses of action. Often, this includes the option of doing nothing.

Returning to the illustrative example of the purchase of a car, executing a cost-benefit analysis is likely to reveal a host of additional considerations that have been omitted in the whole-life cost analysis performed in the previous section. These considerations might include the environmental impact of running the car in terms of energy consumption and CO_2 emissions, its contribution to excessive levels of traffic, the detrimental effect of its emissions on the health of neighbors, the value of the payable taxes to the government, and supporting the jobs at the local car dealership through its purchase.

As indicated in the definition, the goal of cost-benefit analyses is to evaluate how the expected negative impacts, captured in the form of costs, balance against the expected positive impacts. Note here that 'cost' is used in a wide sense which is not only restricted to financial costs but also includes other negative impacts. Although the method is seen to enable informed decisions about the best option, it is important to acknowledge that an exhaustive appraisal of all potential present and future costs and benefits is not normally possible.

The generic method of constructing a cost-benefit analysis typically involves the following steps:*

- define a list of alternative courses of action, policies, or projects;
- make a decision on whose costs and benefits matter and are taken into account;
- decide on categories of impact and select measurement metrics;
- estimate the impacts quantitatively throughout the life of the project;
- convert all costs and benefits into a common metric such as money;
- modify the value of future costs and benefits to reflect their value at present;
- calculate the value of each alternative in the present;
- conduct a sensitivity analysis; and
- recommend a particular course of action, policy, or project.

When conducting such analyses, it is important to note that both costs and benefits are likely to be highly diverse, even in seemingly simple examples, such as whether to buy a car or not. The costs tend to be the more accurately known elements due to the availability of commercial data and market information. Obtaining a clear picture of the benefits arising from a planned course of action is considered to be more challenging. This may be due, for example, to the fact that benefits may incorporate cost savings or may be shaped by the willingness to pay of those affected by the project or their willingness to accept compensation. The general process of evaluating benefits is to identify all parties affected by the project and establish the positive or negative value (usually in financial terms) that they ascribe to the effect on their welfare.

The Stakeholder

An important concept that has emerged toward the end of the 20th century as a cornerstone of modern cost-benefit analyses is the notion of the *stakeholder*. In general terms, any person, group, or organization with a potential interest in a project or activity, known as a *stake*, is a stakeholder. Importantly, this is not limited to individuals or parties that are involved commercially, such as vendors, employees, and

* Adapted from Boardman et al. (2018).

customers. Adapting the definition by Freeman and Reed (1983), stakeholders can be defined as follows:

> **A stakeholder is an individual, group, or organization that has a legitimate involvement with a project or organization. This legitimacy can be the result of ownership, legal rights, moral rights, or being affected.**

IMPORTANT DEFINITION

This means that those in the community that are affected, such as unrelated businesses and members of the wider public, are also stakeholders. To perform a cost-benefit analysis, it will, therefore, be necessary to identify the various stakeholders, understand their needs and expectations, evaluate their attitude (supportive, neutral, or opposed), and be in a position to prioritize the most significant members of the overall stakeholder community to focus the resources available for the investigation.

In systematic assessments of the stakeholder community, known as the *stakeholder management process*, it is usually possible to identify three distinct groups: *primary stakeholders*, *secondary stakeholders*, and *excluded (non-)stakeholders*. While being criticized as simplistic, this approach is nevertheless useful to develop an overview of the stakeholder community. Table 6.7 briefly characterizes the identified groups of stakeholders.

The notion of such a community of stakeholders suggests that, even in seemingly small projects, a large number of people is likely to be affected in some way. Some will be in favor of the project, some will be against it, and a significant proportion is likely to not hold a view at all. Similarly, different stakeholders will make different judgments regarding the relative importance of the elements in the project under consideration. For example, the buyer of a car may consider only the costs involved in owning and operating the vehicle, as modeled in the section on whole-life cost, whereas the government may be primarily interested in the tax income raised. This underlines the importance of defining the scope of cost-benefit analyses carefully.

TABLE 6.7
Different Types of Stakeholders and Their Definitions

Group	Definition	Examples
Primary (internal) stakeholders	Individuals, groups, or organizations engaged in economic transactions with the project or business	Shareholders, customers, suppliers, creditors, employees, and the state (through taxation)
Secondary (external) stakeholders	Individuals, groups, or organizations not engaged in an economic exchange with the project or business	Affected members of the public, parents, prospective customers, and people located nearby
Excluded (non-) stakeholders	Individuals or groups that are unaffected by the project or business	Unaffected businesses, unaffected members, or the public

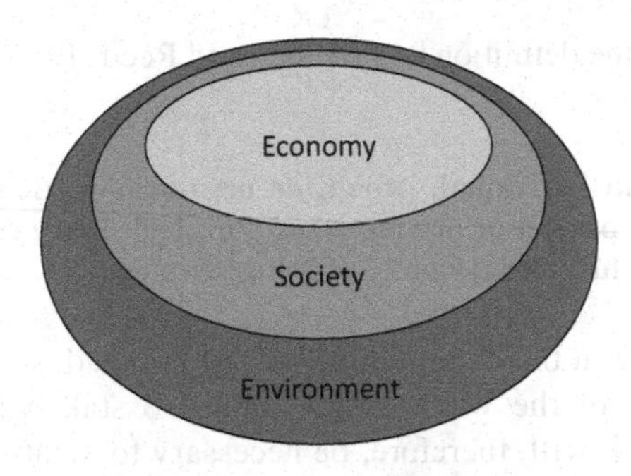

FIGURE 6.4 Components of sustainability

SUSTAINABILITY AND LIFE CYCLE ASSESSMENT

As is evident in the global climate crisis, an increasing focus on the environmental impacts of business, especially with regard to carbon emissions, is required. This is relevant in all areas of the economy but is especially pertinent in sectors generating major impacts such as construction and manufacturing. Importantly, many areas of economic activity produce carbon emissions not only through the processes they employ directly but also through their products which generate additional impacts during their life spans. While the precise nature and extent of these impacts can never be known from the outset, a range of tools are available for their systematic assessment. On the most general level, such analyses assess the *sustainability* of an object or process, which can be defined concisely as follows:

IMPORTANT DEFINITION

Sustainability is the ability of an entity or process to endure over time.

As a concept, sustainability is of particular relevance in the field of *ecology*, which studies the ability of biological systems to exist indefinitely and to sustain a diversity of life forms, including the human population. A common way to view the main aspects involved in sustainability is in terms of the three interrelated aspects of economy, society, and the environment. Also referred to as the 'three pillars' model of sustainability, these aspects are often shown as concentric circles to indicate that they do not exclude each other but are mutually reinforcing and that there is an order of precedence between the environment, society, and the economy. Figure 6.4 illustrates this relationship.

The remainder of this chapter presents the currently dominant method for the analysis of the sustainability of human activity, known as *life cycle assessment (LCA)*. LCA goes beyond the other approaches presented in this chapter by assessing the effects of activities on the natural environment itself, rather than limiting itself to impacts on people. The consideration of environmental sustainability is of interest to managers since it contributes to the long-term viability of business activity in general and is likely to support the corporate social responsibility initiatives of the business, as outlined in Chapter 2. The international standard governing LCA, ISO 14040 (ISO, 1997), defines LCA as follows:

"Life cycle assessment studies the environmental aspects and potential impacts throughout a product's life [...] from raw material acquisition through production, use, and disposal. The general categories of environmental impacts needing consideration include resource use, human health, and ecological impacts."

As with any approach subject to life cycle thinking, the idea behind LCA is the evaluation of impacts occurring over the lifespan of an object, product, or system. The initial step of an LCA is thus to structure the investigated life cycle into distinct phases, or stages. This requires the definition of the *boundaries*, or *scope*, of the overall analysis. The most common boundaries chosen for LCA studies, particularly when assessing manufacturing activities, is to assess the full life cycle of a product or service, ranging from raw material extraction to disposal. LCAs constructed in this way are referred to as *cradle-to-grave* analyses.

A typical cradle-to-grave life cycle with five stages is shown in Figure 6.5. It includes the stages of raw material generation, production, distribution, use phase, and end-of-life, which refers to either disposal, recycling, or re-use. Due to its versatility, this form of LCA can be applied to many industrial products. A further common model summarized in Figure 6.5 is *cradle-to-gate*, which limits the LCA to processes occurring during raw material generation and production. This somewhat constrained approach is frequently chosen if information from the later stages of the life cycle is not available. Another model is *cradle-to-cradle*, which reflects a circular life cycle in which the disposed products form the raw materials required by the

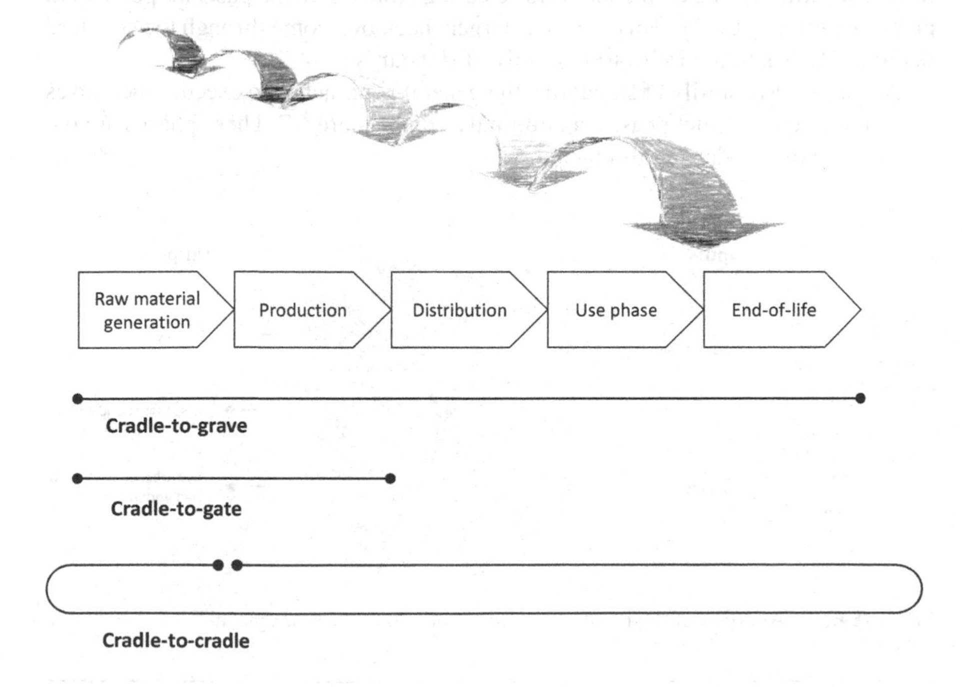

FIGURE 6.5 A typical five-stage product life cycle

following life cycle. This arrangement is not realistic for all kinds of products but is highly desirable from an environmental point of view.

A useful illustrative example for the purposes of this chapter is the application of LCA to the evaluation of the environmental benefits of installing an offshore wind turbine. Apart from the amount of energy that the turbine is expected to produce during its life span, there are a number of additional environmental impacts that are captured by the LCA. These would likely include the following aspects:

- the energy used during the development and manufacturing phases of the project;
- the energy used to process the raw materials used in its construction; and
- the environmental impact of servicing and maintaining the turbine.

Figure 6.6 summarizes several such impacts as inputs and outputs. In this context, a possible motivation to perform the LCA would be to assess if the turbine yields a net benefit to the environment over other forms of energy generation. As shown in Figure 6.6, there is a web of relationships between impacts, which may be of benefit to the environment or impose a cost on it.

An additional difficulty in constructing an LCA is the choice of units. Considering the example of the wind turbine, it is clear that the main output occurs in energy terms. The impacts of the raw materials consumed during construction might also be quantified in energy terms. But how should the disruption to the seabed and marine life be quantified? The confusion from selecting units has in the past hampered comparisons between LCAs. This issue has largely been overcome through the standardization of LCAs in the ISO 14000 family of standards.*

As part of this family of standards, the general approach is to execute such investigations in four distinct phases, as summarized in Figure 6.7. These phases are outlined in the remainder of this chapter.

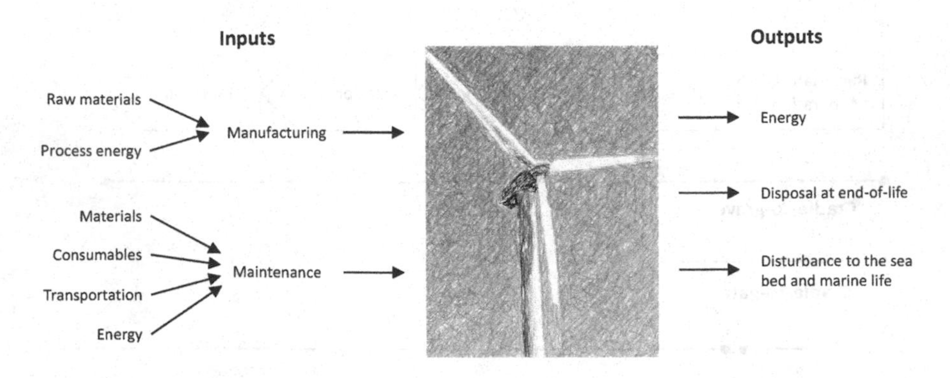

FIGURE 6.6 Analysis of the inputs and outputs of a wind turbine system

* ISO (1997).

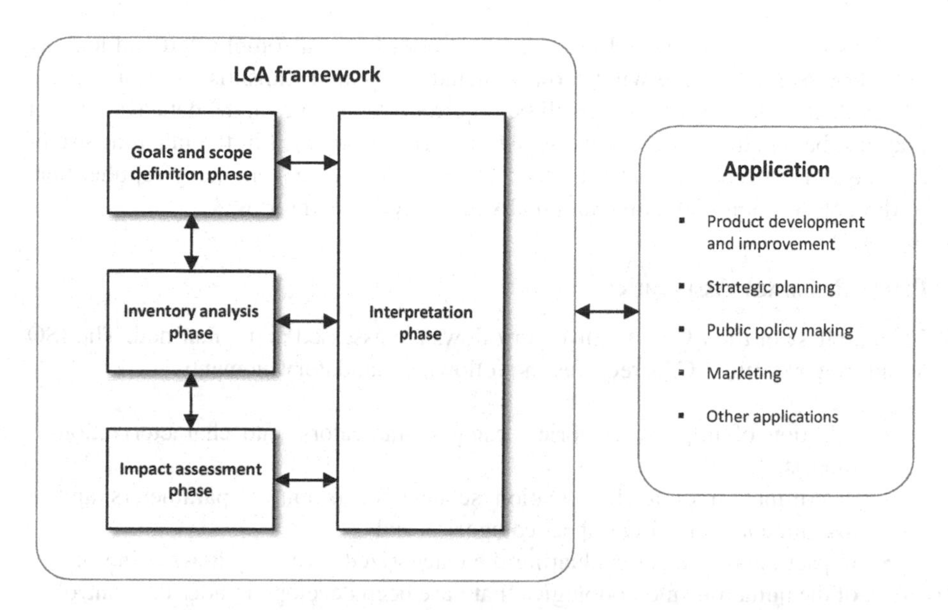

FIGURE 6.7 LCA phases according to ISO 14040 (ISO, 2015)

Phase 1: Goals and Scope Definition

This phase requires an explicit statement of the purpose, goal, functional units, and system boundaries consistent with the application. This will encompass adequate technical information to guide the analysis as well as a definition of functional units. The choice of appropriate functional units is important because the units facilitate communication between different LCA phases. The goals and scope phase will also collate any assumptions or known limitations of the analysis and the categories of impacts that will be considered, for example, CO_2 emissions or net energy generation.

Phase 2: Inventory Analysis

The inventory analysis phase provides the data for the determination of the balance of the environmental impacts within the study. This is done by creating an inventory* of all flows to and from the environment. In the outlined example of the offshore wind turbine, the flows back to nature include the energy generated, the disposal of the wind turbine after it is decommissioned, and the disturbance created in the seabed and for marine life through the wind turbine. Inventory flows from nature include the raw materials, energy, consumables, and resource consumption associated with transportation.

* In this context, "inventory" does not refer to a sum of actual objects, as discussed in Chapter 9. Instead, what is meant by inventory here is an enumeration, or list, of impacts.

Constructing appropriate life cycle inventories forms a complicated and lengthy task. For example, if the wind turbine includes a geared transmission, data on the life cycle inventory of the gears will be required. Depending on the data available, it may not be possible to determine which materials were used in the manufacture of the gears. This indicates that analysts routinely collect data from many supplier tiers in the construction of detailed and realistic life cycle inventories.

Phase 3: Impact Assessment

In this phase of the LCA, the inventory flows are assessed and quantified. The ISO standard governing LCAs requires the following mandatory elements:

- selection of impact categories, category indicators, and characterization models;
- determination of a classification scheme for inventory parameters and assignment to specific impact categories; and
- impact measurement evaluating the categorized inventory flows using one of the numerous methodologies that have been developed under the control of ISO.

The inventory flows are summed up to provide an overall impact category total. The ISO standard is specific as to what can and cannot be included. This allows comparisons between LCAs originating from different studies.

Phase 4: Interpretation

The interpretation phase is a methodical approach to the identification, evaluation, verification, and quantification of the data contained in the life cycle inventory and impact assessment phases. The outcome of this phase forms a set of conclusions and recommendations. Under the ISO standard, the interpretation will include:

- the identification of significant issues based on the inventory and impact assessments;
- the critical evaluation of the study itself considering its completeness and the use of sensitivity checks; and
- a formulation of conclusions, recommendations, and a statement of any limitations of the study.

These elements will highlight important results and determine the level of confidence in the work. The conclusions must be substantive, going beyond simple indications of which course of action is preferable. In the case of the offshore wind turbine discussed in this section, relevant conclusions could highlight the balance between the estimated energy output of the turbine over its life span or the relative significance of the CO_2 released during its transportation and maintenance.

CONCLUSION

This chapter has introduced the perspective of life cycle thinking and various methods available to managers to analyze the life cycles of projects, products, services, processes, and even organizations. The idea of separating the life cycle into discernible stages was presented as a core idea. It has been shown that all such approaches go beyond the maximization of immediate financial gains. One contribution of this chapter is to highlight the broad applicability of the typical cash flow shapes associated with successful business initiatives. It has also been stressed that managerial decision-making will increasingly have to take into account the environmental impacts generated by business activity. As indicated throughout this chapter, such considerations often involve changes of value over time, which relates to a series of techniques introduced in Chapter 15.

The textbook *Engineers in Business: The Principles of Management and Product Design* by Mike Lanigan (1992) provides an effective presentation of profile evaluation and cash flow shape analysis. *Cost-Benefit Analysis* by Anthony Boardman and colleagues (2018) is an authoritative textbook covering cost-benefit analysis. The section on stakeholders and stakeholder management in *Engineering Project Management* by Nigel Smith (ed. 2008) is concise and effective. The literature on sustainability and life cycle assessment is vast. An authoritative textbook covering life cycle assessment with reference to the ISO 14000 family of standards is *Life Cycle Assessment* by Klöpffer and Grahl (2014).

REVIEW QUESTIONS

1. Complete the below statement on life cycle thinking.

   (Question type: Fill in the blanks)

   "The underpinning idea of life cycle thinking is that ____________ exist in time and that it necessary to systematize their ____________ to enable decision-makers to ____________".

2. You are comparing two aspects involved in the development of a new service offering, which are spare production capacity and the requirement to retrain the workforce. Both will have an impact immediately. You have determined a quality rating of very good (equating to a value of "2") and an intermediate level of importance for spare production capacity. You have determined a quality rating of good (equating to a value of "1") and a high level of importance for the requirement to retrain the workforce. Which of the following statements is appropriate?

   (Question type: Multiple response)

   - ☐ The requirement to retrain the workforce has a score of "3"
   - ☐ The requirement to retrain the workforce has a score of "2"
   - ☐ The spare production capacity has a score of "2"
   - ☐ It is not possible to calculate scores because no numerical value is given for importance
   - ☐ Both aspects have the same score

3. Profile evaluation is vulnerable because it is liable to missing the true consequences of the failure of important elements.
   (Question type: True/false)
   Is this statement true or false?
   O True O False

4. Complete the following figure of the typical cumulative cash flow associated with a successful project by inserting the correct labels.
   (Question type: Labeling)

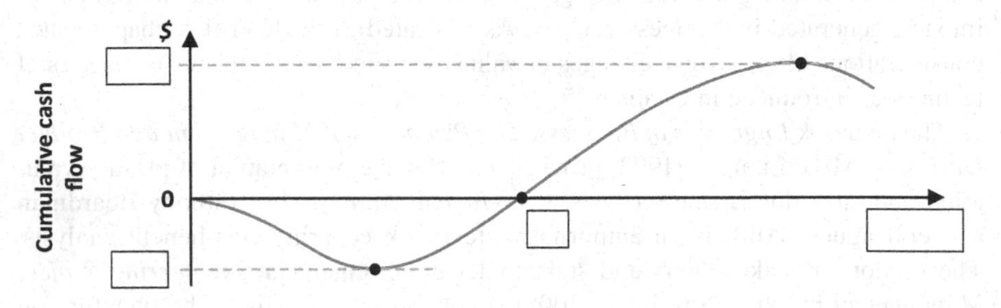

FIGURE 6.8 Insert the correct labels

5. You are investigating the finances of a block of apartments that has fallen into disrepair. The building is still owned by its original investor. The building became cash flow positive on January 01, 1970, and has become cash flow negative on December 31, 2020. The maximum cumulative cash flow obtained during its life cycle was $10 million.
   (Question type: Calculation)
   Calculate the average rate of return.

6. A shipping company is interested in procuring a new container vessel. You are hired to evaluate whole-life costs. The capacity and utilization of container ships are measured in "twenty-foot equivalent units" (TEU), which is based on 20-foot shipping containers. The ship is designed such that a 40-foot container can always be carried in place of two 20-foot containers. It is projected that the average level of utilization is 5,000 TEU. Prior research has shown that the new ship costs $118 million to purchase, and the company will incur total annual operating costs (including crew, insurance, fuel, and port fees) of $20 million. The vessel is projected to depreciate by $3 million each year. The company is planning to sail the ship 180,000 miles per year. The company is expecting to decommission the ship after 30 years releasing its residual value. You are assuming that there is no uncertainty or inflation.
   (Question type: Calculation)
   Calculate the whole-life cost of the vessel per container mile for a 40-foot container.

7. Order the provided steps of cost-benefit analysis in the correct sequence, with 1 being the first, etc.:

   (Question type: Ranking)

   - Convert all costs and benefits into a common metric
   - Calculate the value of each alternative value in the present
   - Define a list of alternative courses of action
   - Decide on categories of impact and select measurement metrics
   - Conduct a sensitivity analysis
   - Decision on whose costs and benefits are taken into account
   - Recommend a particular course of action
   - Estimate the impacts quantitatively throughout the life of the project
   - Modify the value of future costs and benefits to reflect their value at present

8. Match the provided examples of stakeholders to the appropriate stakeholder type.

   (Question type: Matrix)

	Primary stakeholder	Secondary stakeholder	Non-stakeholder
Child of an employee	○	○	○
Owner of preference shares	○	○	○
Debtor	○	○	○
Local government	○	○	○
Employee of an unrelated business	○	○	○

9. Complete the below figure showing the relationship between the so-called "three pillars" of sustainability by inserting the correct labels.

   (Question type: Labeling)

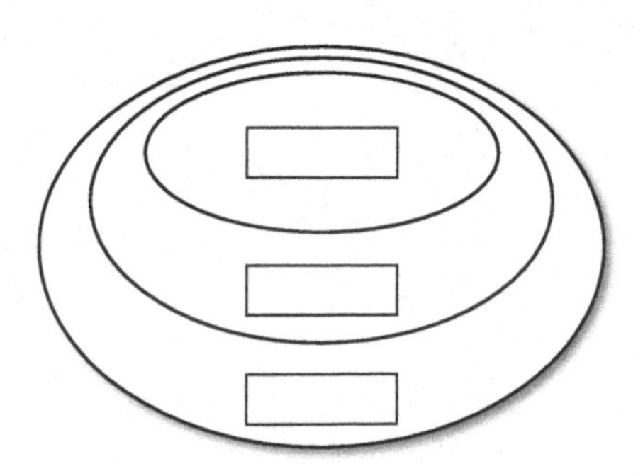

FIGURE 6.9 Insert the correct labels

10. Complete the below figure showing the relationship between the life cycle assessment phases as defined in ISO 14040.

(Question type: Labeling)

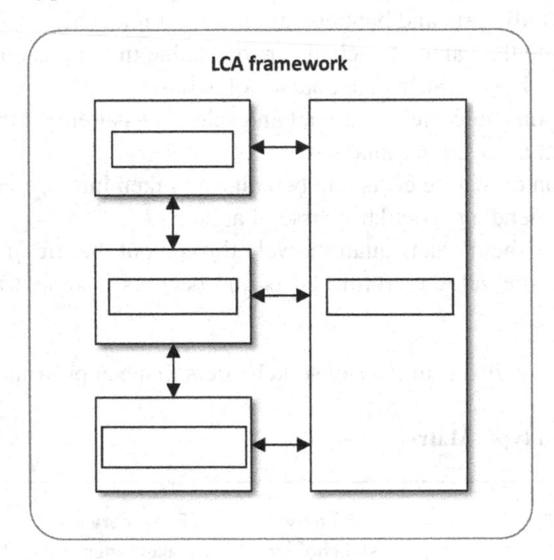

FIGURE 6.10 Insert the correct labels

REFERENCES AND FURTHER READING

Boardman, A.E., Greenberg, D.H., Vining, A.R., and Weimer, D.L., 2018. *Cost-benefit analysis*. Cambridge: Cambridge University Press.

Freeman, R.E. and Reed, D.L., 1983. Stockholders and stakeholders: A new perspective on corporate governance. *California Management Review*, 25(3), pp.88–106.

ISO, 2015. *ISO 14000:2015 Environmental management*. Geneva, Switzerland: International Organization for Standardization.

Klöpffer, W. and Grahl, B., 2014. *Life cycle assessment (LCA): A guide to best practice*. Hoboken: John Wiley & Sons.

Lanigan, M., 1992. *Engineers in business: The principles of management and product design*. Boston: Addison-Wesley Publishing Company.

Smith, N.J. ed., 2008. *Engineering project management*. Oxford: Blackwell.

Part II

The Activity of Managing the Business

7 Business Modeling

OBJECTIVES AND LEARNING OUTCOMES

The objective of this chapter is to provide an understanding of business models and their importance. After introducing major classes of business models, the chapter provides an overview of the main forces shaping them. Additionally, it presents important current methods used for the creation and analysis of business models. Finally, it briefly outlines four important types of business models: vertically integrated businesses, conglomerates, specialized businesses, and platforms. Specific learning outcomes include:

- the ability to appropriately define the term "business model" and to explain why it matters;
- knowledge of underlying classes of business models;
- an understanding of basic drivers that shape business models;
- knowledge of methods available to define and develop business models;
- an understanding of methods used to criticize and analyze business models;
- an understanding of important types of business models, and of the platform business model in particular,
- understanding key terms, synonyms, and accepted acronyms; and
- an appreciation of important thinkers in the field of business models.

WHAT IS A BUSINESS MODEL?

As discussed in Chapter 2, businesses can be characterized from different viewpoints. In many cases, private businesses will be seen from the vantage point of their ability to generate profit, or *earnings*, returned to owners or shareholders, both in the short term and the long term. The underlying idea of how the managers of the business think this profit is generated and, relating to this, how the business conducts its activities, is usually expressed in terms of what is known as a *business model*. Following Osterwalder and Pigneur (2010), a business model can be defined as follows:

> **A business model is a representation of the rationale of how an organization creates, delivers, and captures value, in economic, social, cultural, or other contexts.**

While it is often difficult to accurately and definitively formulate business models before or during the formation of a new business, there is a consensus that having a business model in the first place dramatically increases the chances of a company being successful. For this reason, outside stakeholders will normally focus on the business model when informing themselves about a business. In particular, they will pay special attention to how the business generates its revenue, which is referred to as *monetization*. An important additional function of business models is that they form

DOI: 10.1201/9781003222903-9

the basis for an important document known as the *business plan*, which will be introduced in detail in Chapter 8. Moreover, the business model should not be confused with the *business case*, which is a description of the projected earnings and costs upon which the success of a distinct project, which may well be a business, depends.

The thinking about business models has radically changed in the recent two decades through the spread of information technologies and the resulting rise of the platform business model, as discussed in detail in this chapter. Complicating matters, in many cases, businesses have incentives to hide or obscure their business models from customers or other stakeholders. This could be done to increase their appeal with a customer audience, to increase profitability, to avoid government intervention, or for other reasons. This can make the analysis of business models difficult for outsiders. It is the intent of this chapter to equip the reader with key concepts and methods that can be used to analyze the business models employed by existing businesses and to develop suitable business models for new businesses.

UNDERLYING CLASSES OF BUSINESS MODELS

B2B VERSUS B2C

The term *business-to-consumer*, normally abbreviated using the shorthand *B2C*, denotes the activity of selling products and services from a business to consumers. Businesses that sell primarily to consumers are said to run a *B2C business model*. Moreover, the label B2C can also emphasize that the business trades with the end user directly without involving intermediary businesses such as retailers. Sometimes such intermediary businesses are referred to as *middlemen*.

The emergence of the internet has greatly increased the viability of operating B2C business models and led to disruptive innovation in many industries, as discussed in Chapter 5. In B2C business models, marketing activities are tailored toward consumers, which will normally emphasize eliciting an emotional response in their target audience. This topic is further discussed in Chapter 11. Examples for B2C include most forms of retail, such as shopping malls, restaurants, cinemas, and online retail.

The term *business-to-business*, normally shortened to *B2B*, describes transactions between businesses, such as from manufacturers to wholesalers, or from manufacturers to retailers. The value of individual orders in the B2B setting tends to be far greater than in B2C. Moreover, B2B transactions are typically more complex and varied, frequently involving negotiations of prices or payment terms. In B2B, marketing activities are likely to be focused on demonstrating the value of a product or service. This aspect is discussed further in Chapter 11 of this book. Examples for B2B include most commerce taking place within supply chains and professional services, such as accountancy or consulting.

It is important to note that the distinction between B2C and B2B is not always clear-cut. Operating a B2B business model does not necessarily exclude sales to consumers. Likewise, a B2C business may also have transactions with other businesses. In fact, as will be explored in this chapter, many consumer-facing online service businesses, such as Google and Facebook, operate platform business models that are in their nature hybrids combining both B2B and B2C elements.

There are several other broad classes of business models. These include business models in which consumers transact with other consumers directly, known as *consumer-to-consumer* (*C2C*), for example, selling home-made craft products . A further class that is important in some sectors, such as the defense industry, is *business-to-government* (*B2G*). In this case, businesses sell their products or services mainly to governments and public institutions.

Pure Online versus O2O

Electronic commerce, also known *as e-commerce*, refers to businesses that engage in transactions that occur using electronic systems. An important distinction in business models involving online commerce is between *pure online* and *online-to-offline* (*O2O*) transactions. The initial wave of businesses focusing on the internet in the final years of the 20th century occurred mainly in industries producing information goods, as discussed in Chapter 4. Pure online business models falling in this class are frequently found in software, music, video streaming, and online retail banking. Common subtypes of pure online business models include the following:

- *Online intermediaries*: businesses pursuing this approach act as middlemen between other groups of sellers and buyers. Such businesses do not normally physically process or handle the products or services that are transacted. Monetization is normally done through the payment of commissions.
- *B2C based on advertising*: in this subtype, businesses attract consumers to a website by making free content available. In the process of interacting with this content, visitors are exposed to digital advertising by other businesses that pay a fee to the operator of the site.
- *Online communities*: in this case, businesses operate a digital system or venue through which a community of individual users can interact, based on shared interests. Such businesses are monetized by allowing other businesses or organizations to market their products to users, often in a targeted way and informed by data taken from the members of the community. Most social media businesses and search engine providers operate in this way. Frequently, online communities also generate revenues from the aggregation and sale of community data to third parties.
- *Fee-based*: some businesses are able to charge their users a fee for access to the content provided. This approach is frequently combined with a model offering some form of free access to attract new customers.

As a more recent development, businesses are increasingly adopting approaches that merge online elements with the transaction of physical goods or services. Such approaches are known as *online-to-offline*, or *O2O*, business models. The most basic example of O2O commerce is online retailing, also known as *direct selling*, which involves the online sale of physical goods that are then delivered to the buyer's address or made available for collection in traditional *brick-and-mortar* outlets. Frequently, O2O businesses operate both digital and traditional distribution channels alongside each other.

O2O approaches are deployed in an increasingly wide range of industries, including taxi services, accommodation, food delivery, and health care. For O2O to be appropriate, the nature of the transacted product must be such that the business and the customers benefit from the speed and responsiveness of using information technology in their transactions. For this reason, O2O models are often adopted in industries selling perishable goods, such as hot food, or where capacity is available at specific times only, such as in the hospitality or travel industries.

FORCES THAT SHAPE BUSINESS MODELS

As highlighted throughout this book, many aspects of how businesses work can be traced back to a small number of important forces. While these forces are as old as commerce itself, the way in which they are interpreted and incorporated into business models is subject to trends and fashions. This can, for example, influence how businesses are seen to benefit (or suffer) from being large, whether businesses should concentrate on one particular product or offer a diverse product range, and whether to *outsource* certain activities by engaging other businesses to carry these out. This section briefly discusses two important and fundamental forces that exert great influence over business models and how professional managers think: *economies of scale and scope* and *transaction costs.*

Economies of Scale and Scope

As introduced in some detail in Chapter 3, economies of scale are cost reductions that arise when the scale of an operation or process increases. For business models, economies of scale are of great importance. If a business model relies on an approach or technology that involves significant economies of scale then the business model will normally require that a minimum scale of operations is reached. This may mean that the business will have to capture a certain share of an existing market in a short period of time or will have to grow a new market quickly enough. The importance of scale is so ingrained in thinking about business models that business models are routinely evaluated in terms of their *scalability*, which expresses the ability of a business to benefit from expanding its activities without being limited by its resources or structure. In the context of business models, scalability can be defined as follows:

IMPORTANT DEFINITION

Scalability is a characteristic of a business or business function that describes its ability to tolerate or benefit from an increased scale in activity.

However, it has also been warned that managers should not pursue economies of scale blindly. As expressed in a quote attributed to management scholar Frederick Herzberg, "Numbers numb our feelings for what is being counted and lead to adoration of the economies of scale. Passion is in feeling the quality of experience, not in trying to measure it".* This suggests that managers should not ignore other aspects in their pursuit of business goals.

* Despite its source being unclear, Herzberg's quote is cited in this book since it makes a very clear point about some managers' fixation on economies of scale. Herzberg is encountered again in Chapter 10.

As introduced in Chapter 4, economies of scope form a special kind of economies of scale. Recall that economies of scope arise if the ability to handle a variety of products efficiently permits the production of multiple products or product variants alongside each other at a lower cost than if produced separately. Since this advantage can be used to pursue the goals of a business, the availability of economies of scope has a significant impact on business models. Common approaches targeting economies of scope in business models include offering variety in the design of products, responding quickly to changes in the market, establishing different product lines for customers to choose from, combining products into bundled offerings, and engaging in branding activities. The topic of branding is further discussed in Chapter 11.

Transaction Cost Economics

The effectiveness of markets in organizing economic activity forms a central theme in economics and has been discussed in detail in Chapter 3. From this perspective, the existence of large companies poses a puzzle: if markets are so useful in organizing the allocation of resources, why are businesses the default way of organizing people working together? Wouldn't it be more effective for workers to interact through markets (i.e., to collaborate on their own terms through contracts) than to form businesses? Following this line of thought, the fact that businesses exist suggests that enduring, complex, and hierarchical organizations confer significant advantages. Were this not the case, collaboration between individual people as independent contractors would be far more common.

There is, in fact, an increasing trend to engage contractors rather than to form companies with a paid labor force in the traditional sense. A clear example of this trend is the so-called *gig economy* where large numbers of low-skilled workers are flexibly hired on a contractual basis enabled by advanced software systems, normally by businesses running O2O business models.

Ronald Coase (1910–2013)

This is by no means a new puzzle. In the 1930s, the famous economist Ronald Coase showed that the decision between forming businesses with employees or organizing an alliance of independent contractors can be viewed as a cost minimization problem. Coase stressed the simple intuition that big commercial firms must possess some form of cost advantage over market-coordinated forms of collaboration. He argued that forming businesses carries several advantages, including the lack of expenses of organizing the activity between the collaborators and the absence of the cost of constantly renegotiating contracts. Within a business, decisions can be made using authority without requiring the effort of monitoring and enforcing contracts.

Such questions are addressed by a discipline in economics known as *transaction cost economics*, forming part of the theory of the firm, as encountered in Chapter 3. A basic insight from this field is that market-based arrangements tend to result in lower costs of creating products and services but that businesses tend to have lower costs of coordinating such activities. This includes the costs of setting up operations and running them over time.

In this context, it is relevant that access to cheap information technology and digital networking means that some of the advantages of businesses over market-based collaboration have become less important, at least in principle. For example, by using digital tools and specialized software it is now far more feasible to monitor contracts or to frequently negotiate prices. However, such technologies do not seem to have lessened the importance of businesses in general.

One practical obstacle with market-based arrangements is that writing complete contracts that cover every contingency is impossible. This implies that there will always be rights to make decisions that are not explicitly assigned to another party in a contract. In other words, there will inevitably be *residual control rights* that cannot be captured by contracts, no matter how comprehensive they are. This suggests that running a market-based arrangement that relies on contracts alone to specify who does what is problematic in complex everyday situations. Residual control rights are a good reason to believe that the business is, and will remain, a solution to the problem of organizing people.

METHODS OF CREATING BUSINESS MODELS

There are many useful techniques that help define or refine business models. This section presents three prominent methods. The first two methods, *customer needs analysis* and *value proposition design*, work on the assumption that the characteristics of a product or service should be closely intertwined with the business model adopted. The third method, the *business model canvas*, provides a framework to formally capture this relationship.

CUSTOMER NEEDS ANALYSIS

Customer needs analysis is a technique used to identify what requirements customers have regarding an existing or planned product or a service. Customer needs analysis is used in a variety of product management, brand management, and marketing contexts. It assumes that products or services are means to ends, meaning that customers make decisions based on whether they can achieve a desired state or valuable goal. For example, one customer might buy a particular branded smartphone because they prefer the design of its user interface. Another customer might buy it because its shape is aesthetically pleasing and conveys social status. Both customers are buying the same product but use it for different ends.

Customer needs analysis can be characterized as the process of identifying distinct kinds of interactions between a product and a customer. It concentrates on three different pathways for such interactions. The first pathway is through the features and attributes contained in a product or service, for example the smooth user interface in the above example of the smartphone. The second pathway is through the objective and subjective benefits that a customer obtains through the use of the product or service. An example for this could be the social status derived from using a particular product, such as a luxury-branded item of clothing. The third pathway of interaction occurs through the characteristics of a customer that will allow them to experience those underlying benefits. These traits may be personal, financial, psychological, or social in nature, such as membership of certain social groups. Knowledge of these pathways for interaction allows managers to develop business models that systematically address the needs of the customer.

Alternatively, customer needs analysis can be understood as an attempt to understand the progress that a customer is trying to make in a given circumstance. For this reason, it is also known as *jobs-to-be-done* analysis. Successful innovation within businesses and the creation of business models (which cannot always be separated) can be based on the expertise of helping individual customers solve problems specific to their situation.

Interpreted in this way, four additional concepts and assumptions are associated with customer needs analysis:

- a *job* is a condensed statement of what a person seeks to achieve in a given circumstance. It is assumed that the customer may be incapable of articulating this need;
- the jobs people are trying to address are never reducible to a single physical function or object. Instead, the requirements of customers always have social and emotional dimensions that need to be taken into account;
- the circumstances customers are situated in are more significant than their personal characteristics or the features of the product; and
- commercially successful innovations are marked by the fact that they have solved problems that had only inadequate solutions, or no solutions at all, in the past.

VALUE PROPOSITION DESIGN

In modern management, a *value proposition* is an important concept allowing managers to assess the benefits customers will obtain from acquiring and using a product or a service. Having a clearly stated value proposition helps managers understand which customers should be approached and to find convincing reasons to buy the product or service. It will also allow businesses to set themselves apart from competitors. A value proposition can be defined as follows:

A value proposition is a statement reflecting the entirety of benefits associated with a product or service which a supplier promises a customer will obtain in return for a payment.

The promise of value contained in a value proposition is often expressed as a solution that a product or service supposedly delivers. In this sense, the creation of value propositions, known as value proposition design, is similar to customer needs analysis. However, value proposition design is less targeted on the nuances of human perception and feelings, all of which may have an important effect on the way customers perceive value. Instead, it is more focused on the method and process of generating products and services that are of value to customers.

Osterwalder et al. (2014) developed a framework for the design of value propositions, splitting the process up into the steps of *customer profiling* and *value mapping*. Both elements are brought together to identify if a proposed product or service creates essential *customer gains* and relieves *customer pains*. The objective is to create a situation of *fit* between product features and customer characteristics. There are three kinds of fit that are normally seen to emerge in a staged process in successful value proposition design. These are:

- *problem-solution fit*, which is created when a value proposition is able to address the customers' needs or requirements;
- *product-market fit*, which is created when there is evidence that the product or service is actually creating value for customers by addressing the pains and gains; and
- *business model fit*, which can be demonstrated when there is evidence that the adopted business model is profitable and can be scaled. A variant of this kind of fit is *channel fit*, which is demonstrated if the costs of acquiring an additional customer are lower than the revenue obtained from that customer over time, resulting in *customer lifetime value*.*

The process of value proposition design is frequently supported through a graphical tool, which summarizes and organizes the stages of the process and stimulates discussions among stakeholders about each element. Figure 7.1 shows this graphical tool.

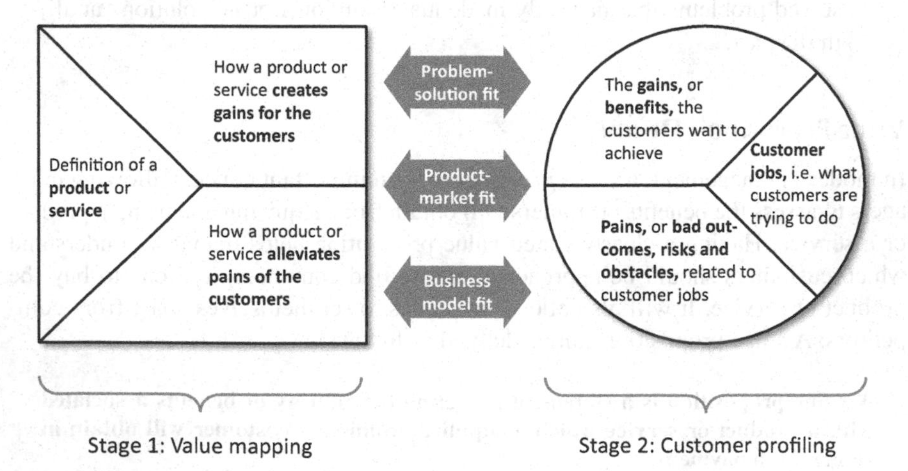

FIGURE 7.1 Graphical summary of value proposition design, adapted from Osterwalder et al. (2014)

* The concept of customer lifetime value (CLV) is introduced in detail in Chapter 11.

Business Model Canvas

Alex Osterwalder (born 1974)

Yves Pigneur (born 1954)

FAMOUS THINKERS

The business model canvas is a framework widely used by managers and entrepreneurs to develop new business models and to analyze existing ones. It is a visual template consisting of an arrangement of business functions that allows characterization of how the different aspects in a business model interact. The business model canvas was proposed in 2010 by business model theorists Alex Osterwalder and Yves Pigneur who are also among the creators of value proposition design, as discussed in the previous section of this chapter. The business model canvas is very widely used because it provides an easily understood standard model of reference. Often, the business model canvas is filled in during dedicated meetings in which a new business is proposed. This forces managers, entrepreneurs, and other stakeholders to consider major aspects of importance when developing business models and to plan how these aspects are aligned. As shown in Figure 7.2, the business model canvas consists of nine elements which can be grouped under four headings: infrastructure, offering, customers, and finances. This section will discuss each of these groups in turn.

Infrastructure: Key Partners, Key Activities, and Key Resources

The "infrastructure" heading contains the *key partners*, *key activities*, and *key resources* elements of a business model. The key partners element is used to determine who the key partners and suppliers of the business are and who will supply important resources. This element also defines which activities the partners will perform and whether there will be alliances or agreements with competitors or non-competitors. The key activities element describes the most important tasks a business needs to execute to realize its business model. This includes major processes involved in delivering the targeted value proposition, as defined in this chapter. The key resources element defines the most important resources required by the business to create value for its customers. These resources may be physical objects and machinery, various forms of intellectual property, required employee skills, and financial resources.

Offering: Value Proposition

The "offering" contains the value proposition, defining the configurations of products or services created by the business to deliver value for specific groups of customers. As introduced in detail in this chapter, a value proposition aims to show

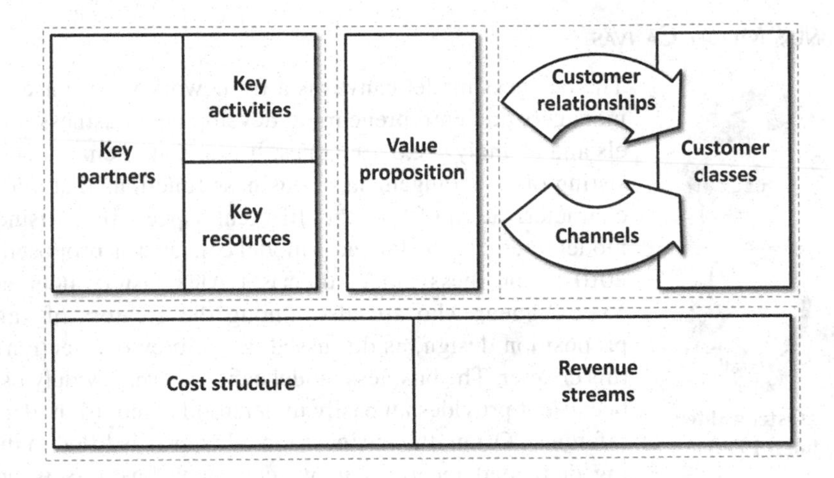

FIGURE 7.2 The business model canvas, adapted from Osterwalder and Pigneur (2010)

how customers benefit through product or service attributes such as novelty, quality, effectiveness, customization, price, design, status, or functionality.

Customers: Customer Classes, Customer Relationships, and Channels

Under the "customers" heading, the *customer classes* element defines various sets of individuals and organizations the business targets with their value proposition.* Customer classes are distinct groups of customers that can be identified by their needs and attributes so that their requirements can be met. Note that not all customers of a company may identify themselves as such, as will become apparent in the introduction of platform business models later in this chapter. The *customer relationships* element defines the relationships between the company and specific customer classes that will allow the acquisition of new customers, maintain the existing customer base and increase revenues from customers. This can involve different strategies such as forming personal relationships between customers and sales staff, self-service systems, or community-based approaches in which potential customers are given the opportunity to interact with each other. The *channels* element states how the business reaches its customer classes by creating awareness of its activities, allowing customers to evaluate the offering, providing points of sales, delivering the product or service, and giving after-sales support. Other businesses, such as distributors, may serve as channels to their customers.

Finances: Revenue Stream and Cost Structure

Under the "finances" heading, the *revenue streams* element defines the revenue the business obtains from each customer class. This requires a statement of how the business is monetized. Revenue can be obtained in different ways, including the sale of products or services, fees for use or intermediation, subscription models, licensing, advertising,

* Customer classes are also referred to as *customer segments* in some texts. To set this apart from the concept of market segmentation, which is a broader concept within marketing, discussed in Chapter 11, the authors prefer the term *customer classes*.

and, increasingly, the sale of personal data obtained from customers. The *cost structure* element reflects the costs arising from the execution of the business model. This may include different forms of costs such as fixed and variable costs, marginal costs, average costs, and sunk costs. To be useful, the statement of the cost structure should provide descriptions of how the costs relate to economies of scale and scope.

METHODS OF ASSESSING BUSINESS MODELS

A group of well-established techniques is available to help managers evaluate the characteristics of existing business models. These techniques are mostly derived from strategic planning where they give managers important cues about what the business should do to achieve its goals. This section will briefly introduce three common techniques: PESTEL analysis, SWOT analysis, and five forces analysis.

PESTEL Analysis

Political, Economic, Socio-cultural, Technological, Environmental, and Legal (*PESTEL*) analysis is a framework of environmental factors that can be used to evaluate the strategic context of a business. This process is known as *environmental scanning*. The assessment of external factors is necessary to capture which broad background factors could have an impact on business models. Knowledge of these factors can be useful, for example, for reasoning about the effects of certain technologies or to understand the viability of a proposed approach to monetization. However, not all identified factors will be of relevance for a given business model. Table 7.1 summarizes examples of environmental factors belonging to each PESTEL category. It is important to note that many factors are interrelated with other factors, such as the speed of technological diffusion and the consumer lifestyles in a specific environment.

SWOT Analysis

An extremely versatile analytical tool that can be applied to business models is *Strengths, Weaknesses, Opportunities, Threats* (*SWOT*) analysis. SWOT analysis aims to provide a systematic overview distinguishing between internal and external environments of a business. In this model, the internal environment is characterized by strengths, which are features of a business that give it an advantage over other businesses, and weaknesses, which are characteristics that place it at a disadvantage. The external environment is determined by opportunities, which are factors in the environment that a business can exploit to its advantage, and threats, which are factors that are likely to create problems. This is readily expressed as a 2-by-2 matrix, as shown in Table 7.2, illustrating SWOT analysis and providing examples for each category.

Five Forces Analysis

Five forces analysis is a method developed by Michael Porter (who first appeared in this book in Chapter 2) to analyze the effects of existing and potential competition on

TABLE 7.1
Examples for PESTEL Factors

Political	Economic	Socio-cultural	Technological	Environmental	Legal
• Government policy	• Economic growth	• Demographic factors	• Available technology	• Carbon emissions	• Employment law
• Public spending	• Interest rates and inflation	• Cultural values	• Technological change	• Climate change	• Company law
• Government stability	• Consumer wealth	• Consumer lifestyles	• Scientific activity	• Availability of natural resources	• Legally enforced standards
• Quality of institutions	• Business cycle	• Cultural practices	• Infrastructure	• Geographical features and location	• Business regulation
• Taxation regime	• Wage levels	• Education levels	• Speed of technology diffusion		
• Available subsidies		• Cohesion in society			
		• Level of corruption			

TABLE 7.2
SWOT Analysis as a 2-by-2 Matrix

	Helpful to the business	Harmful to the business
Internal factors	*Strengths*, for example: • Large established customer base • Growing reputation among customers • Experienced salespeople in the business	*Weaknesses*, for example: • Revenue too low for research and development • Low sales performance of some products or services • Poorly coordinated product lines
External factors	*Opportunities*, for example: • Products or services can both sell to businesses and consumers • Additional markets can be entered with derivative products or services • Known customer interest in additional product variants	*Threats*, for example: • Competitors may rapidly enter the market • A process employed will cause environmental damage and lead to legal action

the profitability of an industry or market.* As the name suggests, this type of analysis is based on the assumption that five forces relating to competition have an effect. These include the threat of new entrants, the severity of rivalry among competitors, the bargaining power of customers, the bargaining power of suppliers, and the threat of substitutes. Figure 7.3 summarizes this model. The rationale behind this approach is that the greater these forces are collectively, the less profitable the industry will be, and vice versa. As with SWOT analysis, the five forces framework can be used to give structure to the analysis of business models. The following sections briefly characterize each force.

Threat of Potential Entrants

In Porter's framework, the threat of potential entrants describes factors affecting how easily new entrants can join an industry. The level of this threat depends on the existence of what is known as *barriers to entry.* As an important concept in management and economics, barriers to entry are obstacles that entrant firms need to overcome to begin operating in a market or industry. They can be defined as follows:

A barrier to entry is an advantage that benefits established businesses in an industry but does not benefit businesses seeking to enter the industry. It thereby deters or prevents entry.

* Porter (1985).

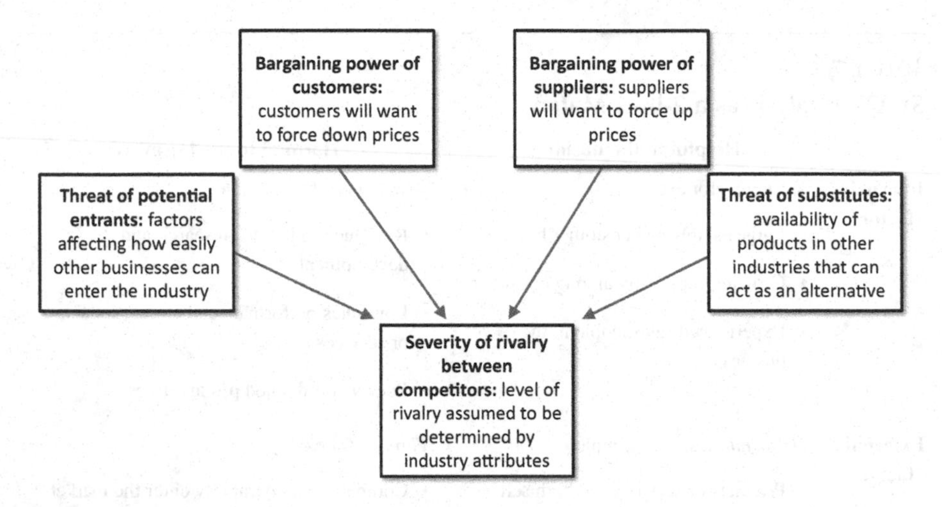

FIGURE 7.3 The five forces framework, adapted from Porter (1985)

There are many kinds of barriers to entry, including economies of scale enjoyed by incumbent businesses, established distribution channels, intellectual property rights, taxes or tariffs imposed on foreign entrants, and the level of customer loyalty.

Bargaining Power of Customers and Suppliers

In the five forces framework, it is assumed that customers will want to force down prices in an industry, seek higher quality products and secure other favorable terms for themselves. Conversely, it is assumed that suppliers to a business will want to charge the highest possible prices for the components required by the business, services, or raw materials and seek other favorable arrangements. The ability to extract favorable terms from customers or suppliers is known as *bargaining power*. It is defined as follows:

The bargaining power of a business is its ability to exert influence over other parties in a transaction in order to maximize its own benefits.

Customers tend to have bargaining power if they are few in numbers relative to suppliers, and vice versa. Bargaining power can be the result of market power, as defined in Chapter 3. Customers have additional bargaining power if there are numerous substitute products to choose from or if they can credibly threaten to switch to an internally produced supply. Suppliers will have extra bargaining power if the product they offer is distinctive or unique and if the costs of switching from one product to another are high. Bargaining power of this kind is also referred to as *lock-in*.

Threat of Substitutes

As discussed in Chapter 4, substitutes are products or services that are similar or indistinguishable or perform similar functions. If substitutes are available to a product or service transacted in a specific industry, then this will increase the competitiveness in that industry. The threat of substitutes will be particularly intense if there are technological innovations that threaten the existing advantages or capabilities of established businesses, if customers in an industry are not loyal to their suppliers, or if established businesses cannot defend their position through legal protections such as intellectual property.

Severity of Rivalry between Competitors

The level of rivalry among the participants in an industry or a market is seen to be determined by a range of factors, including the following:

- *the number of businesses active within an industry*: a large number of businesses is generally seen to make rivalry more fierce;
- *the growth of an industry*: a growing industry is likely to exhibit less rivalry than a stagnating or declining industry in which competitors fight for market share;
- *the relationship between fixed costs and variable costs*: industries with high fixed costs tend to be characterized by a willingness to overproduce and a correspondingly high level of rivalry;
- *the difficulty of exiting an industry*: if special equipment cannot be sold or there are other commitments to remain in the industry, rivalry tends to be high; and
- *the availability of substitutes from other businesses in the market*: if it is possible to interchange products or services from different suppliers, rivalry tends to be high.

MAJOR TYPES OF BUSINESS MODELS

As seen in the previous sections, a range of factors can be used to delineate business models. Each of these aspects carries important implications for the managers of the business, spanning from the development of the original business idea to its day-to-day running. This section outlines four major types of business models, each associated with specific benefits and drawbacks that can be viewed in the light of the concepts and models presented in this book. These types of business models have emerged in a historical process, each forming a response to the circumstances of a particular time. While they are presented in this historical order in this section, they are all still valid today and there are vast numbers of businesses across the globe that fall into each of these categories.

Vertical Integration

As explored in Chapter 1, before the 19th century, the business world was predominantly made up of small firms operating locally. Large and complex business hierarchies did not yet exist. This changed significantly over the course of the 19th century as nationwide markets of millions of customers emerged in some countries, leading to the first truly large private businesses. As also described in Chapter 1, developments in the automotive industry set in motion further advances, such as the manufacturing line, making possible massive, centralized manufacturing operations capable of producing vast quantities of products.

These giant factories obtained their inputs from complex and spread-out networks of suppliers and fed their products into extensive distribution systems. The emergence of such *upstream* (supplier-facing) and *downstream* (consumer-facing) structures further increased the importance of economies of scale. This process is also associated with the formation of large commercial banks and financial institutions to provide the finance required to establish such structures and to facilitate bargaining between powerful buyers and their suppliers.

Figure 7.4 illustrates a typical *supply chain* surrounding a manufacturing business, showing typical upstream structures including various tiers of suppliers and logistics processes as well as downstream processes including finished goods logistics, distribution services, and finally the consumer. Supply chains can be defined as follows:

A supply chain is a system consisting of organizations, people, resources, processes, and information involved in creating a product or service for a consumer.

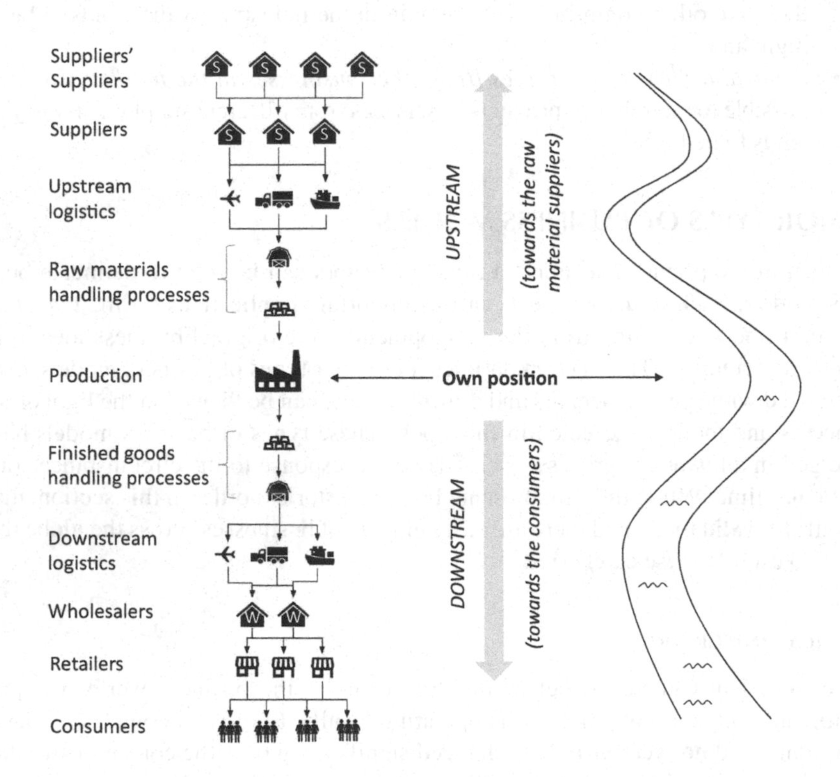

FIGURE 7.4 A typical supply chain structure found in manufacturing

Since a buyer might source inputs from multiple suppliers and sell a product or service to multiple customers, supply chains are also referred to as *supply and demand networks*. It is important to note that the term supply chain should not be confused with the term value chain introduced in Chapter 2. While supply chains are often characterized as *inter-firm value chains*, value chains normally describe processes occurring within a business. In this context, the frequently used terms "upstream" and "downstream" are inspired by the flow of a river in which water flows toward a final destination, as illustrated in Figure 7.4.

One response to the existence of upstream and downstream structures is to own these structures, thereby including them in the business. This process of achieving ownership is referred to as *vertical integration*. Vertical integration can be defined as follows:

Vertical integration is the process by which a business acquires ownership over more than one element of its supply chain or a configuration in which a business owns more than one element of its supply chain.

Following this definition, a business pursuing vertical integration aims to own supply chain elements outside of its original boundaries, located either upstream or downstream. The incorporation of downstream elements of supply chains is known as *forward integration*. Conversely, the integration of upstream elements is known as *backward integration*. The main benefit of vertical integration is that it avoids conflicts between suppliers and buyers resulting from bargaining power and profit-maximizing behavior. A further benefit is that, since the various supply chain elements do not operate through markets, transaction costs are likely to be low.

As discussed in Chapter 1, a famous historical example of vertical integration is the structure of the Ford Motor Company in the 1920s. In this era, Ford owned significant parts of its upstream supply chain. As also evident in the history of Ford, efficiently running a large vertically integrated business is a very difficult managerial task that requires tall hierarchies. To address this challenge, important theoretical contributions were made by practitioners such as Alfred Sloan and scholars such as Peter Drucker. Beyond industry, the emergence of vertically integrated businesses led to new academic institutions such as *business schools* and specialist service providers such as *management consultancies* delivering insight and the necessary expertise to support large vertically integrated organizations.

Conglomerates

Up to the 1950s, significant innovation had taken place in terms of manufacturing methods. These innovations led to considerable economies of scale and the dominance of vertically integrated organizations. At this point industry leaders started looking at other ways in which large size and the ability to run complex structures could be exploited. This gave rise to the *conglomerate business model*. Conglomerates are businesses that are characterized by the ownership of business units that are active in seemingly unrelated markets.

Typically, conglomerates are formed when organizations that are highly proficient and disciplined in accounting and professional management techniques buy underperforming businesses and improve these by applying their sophisticated management techniques and control systems. Owning such a group of business units, also known as a *portfolio*, allows conglomerates to extract profits from some units to fund investment into other units, for example, those in high-growth markets.

Driven by the idea of *diversification*, conglomerates are able to compensate for temporary problems or slumps in some industries through revenues from other industries and to take on large projects by pooling the profits generated by multiple units. Diversification can be defined as follows:

IMPORTANT DEFINITION

Diversification is the activity of entering into multiple markets involving substantially different skills, technology, and knowledge or acquiring different investments to reduce the exposure to any particular risk.

FAMOUS THINKER

Harold Geneen (1910–1997)

During the 1960s, conglomerates involving many different businesses units were considered the embodiment of advanced management thinking. As stated in a quote attributed to the former president of a conglomerate called ITT, Harold Geneen, "telephones, hotels, insurance – it's all the same. If you know the numbers inside out, you know the company inside out". Throughout his tenure as president of ITT, Geneen bought over 400 separate businesses, developing ITT from a telegraph equipment producer into a multinational conglomerate. In this era, it was thought that the units in a conglomerate would generally produce additional synergies in operation, for example, by sharing non-management resources across units. Current thinking among management scholars is, however, that, due to the complexity of managing conglomerates, it is not realistic to obtain large enough benefits from sharing resources across different units. Consequently, toward the end of the 20th-century conglomerates went out of fashion.

The Specialized Business

As the popularity of conglomerates waned in the 1980s and 1990s, it became apparent that other business models are more effective in imposing financial discipline and managerial excellence. This was made particularly clear through the restructuring activities involving dedicated consultancy businesses and *private equity firms*, which specialize in buying underperforming businesses and restructuring them. Following the shift away from conglomerates, the current opinion held by many managers and investors is that companies should concentrate on developing a specific and focused capability and deploy it at scale. Businesses operating in this way are referred to as *specialized businesses*. It is often assumed that businesses adopting this model will be able to acquire more investment and attract high-caliber employees. Specialized businesses can be defined as follows:

Specialized businesses concentrate on the production of a narrow range of products or services in order to bring to bear their expertise, productivity, and other advantages.

In the context of business models, specialization can be thought of as a process in which multiple businesses in a supply chain cooperate and divide work to their mutual benefit. This, of course, echoes Adam Smith's idea of the division of labor, as discussed in Chapter 3. Specialized businesses that focus their efforts and resources on one line of business or product category exclusively are said to run a *pure-play* business model.

FAMOUS THINKER

Jack Welch (1935–2020)

The transition from the conglomerate business model to the specialized business model is frequently illustrated using the example of General Electric. In the early 1980s, CEO Jack Welch completed a drastic reduction in the number of business units from 150 to 15. This process was guided by the simple criterion that for a unit to remain under General Electric's ownership it had to be the leader or in the second position in its market. Additional requirements were for each unit to provide value that no competitor would be able to match and to benefit from General Electric's core expertise in engineering-intensive activities.

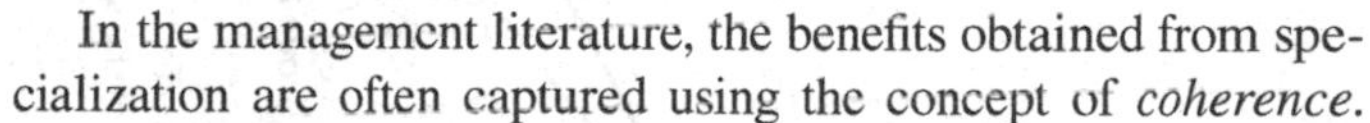

In the management literature, the benefits obtained from specialization are often captured using the concept of *coherence*. The activities of a business exhibit coherence if they form a unified and consistent whole and are free from logical contradictions.

For businesses operating a specialized business model, coherence in all activities has thus emerged as a primary driver of an organization's success. Coherence can be achieved by creating a value proposition that is well-aligned to the internal processes and goals of the business, developing a system of capabilities that is able to deliver the value proposition, and having a matching portfolio of products and services.

A recent example of the challenges faced by large, specialized businesses is provided by the aircraft manufacturer Boeing. In the 2010s, Boeing was troubled by a series of significant problems resulting from a very high, perhaps excessive, degree of specialization. Malone et al. (2011, p.56) summarize Boeing's problems as follows:

> "Looking at today's terrifically complex supply chains, one might think we've already reached the extremes of specialization. Boeing's initiative to build the 787 Dreamliner, for example, was hailed as the epitome of subcontracting – and then proved to have gone a bridge too far when the parts failed to come together as seamlessly as envisioned, and delays ensued. A web page listing just the "major" suppliers of the plane's components contains 379 links. But an aircraft is fundamentally a physical product."

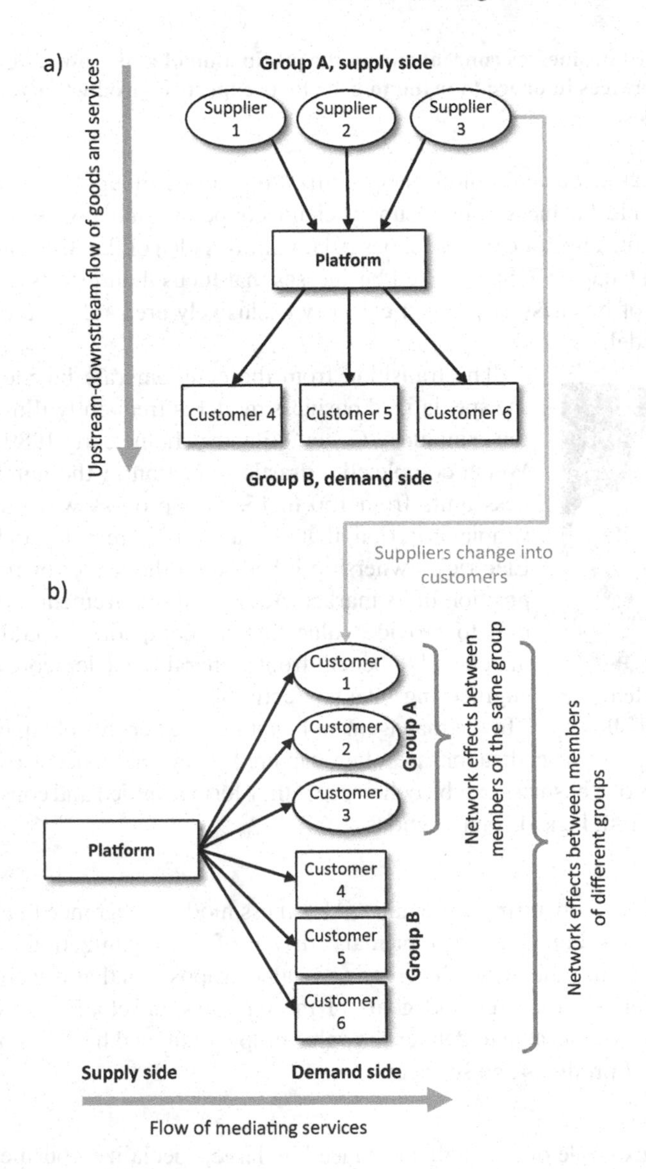

FIGURE 7.5 Reconceptualization of a business from value flow (a) to platform (b)

Platforms

The platform business model has emerged as a very significant challenger to other business models in the recent two decades. The conventional understanding of businesses is that they partake in a linear flow of value from upstream to downstream, as described in the presentation of Porter's value chain in Chapter 2 and in the illustration of the flow from upstream to downstream in Figure 7.4 and in Figure 7.5a. In other words, businesses following traditional approaches aim to create value by controlling a sequence of activities. This kind of linear thinking about business is very old and has been the dominant way of organizing productive activity for many centuries. To highlight this focus on the unidirectional flow of value, business models based on the linear flow of value are sometimes labeled *pipelines* or *pipes*. Businesses manufacturing consumer goods such as packaged food, beverages, toiletries, stationery, and over-the-counter medicines, referred to collectively as *fast-moving consumer goods (FMCGs)*, conform to this idea particularly well.

The platform business model sees the business process in a radically different way. Rather than understanding the business as a process adding value by transforming inputs into outputs, it sees the business as a privately owned venue for interactions between other parties, including suppliers, customers, or both. Instead of perceiving itself as creating value, a platform sees itself as orchestrating the value generation among other parties, thereby adding value indirectly. Following this reasoning, platforms can be defined as follows:

A platform provides a service enabling value-creating interactions between external actors. These actors can be suppliers or customers of products and services, including information goods.

Some authors argue that the platform business model is now the dominant business model and is generally capable of disrupting and displacing other business models.* This argument is underlined by the fact that the five businesses with the highest market valuations in the world (Apple, Microsoft, Amazon, Alphabet, and Facebook) all rely on platform business models, totaling a market capitalization of approximately $7.6 trillion in early 2021.

The reason for platforms to become so important in recent decades can be traced back to the severely reduced transaction and coordination costs enabled by networked information technology. This reduction in costs is, of course, unlocked through the almost-zero marginal cost structure emerging from information technology, as stressed in Chapter 4. Applying ultra-cheap mediating technologies based on digital systems enables a phenomenon known as *network effects*. In this context, a network effect refers to the impact arising to a member of a network or group if another member joins or leaves the network of group. In this sense, network effects are driven by interdependencies or complementarities between members or member groups.

Network effects are powerful because they create *positive feedback loops* that increase the benefits available as the overall number of members increases. In turn, this entices new members to join. Network effects also impose strong opportunity costs on members who wish to leave the platform, thereby creating lock-in. Because network effects in platform businesses are driven by the

* Van Alstyne et al. (2016).

number of members interacting with the business and with each other, they have also been characterized as *demand-side economies of scale and scope.*[*]

The subtlety of the platform business model can be illustrated using the example of a business that offers a digital ride-hailing service. Construed in the traditional, non-platform, way such a business would involve car operators acting as independent suppliers providing a transportation service to the ride-hailing service. The business will then coordinate these services and offer them to the end customers, the passengers. Depicted in the traditional upstream-downstream logic of activity flow, this configuration is shown in Figure 7.5a, featuring two distinct groups, group A (the car operators in this example) and group B (the passengers).

However, the same business looks very different when interpreted as a platform. Through this lens, the car operators are not viewed as suppliers but as an additional group of customers. Doing so construes the platform as providing a mediating service between the two distinct customer groups of car operators and passengers. This view is shown in Figure 7.5b.

The key difference resulting from changing from the pipeline to the platform perspective is that the car operators in group A have switched from the *supply-side* to the *demand-side* of the platform. Crucially, this perspective makes visible the network effects between the different groups, the car operators and the passengers. When an additional car operator joins, the passengers are made better off, for example through shorter waiting times and better coverage. When an additional passenger joins, the result is more revenue for the car operators.

Whether the car operators understand or agree to this logic is not important for the platform. The decisive point is that network effects are created. This applies in the same way to the business model of an internet search engine. Here, the conventional view would see the authors of digital content, for example, web sites, as suppliers (group A) making their content available through the platform to web surfers (group B). This view masks the insight that the suppliers of the content are in truth also customers of the platform.

While appearing somewhat abstract when encountered first, this idea is currently revolutionizing the world of business. This is due to two main reasons. First, some network effects that arise between distinct customer groups, such as group A and group B above, produce positive feedback loops that can lead to explosive growth in platform membership. Apart from creating the potential for very high profitability, at least in the long run, rapid platform growth will mean that it will be difficult for other entrants to challenge a platform once it has established itself. As discussed in Chapter 5, it is thought that such processes may contribute to new, milder patterns of disruptive innovation allowing powerful incumbent platforms to benefit from network effects to remain dominant over time. The second advantage is that if there are multiple customer groups, the platform can take

[*] The disambiguation of different kinds of network effects and mapping them to demand-side economies of scale and scope is conceptually difficult. Gawer (2014) provides a reasonably accessible treatment of this important question.

the revenues generated from one group, known as the *money-side* in the platform world, to subsidize the other group to attract even more new members.

CONCLUSION

This chapter has opened Part II of this book by introducing the topic of business models as the underlying rationale of what businesses do. It has summarized fundamental classes of business models and discussed the main forces shaping these. It has also broached three methods useful in the creation of business models, customer needs analysis, value proposition design, and the business model canvas. This was followed by the introduction of three methods to assess existing business models. In combination, these approaches should equip the reader with a basic understanding of how to approach business models. The remaining part of the chapter presented important types of business models. The chapter closed with a discussion of platform business models. The notion of coherence as a key ingredient of successful business models and the concept of network effects were presented as core ideas in this chapter.

For introductions to widespread methods used to design business models, the books *Business Model Generation* by Alex Osterwalder and Yves Pigneur (2010) and *Value Proposition Design* by Alex Osterwalder and colleagues (2014) are recommended. Many management textbooks present the methods discussed here for the assessment of business models; *Management – An Introduction* by David Boddy (2017) is particularly useful in this respect. The cited article "The Big Idea: The Age of Hyperspecialization" by Thomas Malone and colleagues (2011) explains specialized business models in an accessible way. For a detailed explanation of platform business models, the book *Information Rules* by Carl Shapiro and Hal Varian (1999) is a standard text and highly recommended. For ambitious readers, the article "Bridging Differing Perspectives on Technological Platforms: Toward an Integrative Framework" by Annabelle Gawer (2014) will be rewarding.

REVIEW QUESTIONS

1. Complete the below definition of a business model.

   (Question type: Fill in the blanks)

   "A business model is a description of the __________ of how an organization creates, delivers, and captures __________, in economic, social, cultural or other contexts".

2. Creating complete contracts between contractors that cover all outcomes is possible if done with due diligence. In this case, there will be no residual control rights.

   Question type: True/false)

   Is this statement true or false?

   O True O False

3. Complete the following figure showing the value proposition design process by inserting the correct labels.
   (Question type: Labeling)

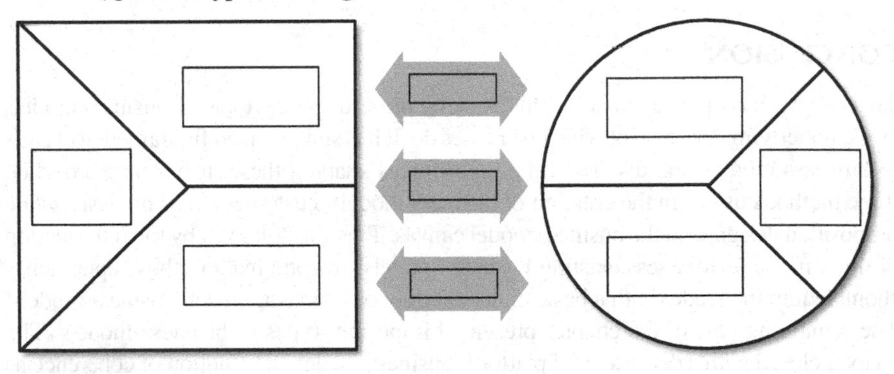

FIGURE 7.6 Insert the correct labels

4. Which of the following statements about value propositions are appropriate?
   (Question type: Dichotomous)

   True False
   O O A successful value proposition captures customers' attention
   O O It is a statement of guaranteed benefits
   O O It often involves a solution to a customer's problem
   O O Value propositions cannot be established clearly

5. Complete the following figure showing the business model canvas process by inserting the correct labels.
   (Question type: Labeling)

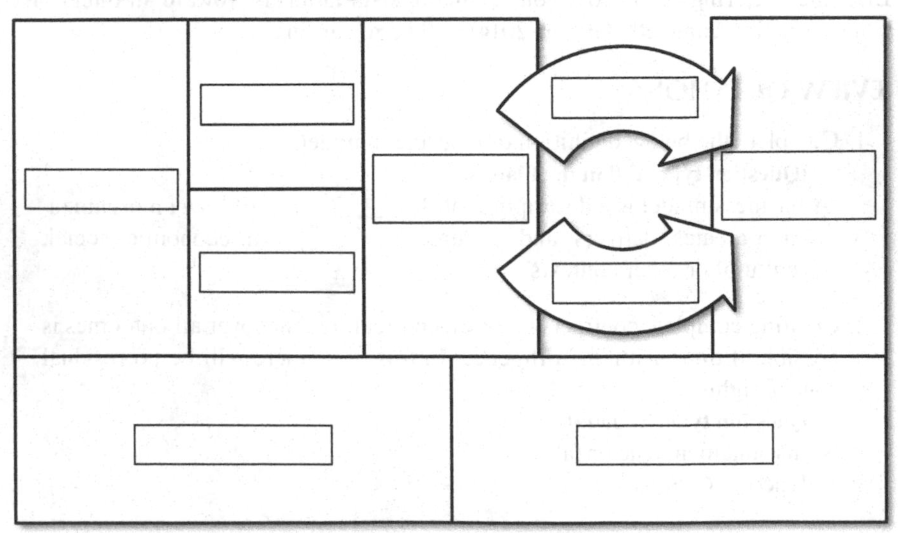

FIGURE 7.7 Insert the correct labels

6. Which of the following is true for conglomerates?
   (Question type: Multiple response)
   ☐ Conglomerates are increasingly considered a successful business model
   ☐ Conglomerates are normally small businesses
   ☐ Conglomerates are superfluous
   ☐ Conglomerates aim to capitalize on economies of scope
   ☐ Conglomerates are specialized technology firms
   ☐ Economies of scale are important for conglomerates
7. Which activities are normally thought to increase the coherence achieved in a business?
   (Question type: Multiple response)
   ☐ Developing matching products and services
   ☐ Increasing scale to achieve relevance
   ☐ Fitting an organization's best capabilities to all products and/or services
   ☐ Creating similar uniform capabilities in each unit
   ☐ Aligning the value proposition to organizational objectives
   ☐ Adopting the same value proposition across business units
   ☐ Creating a complementary capabilities system
   ☐ Overcoming boundaries between units
   ☐ Creating a unique value proposition
   ☐ Obtaining the first or second market position for each product or service
8. Match the provided major types of business models to characteristics of businesses.
   (Question type: Matrix)

	Vertically integrated businesses	Platforms	Specialized businesses	Conglomerates
Presence in diverse unrelated markets	○	○	○	○
Owning multiple elements of a supply chain	○	○	○	○
Facilitating the interaction between different groups	○	○	○	○
Pursuing coherence	○	○	○	○

9. Which of the following statements about network effects are true?
   (Question type: Dichotomous)

   True False
   O O Network effects are always positive
   O O Network effects can create lock-in
   O O Negative network effects are also called demand-side economies of scale
   O O Network effects can create feedback loops
   O O Network effects are driven by complementarities between members

10. In platform businesses, there can be ambiguity about whether a platform member is a customer or a supplier.
    (Question type: True/false)
    Is this statement true or false?
    O True O False

REFERENCES AND FURTHER READING

Boddy, D., 2017. *Management: An introduction*. 7th ed. New York: Pearson Education.

Christensen, C.M., Hall, T., Dillon, K. and Duncan, D.S., 2016. Know your customers' jobs to be done. *Harvard Business Review*, 94(9), pp.54–62.

Gawer, A., 2014. Bridging differing perspectives on technological platforms: Toward an integrative framework. *Research Policy*, 43(7), pp.1239–1249.

Leinwand, P. and Mainardi, C., 2012. The coherent conglomerate. *Harvard Business Review*, June 2012.

Malone, T.W., Laubacher, R.J., and Johns, T., 2011. The age of hyperspecialization. *Harvard Business Review*, 89(7–8), pp.56–+.

McAfee, A. and Brynjolfsson, E., 2017. *Machine, platform, crowd: Harnessing our digital future*. New York: WW Norton & Company.

Osterwalder, A. and Pigneur, Y., 2010. *Business model generation: A handbook for visionaries, game changers, and challengers*. Hoboken: John Wiley & Sons.

Osterwalder, A., Pigneur, Y., Bernarda, G. and Smith, A., 2014. *Value proposition design: How to create products and services customers want*. Hoboken: John Wiley & Sons.

Porter, M.E., 1985. *Competitive advantage: Creating and sustaining superior performance*. New York: Macmillan.

Shapiro, C. and Varian, H.R., 1999. *Information rules: A strategic guide to the network economy*. Boston: Harvard Business School Press.

Sloan, A.P., 1964, McDonald, J. (ed.), *My years with general motors*. Garden City, NY, US: Doubleday.

Van Alstyne, M.W., Parker, G.G., and Choudary, S.P., 2016. Pipelines, platforms, and the new rules of strategy. *Harvard Business Review*, 94(4), pp.54–62.

8 New Business Formation

OBJECTIVES AND LEARNING OUTCOMES

The objective of this chapter is to provide an understanding of the processes and challenges faced during the creation of new businesses. Starting new businesses is an area in which many authors propose simple and unverifiable methods for success. Unfortunately, the odds are stacked against new businesses. This chapter summarizes a range of methods that have emerged to improve the chances of survival of new businesses. It also introduces important parameters involved in the funding of new businesses, the main sources of funding, as well as the personal qualities normally required of founders. The chapter concludes with a brief introduction to business plan writing. This chapter aims to achieve the following learning outcomes:

- an appreciation of the general issue of new business formation and that projects of this kind are extremely risky;
- an ability to critically evaluate proposed methods of starting a business, including knowledge of the breadth of tasks that need to be completed in the process;
- an understanding of important financial parameters involved in starting the business including the level of funding required;
- knowledge of potential sources of funding and of other important prerequisites for starting a business;
- knowledge of the main processes of buying into existing businesses;
- an ability to write a simple business plan; and
- understanding key terms, synonyms, and accepted acronyms.

WHAT IS REMARKABLE ABOUT NEW BUSINESSES?

Starting a new business, for whichever reason, will always require the creation of something new. In other words, creating a new business that customers will be willing to buy products or services from is inevitably an act of progress of some sort. This progress can be *horizontal,* or *extensive*, in the sense that it extends an existing idea or business model that has been shown to work elsewhere to a new context. This might involve a new location, industry, or application. Alternatively, the new business might pursue *vertical* progress, which means that entirely new things are attempted. Vertical progress can be understood as innovative activity in which a new technology* or business model is created.

* As characterized in Chapter 5.

DOI: 10.1201/9781003222903-10

As this chapter will show, starting a new business is always an extremely difficult and risky undertaking for the *founders* of new businesses, also known as *entrepreneurs*, as well as for their financial backers and supporters. It is difficult to successfully achieve by large organizations with extensive resources and it is even more challenging for private individuals on a small budget. As this chapter will highlight, creating a new business is also an intensely social process relying on close communication with others and considerable trust between people. This means that a founder or a *team of co-founders* must work very successfully with other people to achieve their objectives.

Due to the difficulty and the newness inherent to the task of starting the new business, it is necessary that the group involved with the new company is small enough to respond successfully to changing and unforeseen situations. In this sense, a new business has been described as "the largest group of people you can convince to build a different future".* Moreover, successful new businesses are usually built around a deeper insight that is not yet shared by the outside world. Reflecting this, a new business can be fittingly characterized as a "conspiracy" to produce change.

The main objective of a new business is to engage profitably with yet-to-be-convinced customers. This is fraught with uncertainty since the business model the entrepreneurs have created remains to be validated and established.† In this light, a new business can be defined as follows:

A new business is an organization attempting to establish a viable and profitable business model under conditions of extreme uncertainty. New businesses are temporary because either they transition to being an established business or they fail and cease to exist.

In the context of new business formation, a distinction is frequently made between new businesses in general and *startups*. Usually, this is done on the basis that only new businesses with business models that are highly scalable and involve sufficient novelty should be classified as startups. Since both criteria involve arbitrary judgments, this chapter uses the label "startup" synonymously with "new business". Moreover, it is important to note that the formation of new businesses does not only apply to the creation of organizations from scratch. In practice, new businesses are often created from subunits or divisions of existing, successful businesses in a process known as *spinning out*. Alternatively, it is frequently possible for those wanting to acquire ownership of a business to buy into existing businesses. Such aspects are also outlined briefly in this chapter.

* Thiel (2014, p. 10).
† For a detailed discussion of business models, see Chapter 7.

REALITY CHECK: THE VAST MAJORITY OF NEW BUSINESSES FAIL

Starting a business is a surprisingly widespread ambition. In a recent non-representative online survey in the United Kingdom among employed adults, a striking 65% of respondents stated that they wanted to start their own business and an additional 14% of respondents stated that they were not sure.* This is remarkable because the odds of succeeding as founders are very low. Most available studies indicate that a vast proportion of businesses will fail, with the majority estimating failure rates of around 90%. On its own, this number is not particularly informative since it does not establish what business failure actually is and how this relates to the average lifespan of a failed business.

Looking at what constitutes new business failure is interesting since there may be many different consequences for those involved. For most stakeholders, including the founders and the employees, failure will generally mean that the business is unable to continue operating and they will become unemployed. In businesses with unincorporated legal forms, as introduced in Chapter 2, this will simply mean that the business will stop trading and it will attempt to dissolve all contractual obligations against which the founders are personally liable. For incorporated legal forms, the situation is more complex and involves *winding up*, which is a process of voluntary or compulsory *liquidation* in which the objects the company owns are sold to pay for the debts the business has accumulated. The failure of a new business is normally an acrimonious process in which multiple stakeholders lose part or all of their investment. In this sense, the end of ownership marks the failure of the new business.

Where institutional investors are involved, the concept of failure is more sophisticated. As discussed later in this chapter, professional investors often allocate funds to new businesses in a series of *funding rounds* over time. Here, failure can occur if a new business fails to acquire the required funding in any such round and will consequently have to wind up. Intriguingly, however, professional investors can also act as speculators, as discussed in Chapter 2. For these stakeholders, retaining unwanted ownership marks the failure of a new business. This means that failure to exit from a sequence of funding rounds at the appropriate time, for example, through the acquisition of the company by another investor or through the sale of its shares on the stock market, referred to as an *Initial Public Offering* (*IPO*) or a *flotation*, is also considered an investment failure. This can be the case even if the business can comfortably sustain itself and its founders.

As stated in the above, the odds are stacked against new businesses. One study† reports that of 35,568 startups founded in the period 1990–2010 in the United States, only 6,856 achieved an acquisition or an IPO. The same study reports that the average rate of failure to raise additional funds in any given funding round‡ is 67.6%, and the average rate of failure to exit through acquisition or IPO in each funding round is 81.2%. Another study§ reports that only 56% of new businesses survive the first

* Non-representative survey reported by Macnaught (2020).
† Quintero (2017).
‡ The term "funding round" is defined later in this chapter.
§ Mansfield (2020).

four years of their existence. Overall, only 40% of all startups ever become cash flow positive and only 30% ever break even. These sobering numbers suggest that most people have wildly unrealistic assumptions about the level of difficulty and risk involved in setting up a business and the deep, personal consequences failure could have.

PERSPECTIVES ON THE PROCESS OF STARTING A BUSINESS

When surveying the available literature on the process of starting a business, it becomes clear that there are significantly different conceptions of how the process works and what the necessary steps are. This section will present three common characterizations of the process of starting a business: *naïve procedural approaches*, *startup life cycling*, and *funding ladders*.

Naïve Procedural Approaches

Starting a business requires the entrepreneur to confidently handle a breadth of tasks and issues, none of which are likely to lie in the entrepreneur's area of interest or specialization. These include finance, sales, marketing, securing intellectual property, managing human resources, and handling legal issues. This breadth of activities is often described by presenting the process of creating a business as following a tidy sequence of predefined steps. Such frameworks usually assume (falsely!) that it is possible for the founder to establish that the business model and a value proposition are viable before starting the business. Despite this naivety, such frameworks are nevertheless useful because they introduce the novice entrepreneur to the necessary steps involved in setting up the business. They can also serve as an aide-memoire to experienced managers.

There are many competing procedural frameworks of this kind with varying degrees of consistency and realism. Some are available in books and numerous are available on websites. When consulting such frameworks, it would be a grave mistake to assume that following them guarantees the success of a new business. Figure 8.1 summarizes one such framework consisting of ten distinct steps, which do not necessarily have to be completed in the presented order. While this framework is concise and suggests a logical ordering, which may or may not be appropriate, other frameworks are more loosely structured. A framework consisting of 35 steps is summarized in Table 8.1. Despite being introduced as a step-by-step process, this framework is more an enumeration of important (but somewhat obvious) points than a properly defined sequence. Nevertheless, it is instructive in that it introduces the topics and problems new entrepreneurs will have to grapple with.

As can be seen from both frameworks, a multitude of activities must be completed with success to get the business up and running. These range from administrative tasks, such as applying for the required permits, to personal training activities for the founders, such as improving public speaking skills. However, there is no guarantee that even this extensive enumeration is comprehensive. Most likely, there will be many other, important steps to consider. Therefore, the real value of such frameworks lies in indicating general topics the founders should be aware of.

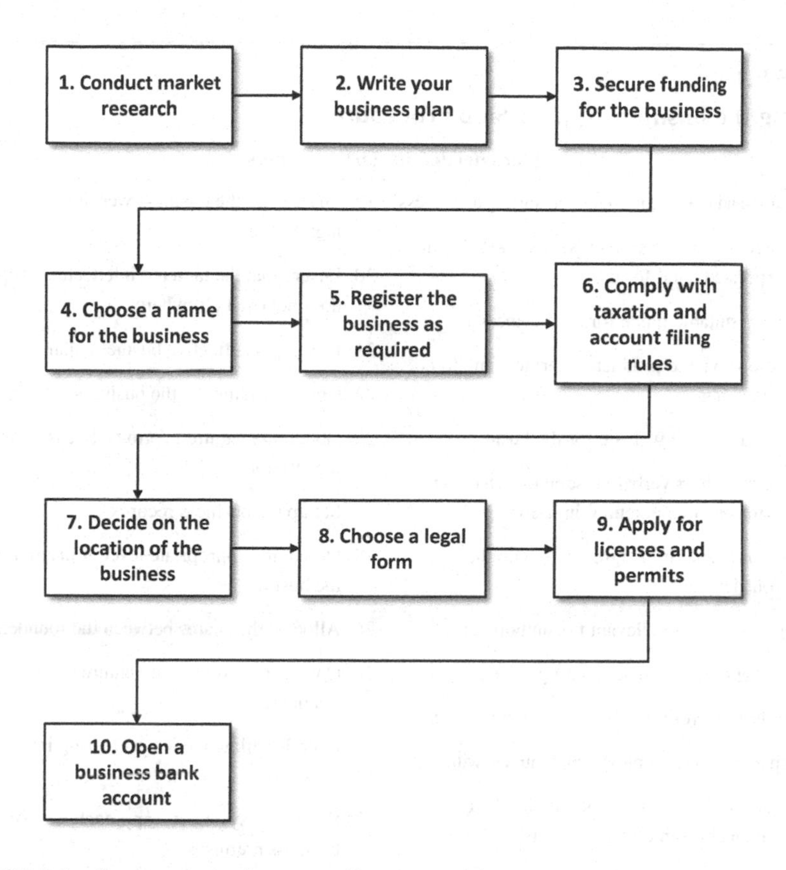

FIGURE 8.1 Starting the business as a 10-step procedure, adapted from US Small Business Administration (2020)

Startup Life Cycling

Instead of attempting to specify the steps that need to be taken when starting the business, other models capture the business formation process in the form of a life cycle. Normally such approaches focus on new businesses that require significant funding. One major model of this type is *customer development*, which was initially introduced as a product development approach in Chapter 5. The advantage of this approach is that it reflects the processes of learning that go on in a new business.

As with life cycling approaches in general, as presented in Chapter 6, the logic of customer development is to break complex processes up into a more manageable sequence of discrete steps. Figure 8.2 shows the full customer development process consisting of four stages: *customer development, customer validation, customer creation*, and *building the business.* The figure also shows how the three categories of fit introduced in Chapter 7 act as *stage gating criteria* for the process to move on to the next stage.

TABLE 8.1
Starting the Business as a 35-Step Procedure

Steps needed to start a business
1. Understand the requirements of starting a business
2. Protect the founders' interests by adopting an incorporated legal form
3. Select a suitable name for the business
4. Develop a viable product or service, but do not take too long
5. Set up a suitable website for the business
6. Develop a short verbal presentation that can be used to convince potential investors
7. Form a clear written agreement with the co-founders
8. Register with the relevant tax authorities
9. Adopt and set up an accounting system
10. Conduct full reference checks for employees
11. Sign appropriate employment contracts with staff
12. Require staff to sign confidentiality and invention assignment agreements
13. Secure the protection of intellectual property
14. Develop personal sales skills
15. Develop an understanding of financial plans and statements
16. Market the business intensely
17. Employ contractors to support the founding team where necessary
18. Develop an effective presentation for potential investors
19. Ensure that the business website is getting high traffic
20. Ensure that the targeted intellectual property does not exist elsewhere
21. Develop an effective business plan
22. Obtain funding for the business
23. Obtain the required permits, licenses, and registrations
24. Set up the business records
25. Obtain the appropriate level of insurance for the business
26. Allocate the shares between the founders
27. Consider the option of venture capital financing
28. Exercise diligence when entering into contracts
29. Exercise a systematic approach to leasing business premises
30. Research the competition in an ongoing way
31. Consider investments from private individuals
32. Consider stock options to incentivize employees
33. Provide excellent customer service
34. Use the services of a suitable business lawyer
35. Develop public speaking skills

Source: Adapted from Harroch (2018).

CORE IDEA

The significance of customer development is that it de-risks the new business by minimizing the need for founders to risk everything in *leaps of faith* with potentially catastrophic consequences. In essence, customer development frames the creation of new business as a learning process, geared toward learning about customers and the overall environment. If a hypothesis held by founders about their product or service, their market, or the customers is shown to be incorrect, the business will adapt its course and search for a new hypothesis.

This often leads to the modification or reconfiguration of business plans by pivoting, as discussed in Chapter 7. If the founders of the new business decide that the original course of action is still appropriate, this is referred to as persevering. Thus, customer development brings together modern thinking about innovation, business models, and new business formation in a coherent whole.

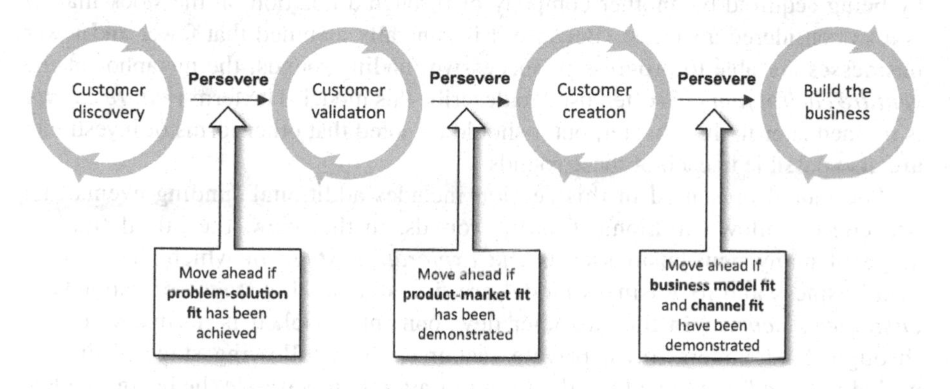

FIGURE 8.2 The full customer development process stage gated by categories of fit, adapted from Blank and Dorf (2012)

There is some evidence that executing pivots has a beneficial effect on new businesses. It is claimed that new businesses that have executed one or two pivots are able to raise on average 2.5 times the amount of funding, have on average 3.6 times more customer growth, and are significantly less likely to begin expanding the business prematurely.*

Funding Ladders

Once a potentially viable business model has been determined and the founders have taken steps to set up the business, they will seek to expand its activities, for example, by selling more products or services to its customers. In many cases, this expansion in activity cannot be funded solely through retained profit. This means that additional funding from investors is normally needed. For this reason, the process of forming a new business is frequently viewed as the sequence of funding events through which the entrepreneurs are able to raise money from

* Marmer et al. (2011).

outside investors, normally in exchange for shares in the business. Collectively, such processes are known as *funding ladders*. Within funding ladders, the distinct events leading up to such funding are referred to as *funding rounds*. They can be defined as follows:

IMPORTANT DEFINITION

A funding round is an event that gives outside investors the opportunity to invest cash in a new business in exchange for shares in the same business.

Viewing the process of starting the business in this way highlights that funding rounds act as filters, selectively determining which businesses are able to continue to operate and which fail. In the eyes of many professional investors, completing a funding round is a success. However, the business's *exit* from the funding process by being acquired by another company or through a flotation on the stock market is also considered a success. Because it is generally assumed that fewer and fewer businesses are able to progress in successive funding rounds, the metaphor of the *venture capital funnel* is often used to describe this model. The term *venture capital* is defined later in this chapter, but it should be noted that other forms of investment are also possible in each of these rounds.

The model presented in this section includes additional funding events that precede or follow traditional funding rounds. In this view, the initial funding stage of many new businesses is the *exploration stage* in which the concept of a business and its business model are first stated. This stage corresponds to *customer discovery* in the customer development model. It is normally funded through the founders' own, private resources. The following stage is the so-called *pre-seed* stage in which the founders attempt to validate the business idea. Corresponding to *customer validation* in customer development, this stage is typically also funded through the founders' own money but usually supported through loans from friends or family or through government grants. In the next stage, the new business begins executing its business model. This is called the *seed stage* and, together with all following stages in the venture capital funnel, forms the *customer creation* and *business building* phases in customer development. The seed stage is normally funded through various kinds of investors, including venture capitalists, business angels, and crowdfunding, which will be introduced later in this chapter. The following stages in the venture capital funnel are funding rounds normally labeled by a capital letter and the prefix *series*. Typically there are up to a maximum of six series funding rounds, ranging from *series A* to *series F*. Figure 8.3 summarizes the venture capital funnel.

Not all new businesses require significant external funding so not all businesses are adequately reflected by this model. Especially small new businesses that do not scale and that do not incur significant costs for setting up may be able to fund themselves through the resources of the founders, their own profitability, and through debt, such as bank loans. This approach carries the advantage that the founders retain all equity in the business. At the same time, this process is disadvantageous because the founders' own money is at stake in the process.

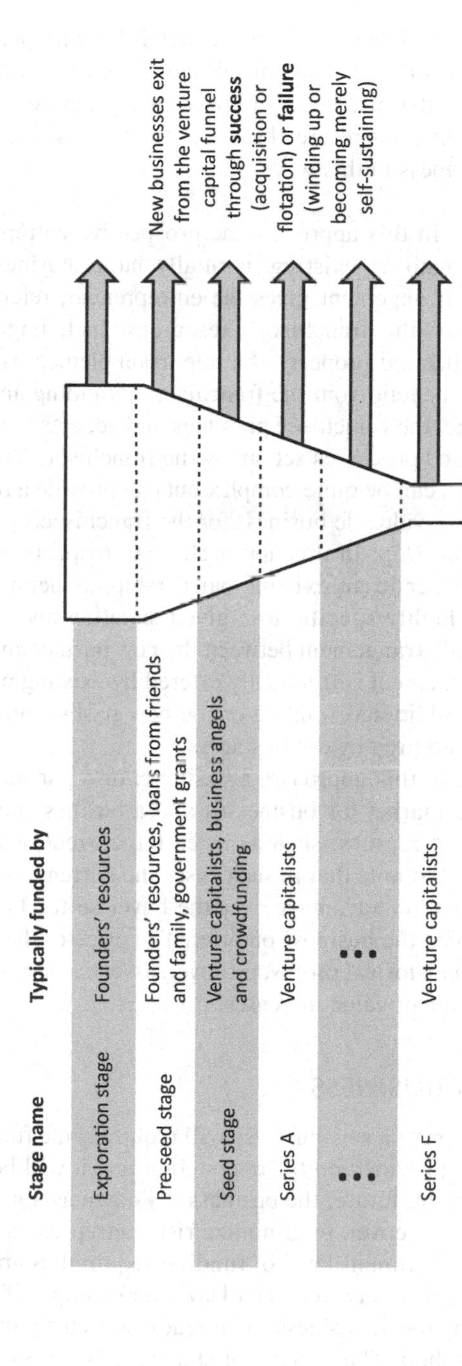

FIGURE 8.3 The venture capital funnel

DE-RISKING THE PROCESS BY BUYING INTO AN EXISTING BUSINESS

An alternative to starting a new business from scratch is for entrepreneurs to acquire an entire existing business or a part or unit of one. This carries the advantage of adopting a business model that has been validated over a period of time and has a history of generating profits. There are three main pathways for entrepreneurs to obtain ownership of a business in this way:

- *Buying a franchise*: In this approach, the prospective entrepreneur enters into an agreement with an existing, normally large, business, known as a *franchisor.* This arrangement gives the entrepreneur, referred to as the *franchisee*, access to the franchisor's resources, including information, processes, and intellectual property. An important element is that franchisees will be able to benefit from the franchisors' branding and advertising activity. In exchange, the franchisee pays fees and accepts the obligation to comply with rules and processes set out by the franchisor. The relationship between both parties can be quite complex but can provide a relatively low-risk way of creating a valuable business for the franchisee.
- *Buying into a partnership*: In this approach, the prospective entrepreneur becomes a new partner in an existing partnership, as defined in Chapter 2. This process is highly specific to a given situation and will normally involve a contractual arrangement between the new partner and the existing partners. This arrangement is frequently offered by existing partnerships if there is a need for additional funding or the new partner offers particular expertise which is required by the business.
- *Buying a business*: In this approach, a business, or a part of it, is offered for sale. There is a market for businesses and a business may be offered for clear and obvious reasons, such as when the current owner wishes to retire. It is important to note that in such cases the current owner possesses a significant information advantage over the buyer so the buyer will have to carefully scrutinize the business on offer. The price of the business will usually be based on historical profits, turnover, assets, stock, and the expectation of future profits or value increases.

FUNDING THE NEW BUSINESS

In most cases, the formation of a new business will require some funding from external sources. If funding is provided on the assumption that it will be repaid through the revenues generated in the future, the business is known as a *private venture*, as introduced in Chapter 6. Moreover, to minimize risk, entrepreneurs will plan to set up the business with the minimum level of funding required. Sometimes it is possible to establish a new business at a very small cost, for example, if the entrepreneur converts an existing hobby into a business and already owns the required equipment, such as tools and a workshop. The process of starting a business from the ground up with minimal or no outside investment is known as *bootstrapping*. In such cases, all that may be required before the business begins operating is to engage in some marketing activities, as introduced in Chapter 11, to register the business with the authorities, and to secure the necessary permits.

In other cases, businesses require significant upfront investment before they can start operating. If this form of expenditure is not recoverable, it forms a *sunk cost*. New restaurants are frequently cited examples for businesses with high upfront costs before the first customer can be served. In engineering businesses, significant research, design, and development activity is often required, with uncertain outcomes, before the business can begin trading.

Major Financial Parameters in the Process of Starting a Business

After founding the new business, cash is spent on materials and equipment, hiring the first employees, and initial marketing in order to begin generating revenues. If the new business is viewed as a private venture, the overall amount of money required during the formation of the business corresponds to the point of maximum investment I_{MAX}, as shown in Figure 6.2 in Chapter 6. Unfortunately, there is no useful rule to determine the level of funding needed to start the business, since this depends on the nature of the business and on its product or service offering.

An important additional question relating to the amount of funding needed is how much cash the business will require in its operation. For example, if the business will be operating in property development, it will require large pools of cash to buy properties to renovate and sell with a profit. Chapter 15 presents a range of tools used by managers to assess such requirements.

The following general points give an orientation to the level of funding required:

- Once the nature of a product or service offering has been established and the activities of the business are clear, financial planning can be performed, resulting in a *profit and loss budget* and a *cash flow forecast*. These methods are introduced in Chapter 14.
- Generally, it is thought that entrepreneurs should attempt to secure as much funding from outside sources as possible since a higher level of funding gives the business more time and resources, which may be needed to correct potential errors.
- Entrepreneurs must include their personal living expenses when planning the new business. A business that does not sustain the founder or co-founders is not viable. However, many prospective entrepreneurs are willing to sacrifice some of their personal income when starting their businesses.
- Entrepreneurs should plan for the unexpected and use risk management techniques. Chapter 12 introduces basic risk management methods and shows how contingencies can be quantified financially.
- An adequate budget should be allocated toward professional fees and services, such as an accountant and legal advice.

If the company has received funding from professional investors, the *burn rate* is normally used to track the amount of cash that a company spends every month before it starts generating its own income. In terms of the cash flow model presented in Figure 6.2, the burn rate corresponds to the negative cash flow, measured monthly, before the business starts generating revenue at time t_{FS}. The burn rate is also used to determine the time a new business has before it runs out of money. This is known as *runway*. The runway is calculated using the business' overall initial funding and the burn rate as follows:

$$\text{Runway} = \frac{\text{Total initial funding secured}}{\text{Burn rate}} \tag{8.1}$$

For example, if a company has received $100,000 in startup funding and it spends $10,000 a month, its burn rate is $10,000 and its runway is 10 months. If the burn rate exceeds the forecasts made during the financial planning stage or if revenues are generated later than expected, the usual response is to reduce the burn rate, irrespective of the level of remaining funds. In this situation, appropriate actions include reducing the number of staff and cutting expenses for office lease, technology, or marketing activities.

Sources of Funding

As summarized in Figure 8.3, several sources of funds can be approached to acquire the funding needed. It is assumed that the initial exploration of the viability of the business will be supported by the founders' own resources. Beyond this activity, which is normally performed alongside a full-time occupation such as a job or education, the time requirements of the new business normally dictate that additional funding must be secured. The following briefly characterizes the most important sources of external funding.

Friends and Family

During the early parts of the business formation process, the most important source of funding will be from relatives and close personal connections, referred to as *friends and family*, or alternatively, *friends, family, and fools* (*"the three Fs"*). While this way of securing funds may be the least costly in terms of interest,* it is important to make sure that all parties involved understand the nature of the business and the risks involved. Everyone involved should have a clear appreciation of the consequences for the personal relationships with the founders should the business fail. Moreover, it is advisable to create a legal basis for such funding by forming clear contractual arrangements between those involved and to understand any tax implications.

Loans from Retail Banks and Commercial Banks

For credible entrepreneurs with good credit scores and otherwise good financial records, it will normally be possible to arrange loans from commercial lenders such as retail banks to fund a new business. Such lenders will charge a level of interest

---

* See Chapter 15 for an in-detail introduction to interest rates.

that reflects the perceived risk of the investment. To demonstrate the commitment to the business, the applicant will normally be required to invest a significant amount of their own money into the business, which is referred to in the investment community as having *skin in the game*. Additionally, the lender will often want some form of security in case the business fails, which will usually be the applicant's house. This security is known as *collateral*. Again, the founder should carefully consider if the risk of losing their home is acceptable.

Business Angels

Business angels, also known as *angel investors* or *private investors*, are wealthy individuals who invest in new businesses by providing their own funds, usually in exchange for equity. The level of equity demanded by such investors may be significant. Moreover, such investors may offer other forms of support to the founders, for example through mentoring or by introducing valuable new business contacts. Business angels tend to have ownership stakes in several businesses, known as a *portfolio*. In the venture capital funnel model described in this chapter, business angels are typical providers of seed-stage funding.

Venture Capital

Venture capital (*VC*) is a form of funding that is provided to new businesses by specialized businesses called *venture capital firms* or *venture capital funds*. The typical point at which venture capital is sought is after the seed stage, as shown in Figure 8.3. Venture capital firms provide funding in exchange for equity in the new business. As described in this chapter, the aim of such investors is to gain very large returns from successful investments through exit events such as an acquisition by another business or through a flotation on the stock market. High returns are required by venture capital firms because most new businesses in their portfolio are likely to fail, meaning that most investments cannot be recovered.

Government Grants and Loans

Many new businesses are able to secure funding through *government grant schemes*. Such schemes may be national or local, aiming to support new business activity through grant funding. The great thing about grants is that they do not have to be paid back. Frequently, specific funding can be applied for at government agencies to complete specific projects, for example, to develop or commercialize new technology. In other cases, loans are offered at favorable conditions such as low interest rates or with lenient repayment terms. Additionally, government grants are often offered to promote collaboration between groups of organizations, such as businesses and universities. Such groups of organizations are known as *consortia*. Due to the low risk inherent to this source of funding, government grants and subsidies should be sought wherever possible.

Crowdfunding

Crowdfunding is a source of funding in which a new business raises money from a large number of amateur investors, normally through a service provided by an internet platform, as described in detail in Chapter 7. The crowdfunding platform thus brings together groups of entrepreneurs seeking investment and groups of amateur investors willing to provide or donate funds. If funded in this way, new businesses normally do not sacrifice an equity stake. Instead, the resulting product, or a special version thereof, is typically offered to amateur investors as a reward. Crowdfunding has been used for a wide range of for-profit businesses, including artistic or creative projects and toward the development of new technologies. Crowdfunding can be ethically and legally problematic if the investors are misled or if the nature of the product or service offering is fraudulent. Since the potential funders are amateurs, they will not normally be in the position to exercise due diligence.

QUALITIES OF THE FOUNDERS

Given the high degree of uncertainty and the relatively low probability of success, the founders behind any new business must be fully committed. Frequently, the reason to start a new business is the dislike of a current job or of employed work altogether. On its own, this rather negative motivation does not bode well for the new business. Another frequently cited reason to start a business is the expectation of large financial rewards. Again, given the small prospects of success, such expectations should be tempered with realism and caution. Arguably, the best reason to start a new business is when qualified and available people serendipitously recognize an opportunity to offer a solution to an unmet need or to make a change that will benefit potential customers. Such a situation is rare and almost impossible to plan, however.

The founders must understand the advantages and disadvantages of running their own business. The advantages include control over one's own activities and fortunes, the ability to manage one's own time, and the satisfaction of growing and developing a business. A major drawback is that the new business is likely to require a high level of effort and responsibility, especially if staff are employed early on. Founders will normally face long working hours, little time off, uncertain holidays, and, at least initially, lower pay than in employed work. Considering these sacrifices, it is essential that the founders' personalities and personal circumstances match the aspiration of starting a business.

BUSINESS PLAN WRITING

One of the most important documents for any business is its *business plan*. A business plan is a formal written document describing a number of aspects, including the aims of the business, the methods that will be adopted to attain these aims, and the timescale of the formation of the new business. It also provides detailed insight into the nature of the business, its business model, and various other important items of background information. Ideally, a business plan is a document that serves both as a tool to present a business to outside stakeholders, such as the providers of funding, as well as a roadmap for the managers running the businesses.

The content and format of a business plan are determined by the goals of the founders and the intended audience, emphasizing aspects of particular interest. For example, a business plan written for a venture capital firm will attempt to be as convincing as possible regarding the viability of the business in order to justify the proposed investment. It will also clearly spell out the expected return on investment and when the exit event is planned to take place.

Writing business plans draws on a wide range of information about the proposed business and involves several disciplines in management, most of which are introduced in this book: innovation and intellectual property (Chapter 5), business modeling (Chapter 7), operations management (Chapter 9), human resource management (Chapter 10), marketing (Chapter 11), and financial planning (Chapter 13). Since the format of the business plan depends on the context of its use and presentation, it is customary for new businesses to generate their business plan in a range of alternative formats.

The business plan is often supported by two additional pre-prepared methods of communication. The first is the *elevator pitch*, which is a short verbal presentation of the founder's vision. The elevator pitch is normally based on the executive summary contained in the business plan. It is called an elevator pitch because it is designed to be deliverable in a very short time during chance encounters with potential supporters of the projects, for example, while riding an elevator with wealthy individuals. Durations vary, with a standard length being around 1 minute. The second method of communication is the *pitch deck*, which is a brief digital slide show and oral presentation, again based on the executive summary of the business plan. Pitch decks are often supported with additional information to further pique the interest of potential investors. They should be carefully crafted and make effective use of figures and drawings to present the business opportunity in a favorable and accessible way.

While there is no single correct format for business plans, the following describes a widely accepted standard format that provides a good basis for further adaptation.* In any case, the business plan should be visually pleasing and prepared with care. Its overall length will depend on the amount of funding required and the complexity of the proposal. The typical length of a business plan is approximately 20–30 pages. An important prerequisite to writing convincing business plans is clarity about the business model pursued by the entrepreneurs. A fully developed business models canvas, as introduced in Chapter 7, forms an excellent basis for writing the business plan.

Cover Sheet and Executive Summary (One or Two Pages)

The business plan begins with a cover sheet identifying the business name, the founders' names, the current business address, and contact details. This is followed by an executive summary highlighting the attractions of the business idea as described in the business plan. It should very briefly answer the following questions:

- What is the business about?
- What is the target market?

* The format of the business plan described in this section is based on the template provided by the Prince's Trust, which is a British charity with the mission of promoting entrepreneurship among disadvantaged young people (Prince's Trust, 2020).

- What are the aims of the business and what is its growth potential?
- What are the anticipated profits?
- What level of funding is sought?
- What are the prospects for the investors?

Founder Background and Team (Two or More Pages)

This section should explain the founders' motivations, past employment, business record, and relevant achievements. It should do so for any additional member of the team supporting the business who is not a founder. The section should show that these individuals hold the necessary expertise, skills, and qualifications. If there are weaknesses in the skills of the team, this section should present a mitigation plan.

Description of the Product or Service (Two Pages)

This section provides a simple description of the product or service that will be offered by the new business, supported by relevant explanations. It is important that this section is concise and does not use technical jargon. It should set out that the targeted product or service is unique and explain its value proposition (as described in Chapter 7). It should also describe if the product or service will be offered in different variants. A further important point is to disclose if the new business holds relevant intellectual property or if this will arise in the future (as outlined in Chapter 5).

The Target Market (Three or Four Pages)

This section summarizes the market for new products or service offerings. It summarizes the most important market parameters including market size and historical and projected market growth. It should also define relevant market segments and the needs and attributes of relevant customer classes and how these customers will be reached. Relevant methods to structure and obtain such information are introduced in Chapters 7 and 11. This section should also list existing customers or orders and indicate where relevant organizations or individuals have expressed the intent to place orders in the form of letters of intent.

Market Research (One or Two Pages)

The purpose of this section is to present evidence from desk-based and field-based market research, for example from surveys of potential customers. It should make clear what methods have been used to collect these data. If any test trading has been undertaken or minimum viable products have been deployed, as introduced in Chapter 5, this should be presented in this section. It is possible to combine this section with the previous section.

Marketing Strategy (One or Two Pages)

This section presents the approach to marketing and sales, making clear how the product or service will be sold. It is useful to structure this section along the concepts of customer relations and channels as part of the business model canvas introduced in Chapter 7. It should identify the structure of the sales team and explain how the value proposition will be communicated. This should include an outline of the planned pricing strategy. This section should also identify how much the planned marketing activities will cost. It can be developed using the methods and models presented in Chapter 11 of this book.

Competitor Analysis (Two or More Pages)

This section summarizes the competitors in the market for the targeted product or service or in the market for related products or services. It should identify who the main competitors are, their size, their pricing approach, their relative advantages and weaknesses, and their likely response to the new business. The section should also make clear the nature of the competitive environment in the industry. Further, this section should outline how the proposed new business relates to the competitors. Often this is done using a SWOT analysis or a five forces analysis, as introduced in Chapter 7. Finally, the section should identify the *unique selling proposition* (*USP*) that helps the targeted product or service stand out from the competition.

Operations and Logistics (Two or More Pages)

This section presents the structure of the operations of the new business. It is useful to draw inspiration from the infrastructure elements of the business model canvas approach presented in Chapter 7, including an overview of key partners, key activities, and key resources. It should explain how the product will be manufactured or the service will be generated and what items of equipment will be used. This section should also state the location of the premises of the new business. Further aspects to describe in the section include transport arrangements, legal and insurance requirements, and operations staffing and management. This section can be developed using the operations management methods presented in Chapter 9 of this book.

Costs and Pricing (One or More Pages)

This section explains the costs and pricing approach of the new business. This can be informed by the key resources, key activities, cost structure, and revenue streams elements of the business model canvas. It should describe in detail the planned revenue streams of the business, its approach to monetization and ideally also explore the willingness to pay of the targeted customers, as introduced in Chapter 3. Where applicable, the pricing strategy of competitors should be discussed.

Financial Planning (One or Two Pages)

This section presents sales and cost forecasts as part of á *profit and loss budget* and a *cash flow forecast* for the proposed business, as introduced in Chapter 13. A suitable planning horizon should be chosen; normally three years is seen as appropriate. The forecasts normally reflect monthly or quarterly periods. This section should clearly identify or estimate when the business expects to break even and starts returning a profit to the investors.

Prospects and Contingency Plans (One or Two Pages)

The final section of the business plan should summarize the objectives of the business, both in the short term and in the long term. It should clearly state the proposed investment and justify this requirement convincingly. If appropriate, this section should also suggest shareholdings among future business owners. This may mean that existing shareholders will have to be disclosed in this section. Moreover, it should clearly outline the founders' expected valuation of the company after it starts operating and identify possible exit events, such as potential points of acquisition or even a stock market flotation. Finally, this section should present contingency measures if starting the business does not go according to plan. Chapter 12 in this book introduces risk management techniques that can be used to design contingency measures.

CONCLUSION

This chapter has introduced the activity of forming new businesses. It highlighted the core idea that the creation of a new business is in itself a process of innovation and is normally based on some form of minority opinion held by the founders that is not shared by the wider public. It has also stressed the high levels of risk and extreme uncertainty in the process. Modern approaches to new business formation such as startup life cycling, also known as customer development, attempt to systematically de-risk this process by treating it as an *epistemic*, knowledge-generating, process in which new knowledge is created as efficiently as possible. This was presented as the second core idea of this chapter. To start operating a business with a lower level of risk, it may be possible to buy into existing businesses. The chapter has provided an overview of sources of funding available to founders and described the content of the main document required to obtain such funding, the business plan.

The theory and practice of starting new businesses have changed dramatically over the recent two decades. Modern approaches aim to systematically reduce the severity of the risk to founders and investors. Important and seminal books on this topic are *The Startup Owner's Manual* by Steve Blank and Bob Dorf (2012) and *The Lean Startup* by Eric Ries (2011). As evident by the many references to other chapters, forming a new business requires a wide variety of management techniques. For this reason, many of the other texts recommended throughout this book will be of relevance.

REVIEW QUESTIONS

1. Complete the below definition of a new business.

   (Question type: Fill in the blanks)

   "A new business is an ________ attempting to establish a viable and profitable ________ under conditions of ________. New businesses are ________ because they either transition to being an established business or they ________."

2. New businesses are more likely to succeed than to fail and are likely to return any investment to their funders.

   (Question type: True/false)

   Is this statement true or false?

   ○ True ○ False

3. Identify which of the following statements about the customer development approach are true or false.

   (Question type: Dichotomous)

True	False	
○	○	It involves the decision to persevere, pivot, or reject.
○	○	It increases the chances of successfully starting the business to over 50%.
○	○	Its purpose is to identify the needs of the customer.
○	○	Its goal is to test the hypotheses held about the business by the founders.

4. Order the provided steps in the startup life cycle.

   (Question type: Ranking)

   - Customer creation
   - Build the business
   - Customer discovery
   - Customer validation

5. Which of the following are normally considered successful outcomes by a venture capital provider?

   (Question type: Multiple response)

   ☐ The new business becomes cash flow positive and sustains its owners indefinitely.
   ☐ The new business shuts down, but many important lessons have been learned.
   ☐ The new business is acquired by a competitor and shut down immediately.
   ☐ The new business is acquired by a competitor and grows further.
   ☐ The new business is has invested substantial amounts into research.
   ☐ The new business is floated on the stock market.
   ☐ The new business has led to exciting opportunities.

- ☐ The new business is based on an excellent business plan.
- ☐ The new business has wound up very efficiently.

6. The franchisee supports the franchisor with resources, including access to branding, processes, and intellectual property.
   (Question type: True/false)
   Is this statement true or false?
   ○ True ○ False

7. A new business has secured $2,000,000 in funding from investor A and $500,000 from investor B. It is showing a burn rate of $125,000 per month:
   (Question type: Calculation)
   Calculate the runway of the new business.

8. Match the provided major sources of funding for new businesses to the given characteristics.
   (Question type: Matrix)

	Government grants	Venture capital	Crowdfunding	Bank loans	Friends and family	Business angels
Potential negative impact on close personal relationships	○	○	○	○	○	○
Are likely to require collateral	○	○	○	○	○	○
Often offered for specific projects	○	○	○	○	○	○
Typically sought after the seed stage	○	○	○	○	○	○
Funded by wealthy private individuals	○	○	○	○	○	○
Money from amateur investors	○	○	○	○	○	○

9. Business plans should not contain cash flow forecasts because these are uncertain.

   (Question type: True/false)

   Is this statement true or false?

   O True O False

10. Which of the following is not a part of the business plan?

    (Question type: Multiple choice)

    O Cover sheet
    O Section on operations and logistics
    O Section describing market research
    O Explanation of contingency planning
    O Invention disclosure

REFERENCES AND FURTHER READING

Blank, S. and Dorf, B., 2012. *The startup owner's manual: The step-by-step guide for building a great company.* Pescadero: K&S Ranch.

Harroch, R., 2018. The complete 35-Step guide for entrepreneurs starting a business. Available at: https://www.forbes.com/sites/allbusiness/2018/07/15/35-step-guide-entrepreneurs-starting-a-business/#341395c184b5 [Accessed October 16, 2020].

Macnaught, S., 2016. Startup statistics 2020. *Micro Biz Mag.* Available at: https://www.microbizmag.co.uk/startup-statistics/#:~:text=672%2C890%20start%20ups%20were%20founded,Or%2076.8%20per%20hour [Accessed October 16, 2020].

Mansfield, M., 2020. DDStartup statistics – The numbers you need to know. *Small Business Trends.* Available at: https://smallbiztrends.com/2019/03/startup-statistics-small-business.html [Accessed October 16, 2020].

Marmer, M., Herrmann, B.L., Dogrultan, E., Berman, R., Eesley, C., and Blank, S., 2011. Startup genome report extra: Premature scaling. *Startup Genome*, 10, pp.1–56.

Quintero, S., 2017. Dissecting startup failure rates by stage. *Medium.* Available at: https://medium.com/journal-of-empirical-entrepreneurship/dissecting-startup-failure-by-stage-34bb70354a36 [Accessed October 16, 2020].

Ries, E., 2011. *The lean startup.* New York: Crown Business.

Thiel, P., 2014. *Zero to one: Notes on start ups, or how to build the future.* New York: Random House.

U.S. Small Business Administration, 2020. *10 steps to start your business.* Available at: https://www.sba.gov/business-guide/10-steps-start-your-business [Accessed October 16, 2020].

9 Managing Operations

OBJECTIVE AND LEARNING OUTCOMES

The objective of this chapter is to introduce the field of managing operations to the reader. After defining operations management and presenting a brief overview of its main functions, the chapter characterizes the most important classes of operations systems. This is followed by a brief summary of the topics of supply chain management, demand forecasting, and enterprise resource planning (ERP) systems. Since Lean arguably forms the dominant operations paradigm of the present and a universal method for process improvement, the chapter builds on Chapter 1 in this book by further characterizing Lean through an overview of its main principles, a discussion of the focus on the elimination of waste, and a brief summary of major Lean methodologies and tools. This chapter aims to achieve the following learning outcomes:

- an appreciation of the general field of managing operations;
- the ability to define operations management and to characterize operations as a transformation system using operations processes;
- an understanding of the main functions of operations management;
- knowledge of important operations systems and the role of available technology in constraining these;
- an understanding of the topics of supply chain management, demand forecasting, and enterprise resource planning systems;
- knowledge of Lean, including important principles and major methods employed by it; and
- understanding key terms, synonyms, and accepted acronyms.

A SYSTEM OF ACTIVITIES DELIVERING A PRODUCT OR SERVICE

The creation and running of effective systems that support the activity of the business is the task of *operations management*. Its goal is to implement and run processes in a business that can repeatedly, consistently, and reliably deliver the products and services offered by the business. This goal, crucially, includes developing, measuring and improving processes over time. Since the business is likely to be in a competitive situation, managing operations efficiently is often a vital requirement for success. Usually this involves the management of uncertainty and unforeseen events. A further operational criterion is that processes must comply with relevant frameworks of rules and regulations. Operations management can thus be defined as follows:

DOI: 10.1201/9781003222903-11

IMPORTANT DEFINITION

Operations management is the set of activities, decisions, and responsibilities relating to the management of the processes used to deliver products and services.

CORE IDEA

A useful lens through which the operations of a business can be viewed is as a system. The universal features of any system are that it exists in an environment, has a boundary, an internal structure, and a purpose. Expressed in a general way, the operations system run by a business takes inputs, such as capital, equipment, materials, labor, and information and processes these into outputs such as products and services for customers. Note, however, that any such transformation system will create unintended and, most likely undesirable, externalities as outputs, for example, in the form of carbon emissions. Figure 9.1 summarizes the broad view of the operations in a business as a transformation system.

The field of operations management is usually associated with five distinct topics that must be addressed when setting up and managing operations: capacity management, setting and maintaining standards, inventory management, scheduling, and controlling operations. Figure 9.2 summarizes these functions, highlighting the central role of controlling operations. This section will introduce each function briefly.

Capacity Management

Any operation in a business has an upper limit for the production of outputs, be they products or services. This limit is known as the *capacity* of an operation. The capacity is limited by the availability of the resources used to generate the product or service sold to the customers, which typically include workers, equipment, materials, and finance. Capacity can only be changed within constraints, and differently so in the short term and long term.

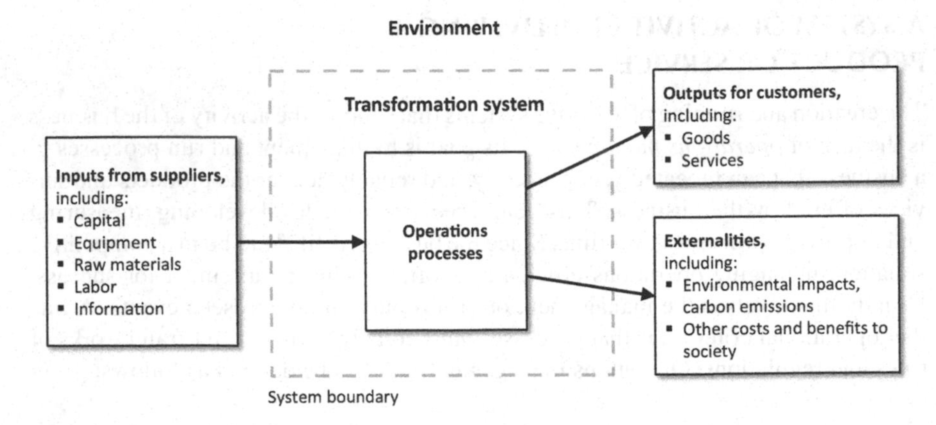

FIGURE 9.1 Operations as a transformation system, adapted from Naylor (2002)

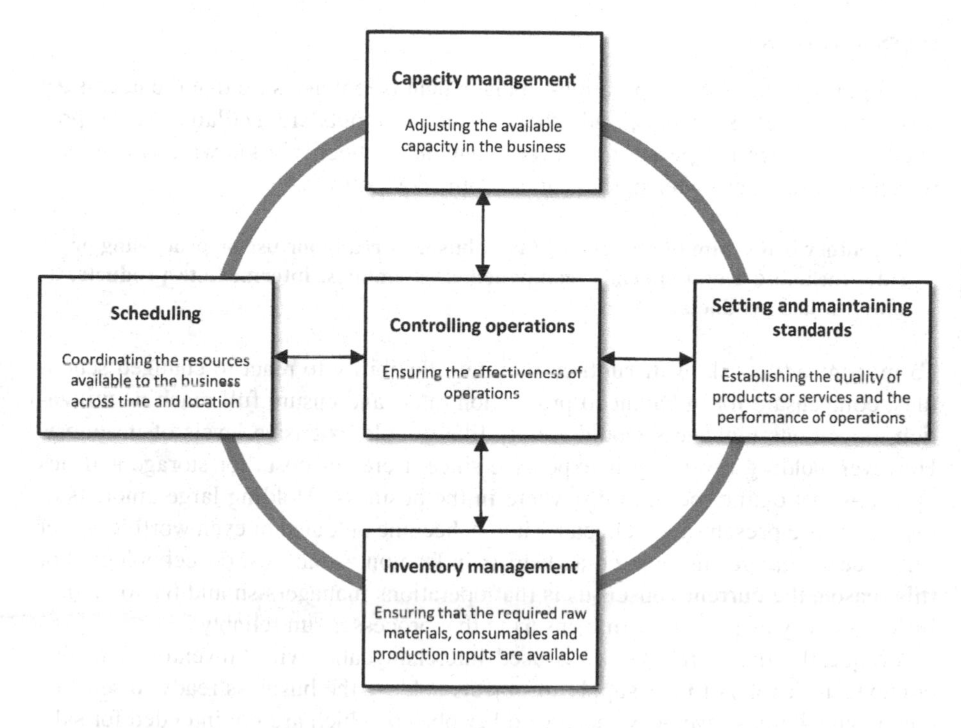

FIGURE 9.2 The functions of operations management

The utilization of capacity in an operation may be at the level it was designed for, known as its *nominal capacity*. In operations, it is possible that actual capacity utilization exceeds the nominal capacity for a short period of time. However, it is problematic to maintain high or excessive levels of capacity utilization for too long. Therefore, in the medium or long term, it is necessary to adjust the available capacity of a system by extending or reducing it. Since making decisions about capacity has consequences in the future, managing capacity appropriately is an important determinant of the ability of a business to achieve its goals over time.

Setting and Maintaining Standards

The standards set by a business in its operations relate to either the quality of the products or services created for the customers or to the performance standards of the processes employed by the business. The management of quality standards is discussed as a separate topic in Chapter 12. From an operations management perspective, performance standards normally capture how long the completion of the various processes carried out in the business should take and how well these are performed. This informs decisions regarding capacity and planning activities.

Inventory Management

An important function of operations management is to make sure that the necessary raw materials, consumables, and other production inputs are available for the production processes in operations. These inputs are collectively known as *inventory*, which is an accounting term, that can be defined as follows:

Inventory is the sum of objects held by a business ready for use or processing or sale, including raw materials, components, assemblies, intermediate products, and finished products.

To minimize the risk of disruption, maintain flexibility to react to changed schedules, compensate for unbalanced production rates, and ensure full machine utilization, operations managers would ideally like to hold extensive levels of inventory. However, holding inventory is expensive since it creates costs for storage and ties up funds that cannot be used elsewhere in the business.* Holding large amounts of inventory also presents a risk because it may become outdated or even worthless over time due to changes in demand and shifts in fashion, standards, or technology. For this reason, the current consensus is that operations managers should try to hold as little inventory as possible while ensuring that processes run reliably.

Frequently, the word "stock" is used interchangeably with "inventory". In this context, stock refers to the supply of objects held by the business ready to sell to a customer, whereas inventory includes other objects which are not intended for sale. Strictly speaking, all stock is inventory but not all inventory is stock. However, as noted in Chapter 2, stock can also refer to shares, which is an entirely different meaning. The multiple meanings of the word "stock" are a frequent source of confusion and ambiguity, even for seasoned managers.

Scheduling

The activity of coordinating the operational resources available to a business across time and location is referred to as *scheduling*. Scheduling takes into account information about the *order flow* received by the business and matches this flow to the available capacity. To avoid disruption through unexpected events, necessary spare capacity is maintained and queueing processes between activities are controlled. In summary, scheduling determines the sequence, timing, and speed with which all tasks required to generate the business's products or services are carried out.

Scheduling activities take place according to different time frames, known as *time horizons*. *Aggregate scheduling* is implemented for longer time horizons, usually ranging up to five years into the future, and usually forms the basis for the planning of capacity. *Master scheduling* relates to the scheduling activity with a time horizon of a few months. The *dispatching* function in scheduling is concerned with the immediate coordination of resources.

* A frequently cited and unverifiable rule of thumb is that the annual cost of holding an object as inventory amounts to approximately 25% of its value.

Additionally, it is important to note that scheduling systems can operate internally within the business or reach across the boundaries of the business. For example, it is possible for a scheduling system to automatically generate orders for components needed from outside suppliers.

Controlling Operations

The control function in operations management monitors and controls the operations activities in the business. Without such control, the effectiveness of the other operations functions would be severely diminished and could not improve over time. The general control cycle for operations processes consists of four phases:

- setting the criteria, standards, and objectives for operations processes;
- measuring the state and performance of individual processes;
- comparing the measured data to expected outcomes; and
- acting decisively in different time horizons to bring the process in line with objectives.

DESIGNING OPERATIONS PROCESSES

A task of significant practical importance faced by senior operations managers and the leadership of the business is to make decisions on how the operations functions required by the business should be structured. In particular, the ability to deliver the value proposition formulated in the business model and meeting the three criteria of fit (problem-solution fit, product-market fit, and channel fit) described in Chapter 7 depend on specifying and running appropriate operations systems.

Prior to the determination of the structure of its operations, however, the business must determine its *span of operations*, which is the overall extent of operations it performs. This topic was discussed in Chapter 7, where the opposing business models of vertical integration, with a large span of operations, and the more modern approach of the specialized business, with a minimal span of operations, were contrasted. In either configuration, the span of operations is determined by the business model that has been chosen for adoption.

Defining Production Systems

If the pursued business model requires the production of material, or *tangible*, products, the operations process is based on a design for a production system. Typically, this takes into account the capacity of the required production system and its flexibility in terms of the ability to cater to different volumes and varieties of products. It should be emphasized that such decisions are heavily dependent on the characteristics of the available technology. From the perspective of economics, an important additional criterion is whether the production system is capable of delivering economies of scale (introduced in Chapter 3) and economies of scope (introduced in Chapter 4).

The *product-process matrix* is a traditional model used to structure such decisions*. This model suggests that production systems cannot be set up to simultaneously deliver products in high volumes and in many varieties. This means that the available production technology imposes a trade-off between volume and variety: if many product units are required at low cost, then the production system can produce few or only one variant of this product. Conversely, if multiple variants are required, then the production system must be set up in a more flexible way, geared less to the production of large quantities at the lowest possible cost. In operations management, this is known as the *volume-variety trade-off.* In the traditional version of the product-process matrix, four archetypal production systems are considered: *project operations systems*, *job-shop operations systems*, *batch operations systems*, and *line and continuous operations systems.* Figure 9.3 shows the product-process matrix in a graphical form.

This section briefly characterizes each operations system reflected in the product-process matrix, with the addition of *digital manufacturing operations systems*, which constitute a novel approach based on existing production systems that offers new possibilities, as discussed in the below. It is important to note that the presented archetypal production systems are not fully distinct so there are operational systems that combine the presented categories. Examples for each kind are provided in Table 9.1.

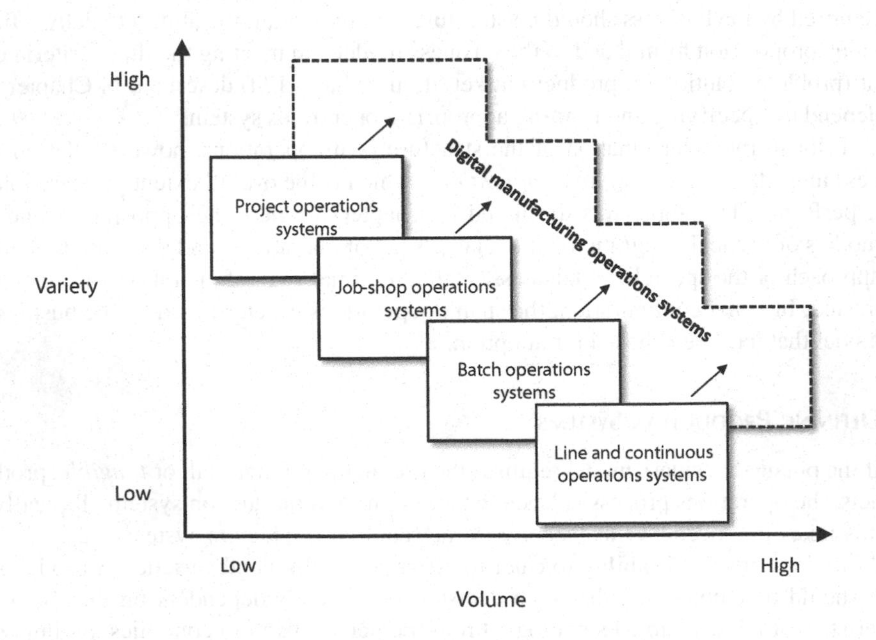

FIGURE 9.3 The product-process matrix, adapted from Boddy (2017)

* Hayes and Wheelwright (1979).

TABLE 9.1
Summary of Production and Service System Designs

Archetypal product or service system	Example industries	Technological characterization	Scale and variety of output	Specialization of workers	Relative complexity of scheduling
Project operations systems	• Construction • Marine engineering • Film-making	• Specialized equipment	• Very low volumes, down to a single unit, high variety	• Highly specialized and skilled workers	• High level of complexity
Job-shop operations systems	• Industrial equipment • Printed products	• General-purpose tools	• Small volumes, different varieties	• Skilled workers	• High level of complexity
Batch operations systems	• Molded car components • Office furniture	• Special tooling, often using flexible manufacturing systems	• Low to high volumes and some varieties	• Mixture of skilled and unskilled workers	• Reduced level of complexity
Line and continuous operations systems	• Automobiles • Consumer electronics • Chemical industry	• Dedicated processes, high degree of automation, long-linked technologies	• High and very high volumes, low variety (if any)	• Mainly unskilled	• Low level of complexity
Digital manufacturing operations systems	• Aerospace products • Defense • Medical devices	• Flexible technology, linking of physical and digital processes	• High volumes and high variety	• Highly skilled	• Highly complex automated scheduling

(Continued)

TABLE 9.1 (CONTINUED)
Summary of Production and Service System Designs

Archetypal product or service system	Example industries	Technological characterization	Scale and variety of output	Specialization of workers	Relative complexity of scheduling
Professional service systems	• Legal services • Medical practices • Beauticians	• Non-dedicated equipment, not normally automated, intensive technologies	• Small scale, personalized service	• Highly skilled, often professional	-
Service shop system	• Estate agents • Car repair shops	• Intermediate levels of automation	• Small to large scale, intermediate standardization	• Limited skills and training	-
Service shop system and service factory systems	• Hotels • Resorts and recreation • Logistics	• Low levels of interaction and customization	• Small to large scale, mostly standardized services	• Limited skills and training	-
Mass service systems	• Consumer-facing internet services • Mass transportation • Supermarket retail	• High land highest levels of automation, frequently based on or involving mediating technology	• Large scale, standardized service	• Minimally skilled	-

Project Operations Systems

Project operations systems produce goods in small numbers, down to a single unit. Usually, they are organized to carry out individual tasks and processes in parallel over a sustained period of time. An important characteristic of project operations systems is that the products are not normally moved during production. Project operations systems are often characterized by the involvement of different workers and types of equipment. Organizing such systems entails a significant and complex scheduling task. This complexity is often compounded through the involvement of highly specialized workers and equipment, which are likely to be costly and of low availability. A group of well-known techniques frequently employed by managers to plan and monitor project operations systems is introduced in Chapter 16.

Job-shop Operations Systems

A job-shop operations system is designed to be flexible enough to allow a wide variety of tasks while producing a greater volume of output than a project operations system. Overall, the volume of production is still relatively low, however. Job shops typically deliver specialized products with the same design more than once. This style of operations usually makes use of general-purpose tools and is characterized by highly skilled workers. The patterns of demand served by job shops are marked by unpredictability and small order sizes, so facility planning and business structures are not normally adapted fully to any particular product. Planning job shops is challenging and complex due to the irregularity inherent to this form of operations.

Batch Operations Systems

Batch operations systems are production systems that use special tooling, often employing flexible manufacturing systems, as introduced in Chapter 4. They generate higher volumes of products than job shops. Batch operations systems are characterized by pre-planned and documented techniques, usually breaking down worker activities into small, incremental steps. The processes employed in batch operations are typically carried out by a mix of highly skilled, semi-skilled, and unskilled workers. Compared to job-shop operations systems, the reduced variety in batch operations systems also reduces operational and planning complexity. Volumes in batch operations systems can vary strongly but are not high enough to warrant line operations systems featuring more dedicated systems.

Line and Continuous Operations Systems

Line operations systems are adopted when the production volume is large enough to justify the installation of dedicated manufacturing systems with high levels of automation. This enables very high production volumes with small numbers of workers, mainly employing an unskilled workforce. Generally, line operations systems exhibit low flexibility and variation and feature low planning and scheduling complexity. They are, however, highly sensitive to disruptions in the supply of raw materials and other inputs.

While line operations systems, also known as manufacturing lines, as discussed in Chapter 1, are associated with *integral products*, which are products that form countable and discrete units, continuous operations systems process a flow of bulk products that can only be measured in terms of a dimension, such as weight or volume. To reflect this difference, such products are also known as *dimensional products*. Like line operations systems, continuous operations systems are normally geared for high production volumes, utilizing dedicated equipment and a high degree of automation. Both line and continuous operations systems are clear examples of long-linked technologies in Thompson's influential classification of technologies presented in Chapter 5.

Digital Manufacturing Operations Systems

As the result of the embedding of information technology in manufacturing, digital manufacturing refers to new manufacturing approaches combining digital technologies with production operations systems. The idea underlying this is that digital manufacturing, also referred to as *Industrie 4.0* or the *smart factory*, allows the modeling, analysis, simulation, and control of all equipment used in the factory to optimize efficiency, while at the same time retaining flexibility. This is made possible by extensive real-time data exchange between a range of technologies and processes, including cyber-physical systems, the Internet of Things (IoT), cloud computing, artificial intelligence, robotics, and novel manufacturing processes such as additive manufacturing. This allows digital manufacturing operations systems to tolerate both high volumes and high variety, thereby escaping the traditional trade-off associated with the product-process matrix, as shown in Figure 9.3. Digital manufacturing operations require highly skilled workers and their complex scheduling tasks are addressed by advanced automated planning systems.

Defining Service Systems

Operations used to generate services can be thought of in a similar way to production systems. Again, not always clearly distinguishable into different categories, four broad archetypes for service systems are normally delineated.

Professional Service Systems

Professional service systems are operations in which customers interact intensively with the business, usually over a prolonged period of time. Service systems of this kind are designed to deliver high degrees of customization and adaptation to the customer. To discharge such a service, highly skilled, usually professional, workers are required. Professional service systems do not normally rely on dedicated equipment and high degrees of automation. Interpreting such service systems as technologies in their own right, they are represented by intensive technologies in Thompson's classification.

Service Shop Systems

Service shop systems are designed for intermediate levels of interaction with customers and customization. This allows such systems to efficiently serve relatively large numbers of customers. Normally service shop systems are based on a standardized service that can be adapted to individual customers. The workers involved in dispensing such services tend to have limited skills and training, usually adhering to fixed and pre-defined service procedures.

Service Factory Systems

Service factory systems are service systems operating at a high volume with a low degree of customer interaction or customization. Service factories often require that an element of the service provided is generated away from the customers. Typically, the element generated remotely involves a higher level of skill or training.

Mass Service Systems

Mass service systems are implemented to deliver a standardized service in very large volumes. This mode of service provision features very limited interactions with the customers and very little, if any, customization. Staff will be minimally skilled and rigidly execute processes that are pre-defined in a detailed way. In addition, mass service systems will be automated wherever possible. A further characteristic of mass services is that they often involve groups or even crowds of customers.

Interestingly, virtually all consumer-facing digital platform businesses, as characterized in detail in Chapter 7, are mass service systems.* This includes search engines, such as Google, or social media businesses, such as Facebook. While not all mass service systems are based on mediating technology (following Thompson's useful classification, as presented in Chapter 5), most businesses relying heavily on digital mediating technology operate mass service systems.

As for production systems, the distinction between the aforementioned types of service systems is not always possible. Many hybrid configurations can be found. Table 9.1 summarizes the manufacturing and service operations systems outlined in this section and provides examples of industries in which these operations systems are typically found.

Once a business has settled on an appropriate type of operations system, a number of subordinate operational decisions must be made. These include the determination of the best location for the operations of the business, the specification of an optimal facility layout, and the selection of a scheduling process to control the timing of activities performed. These decisions will depend largely on the nature of the business and its operations type.

The Role of New Technology in Reshaping the Volume-Variety Trade-off

As evident from the brief history of management presented in Chapter 1, manufacturing operations are a hotbed of innovation. This has over time yielded new options in the volume-variety trade-off, both in the manufacturing sector as well as in the service sector. As outlined in this chapter, the process of overcoming operational trade-offs is currently occurring in manufacturing through the emergence of digital manufacturing.

* In the original delineation of service systems by Schmenner (1986), mass service systems are identified as being labor intensive. This is obviously not the case with digital platform businesses, in which the mediating work is carried out in an automated way through digital systems.

It is important to note that the distinction between different operations systems is itself largely due to the characteristics and limitations of the technology available at a given time and place. For example, the significant decline in the costs of information technology has enabled the rise of extremely influential digital mass service operations based on the provision of standardized and fully automated, yet highly personal, services, such as social media or internet search engines. It is expected that the availability of practically useful artificial intelligence systems will further reshape existing volume-variety trade-offs by allowing entirely new operational structures.

There are several examples of how technological and managerial innovations have allowed businesses to overcome historical trade-offs in their operations. Among these are:

- the emergence of flexible manufacturing systems in the second half of the 20th century, as outlined in Chapter 4, which have opened up many new applications for batch operations systems; and
- the implementation of *mass customization systems* allowing the manufacture of mass-produced goods that can be modified to meet specific customer preferences. This was achieved through a process of *modularization*, allowing the flexible combination of mass-produced modular elements into products. A great example for mass customization is provided by the automotive industry, where customers can specify vehicles to their requirements and preferences.

SUPPLY CHAIN MANAGEMENT

Most specialized businesses, as introduced in Chapter 7, will rely on other companies to provide raw materials or bought-in components. While processes occurring in other businesses cannot be controlled directly, they may nevertheless be of vital importance for the success of a business. This challenge is addressed through the field of *supply chain management*, which is often characterized as a topic within operations management. It can be defined as follows:

Supply chain management is the set of activities, decisions, and responsibilities that coordinate suppliers, manufacturers, and distributors so that products or services are produced and distributed efficiently, whilst satisfying the requirements of the members of the supply chain.

The activity of organizing the flow of products, services, and information across different processes and boundaries is commonly seen to involve multiple important business functions. Among these are:

- *Supply chain planning*, which involves the allocation, deployment, and consumption of the resources available to one or more businesses to meet projected or actual demand in the long term.

- *Supply chain scheduling*, which refers to the allocation of specific resources to activities across time periods in the short term.
- *Supply chain control*, which ensures that plans and schedules are met or adhered to in the present. If a deviation is identified, action will be taken to mitigate risks and avoid further problems.

Classification of Different Supply Chain Structures

An important topic in supply chain management is the classification of different supply chain structures. This aspect is important since the structure of a supply chain has important implications for its planning, scheduling, and control. A common delineation is between *make-to-stock*, *assemble-to-order*, *make-to-order*, and *engineer-to-order** supply chains.

Make-to-stock (MTS) supply chains are structures in which all activities feed into an inventory of finished products from which the customer, who forms the end of the supply chain, is served. This conforms to the traditional approach to mass-production, as introduced in Chapter 1. While easy to plan and very cost-efficient, this approach is highly inflexible. This causes problems if demand is volatile or customers require customized products.

A more flexible approach is assemble-to-order (ATO), in which the assembly of the final product is delayed until a customer order is placed. This approach combines the low cost of the fabrication of standard components with the ability to serve customers more flexibly. This structure supports many mass customization approaches.

An even more flexible structure is make-to-order (MTO), also known as *build-to-order* (*BTO*) in which fully developed products or components thereof are fabricated, assembled, and shipped to the customer upon receipt of an order. Operations systems based on digital manufacturing, as characterized in this chapter, form an important enabler for such structures.

The most flexible structure is engineer-to-order (ETO), in which product development and design activities are triggered by the receipt of a customer order. In engineer-to-order, there is in principle no requirement for inventory at all. Figure 9.4 graphically summarizes the introduced supply chain classes.

DEMAND FORECASTING

An important role in the planning of operations and supply chains is played by the *forecasting* of future demand levels. Generally, forecasting refers to the prediction or anticipation of future situations, events, or conditions. A particularly important activity in forecasting is the prediction of change over time. Of course, forecasts are always uncertain so there can never be a guarantee that a forecast is correct. However, it is accepted among operations managers that forecasting is necessary. Without forecasts, businesses would have no choice but to react passively as events

* While the structures introduced in this section are often introduced as supply chain configurations, they are equally applicable to the operations within a business.

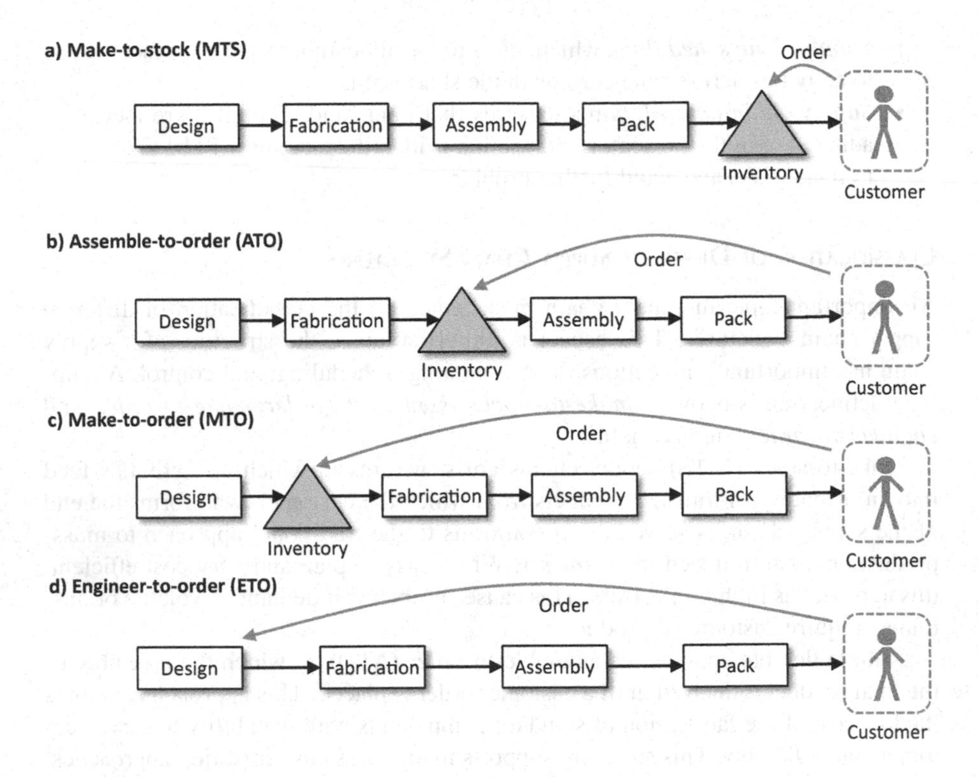

FIGURE 9.4 Illustration of different supply chain structures

unfold. It is also accepted that relying on outdated or otherwise low-quality forecasts is detrimental to the performance of operations.

Forecasts of demand can be made at different levels of analysis known as *levels of aggregation*. For example, a forecast might be constructed for an entire product category, such as "dairy products", or for specific items for sale, such as "semi-skimmed milk in one-pint cartons". In operations management, a specific item at this level is referred to as a *stock keeping unit*, mostly abbreviated to *SKU*. Forecasts that predict changes in aggregate variables, e.g. in product categories, are generally seen to be more accurate than forecasts for specific SKUs.

As with scheduling, time horizons are an important issue in forecasting. Typically, *long-term forecasts* cover time periods ranging from 2 to 20 or more years in the future. *Medium-term forecasts* typically cover periods from one to three years in the future. *Short-term forecasts* typically span from the present up to one year in the future. These time periods overlap, which is intentional, showing that it is not always possible to categorize a specific time horizon.* Short-term demand forecasts are seen as more accurate than long-term forecasts.

As part of the function of capacity management in operations management, demand forecasting serves four main purposes:

* Naylor (2002).

1. *Informing strategic decisions to enter or remain in a market*
   High-quality forecasting allows the leadership of the business to make informed strategic decisions about whether to enter a new market or remain in an existing market. Entering into a new market is a risky and often expensive step to take for a business so forecasts must indicate sufficient expected profits from the new market. Remaining in an existing market that has insufficient demand to warrant the available capacity is equally costly and should be avoided.
2. *Determining the capacity needs of the business*
   High-quality forecasts of long-term developments in demand will allow the leadership of the business and senior managers to make better decisions on how to develop facilities and infrastructure. They will also allow the cultivation of long-term relationships with important suppliers.
3. *Adapting resource levels*
   An important function of medium-term forecasts is to allow operations managers to coordinate resource levels in the business. This could mean recruiting additional employees or making employees redundant. This could also entail balancing work across different sites or locations in the business and ensuring that other partners in the supply chain are able to expand or contract their activities to support the business.
4. *Enabling cost-effective, responsive operations*
   Short-term forecasting plays an important role in allowing the operations system of a business to adjust to impending changes. Good short-term forecasts increase the overall quality of the responses of operations systems to changing circumstances. In this way, short-term forecasting supports multiple functions of operations management, particularly inventory management and scheduling functions. In this, accurate forecasts greatly aid the effective scheduling of staff, materials planning, stock control, and maintenance planning.

Qualitative and Quantitative Forecasts

Operations managers generally distinguish between two broad kinds of forecasting, *qualitative forecasting* and *quantitative forecasting.* Qualitative forecasting refers to a group of techniques aiming to collect information, knowledge, experience, and subjective opinions from decision-makers, specialists, and other experts. These are used to form speculative descriptions of the future. Typically, qualitative forecasting anticipates long-term developments and changes. There are several approaches to qualitative forecasting; the following briefly outlines three common techniques:

- The *Delphi method*, also known as the *jury method*, assembles a panel of approximately ten experts for a survey, asking them to speculate about changes to their specialist area in the long-term future. If there are significant differences of opinion in the panel, the outcomes of the panel are circulated and the experts are asked to give their specialist opinion again.

Coupled with other information, such as statistical models, the predictions made by the panels are frequently used to support or critique the planning activities of senior managers.

- In the *sales force survey*, sales data are extrapolated over time to construct forecasts of demand. The resulting model is enriched through the addition of detailed information and opinions held by the salespeople and their network of contacts, which should include customers. This information is usually collected from salespeople via focus groups and interviews.
- A further way to build qualitative forecasts is by conducting a *customer survey*. Customers can be surveyed directly for their assessment of the future by asking questions such as "how much money will you spend on our products next year?". Since customers are invited to speculate about their own future behavior, forecasters need to exercise caution and judgment when using this kind of information. This information can be collected from customers through interviews or surveys. Increasingly, web-based questionnaires are used, for example, utilizing social media platforms.

The other broad branch of forecasting, quantitative forecasting, uses numerical and statistical techniques to gather data and build models of relevant relationships. Within quantitative forecasting, two main approaches are used: *time series modeling* and *causal forecasting*.

Time series models are based on the assumption that the future behavior of the subject under investigation is in some way shaped by its past behavior. Forming its own discipline within statistics and economics, time series modeling investigates variables that change over time. Complicating matters, the existence of trends in unrelated variables can lead to false conclusions if incorrect statistical methods are applied. This makes time series modeling a non-trivial and highly technical activity.

Despite the complexity inherent to the matter, demand forecasters in operations management often seek practical insight by attempting to decompose the behavior of variables over time into component factors that cause this behavior. Figure 9.5 shows four common patterns of behavior in time series variables commonly associated with the demand for products. Figure 9.5a illustrates constant demand for a product with random variation, and Figure 9.5b shows linearly growing demand for a product with random variation. Figure 9.5c depicts the growth of demand at an increasing rate, also with random variation and Figure 9.5d illustrates the case of demand for a product that follows a linear growth trend but is also characterized by cyclical variation. This pattern of regular fluctuation, known as *seasonality*, is particularly important in demand forecasting since the demand for some products changes in a regular fashion according to recurring seasons or other cycles. Consider, for example, the fashion industry, where demand for hats and scarves will be greatest in autumn and winter and will be subdued in the warmer months.

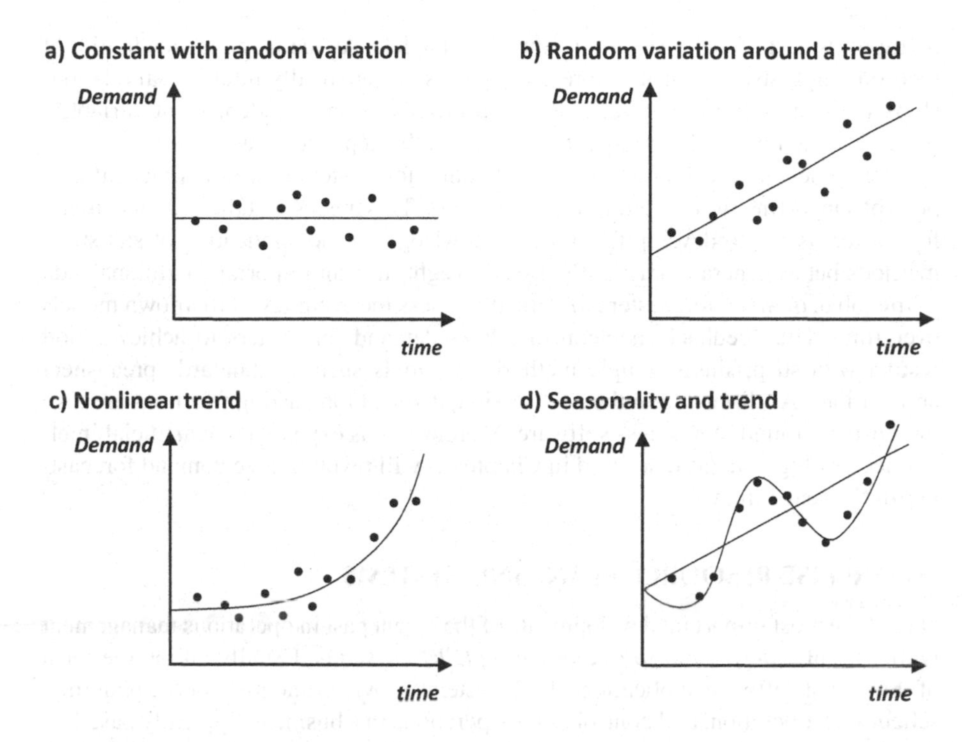

FIGURE 9.5 Common patterns of time series variables

Apart from decomposing historical time series data, two practical approaches are frequently used by demand forecasters to construct simple models using such data:

- *Trend lines* of different types can be fitted to historical demand data using many software packages, including the most common spreadsheet applications. These applications are generally capable of also returning the trend line as a mathematical function. To forecast demand into the future, the identified trend is numerically or graphically extended into the future. This approach may combine multiple components to form a more complex demand pattern, as illustrated in Figure 9.5.
- Where random variation and fluctuations are excessive, moving averages can be calculated to remove some of the random variation. This can also be done easily using common spreadsheet packages. This method allows a clearer identification of the supposed underlying behavior and its components, again with the goal of identifying a trend that can be projected into the future.

Causal forecasting is similar to time series modeling but uses variables that are assumed to be invariant over time. This makes the statistical approaches used in causal modeling less complex; it permits the use of relatively simple regression methods as part of *cross-sectional* statistical models. As the name of this family of methods suggests, this form of forecasting aims to statistically infer causal relationships between a dependent variable and a cross-section of independent variables, which can then be used to predict the change in the dependent variable.

The models and techniques used by demand forecasters serve the practical purpose of supporting the operations of the business. This means that the goal of demand forecasters is not to develop theoretical knowledge on the application of statistical methods but to generate practically useful insight. It is an important additional side of the job of demand forecasters to critically assess the accuracy of their own models over time. This feedback mechanism allows demand forecasters to achieve good results with surprisingly simple methods and tools such as standard spreadsheet applications. As discussed in the next section, demand forecasting functions are also increasingly found in business software. Moreover, it is expected that artificial intelligence and big data (as discussed in Chapter 4) will revolutionize demand forecasting in the near future.

ENTERPRISE RESOURCE PLANNING SYSTEMS

One of the most important developments of the recent past in operations management is the rise of *enterprise resource planning (ERP)* systems. Usually taking the form of integrated software applications, ERP systems provide functions for the planning, scheduling, operation, and control of the operations in a business. Typically based on the use of a central database storing all relevant information, ERP systems monitor many activities and resources present in the business. This includes the broad operations functions as shown in Figure 9.2 but also the management of other aspects, such as cash levels, orders, staff, and customer data. In the present, the vast majority of large manufacturing and service businesses operate ERP systems. In these businesses, ERP systems, such as SAP's S/4 HANA or Oracle's ERP Cloud, effectively serve as the backbone for information and data exchange.

The concept of ERP rose to popularity in the 1990s as an evolution of operations-focused software systems historically found in manufacturing businesses. Important predecessors of ERP systems are known as *materials requirements planning* (*MRP*) systems and *manufacturing resource planning* (*MRPII*) systems. Driven by the desire to increase the performance and efficiency of the entire business by integrating different activities and sharing data, ERP forms a wider-reaching concept that, at least in principle, aims to address all important business functions. Recent generations of ERP systems reach beyond the boundaries of the business and exchange information with other businesses and outside stakeholders, which is particularly important for supply chain management.

While ERP systems are frequently tailored to specific businesses or are custom-built, a typical ERP system will exhibit the following general characteristics:

- it is designed as a seamlessly integrated system, rather than a collection of distinct software packages. This is done so that functions can be integrated with each other and data can be shared with ease;

- it is able to respond rapidly to new information and data, allowing immediate changes or responses to changes. This ability is known as *real-time* operation;
- it contains a powerful central database architecture that receives data and feeds data to all other functions of the ERP system and beyond;
- it is designed to provide a consistent user experience so that users can easily navigate from one function to another. This is important because ERP systems can contain a large and diverse group of functions for use by a wide group of employees; and
- such systems are normally specified and installed in large and very expensive projects with support from external specialists and consultants. To protect such an investment, ERP systems are often supported and maintained by dedicated information technology teams within the business after installation.

It is important to note that the evolution of ERP systems is far from complete. Additional functions and modules are still being added. This is likely to increase the importance of ERP systems even further, elevating them to a major factor for competitive advantage. Current themes in ERP integration include concepts such as artificial intelligence, highly distributed sensing, Internet of Things, and the use of mobile devices.

Table 9.2 summarizes functionalities typically covered by current ERP systems, highlighting that ERP systems have already become the nexus of information flows in many modern businesses.

LEAN

As discussed and defined in Chapter 1, Lean is a principle-based management philosophy originating from Japan which has its roots in the Toyota Production System. It draws on manufacturing systems, operations, supply chain management, and human resources in an organization to establish a culture of learning and continuous improvement. For many managers, Lean forms the most prominent and most universal process improvement method. Due to this importance, basic aspects of Lean are covered in the remainder of this chapter.

An important tenet of Lean is that valuable knowledge of the operations of a business resides with the workers involved in production or the generation of services. To exploit this knowledge, Lean stresses that workers should take personal responsibility for the activities in the business and be empowered to improve processes. This not only necessitates a change in culture among the workers but also requires a new mindset of managers who must be prepared to empower all employees, regardless of their role in the business. The inherent concern for human agency and respect for people marks Lean as a *humanist* approach. This reflects a general concern for people and their wellbeing, setting it apart from other management schools focused more narrowly on the achievement of growth and financial objectives.

The characterization of Lean in this chapter goes beyond the brief outline of Lean presented in Chapter 1. It introduces the basic principles of Lean and shows how they

TABLE 9.2
Functionalities Covered by Current ERP Systems

Function	Description
Sales administration	Functions allowing placement of orders, scheduling of orders, coordination of shipping, logistics, and invoicing of customers
Customer relationship management	Functions that are designed to manage the relationships with customers and extract information on these. Customer relationship management systems are further discussed in Chapter 11
Business intelligence	Methods and technologies used to analyze business information to support decisions by managers
e-commerce	Systems for the facilitation of transactions using online services or over digital networks
Procurement	Administration of the acquisition of goods, services, or information from outside sources and the associated inbound logistics
Manufacturing resource planning and product life cycle management	Running the operations system within the business as introduced in this chapter. This includes managing the product life cycle (PLM), as introduced in Chapter 6
Distribution and outbound logistics	Control of warehousing, movement of goods, and logistics
Accounting	Feeding in and extracting information from the accounting function in the business, ensuring compliance with regulations. An important function in this is running the general ledger of the accounting system, as introduced in Chapter 13
Enterprise asset management	Management of resources, including equipment and infrastructure, controlled by the business with the objective of minimizing life cycle costs, using methods introduced in Chapter 6
Supporting human resource management	Scheduling staff and supporting the human resources management function of the business, as introduced in Chapter 10
Corporate performance and governance	Providing insight into the performance of the business for senior managers, for example, using methods presented in Chapter 14. This includes facilitating the oversight of the activities of managers at various levels, introduced as corporate governance in Chapter 2

flow into a group of well-known methods associated with it. As stressed in Chapter 1, Lean has emerged from manufacturing operations. Due to its overwhelming effectiveness in improving processes, however, it is now used by businesses in all sectors and even in non-profit and governmental organizations.

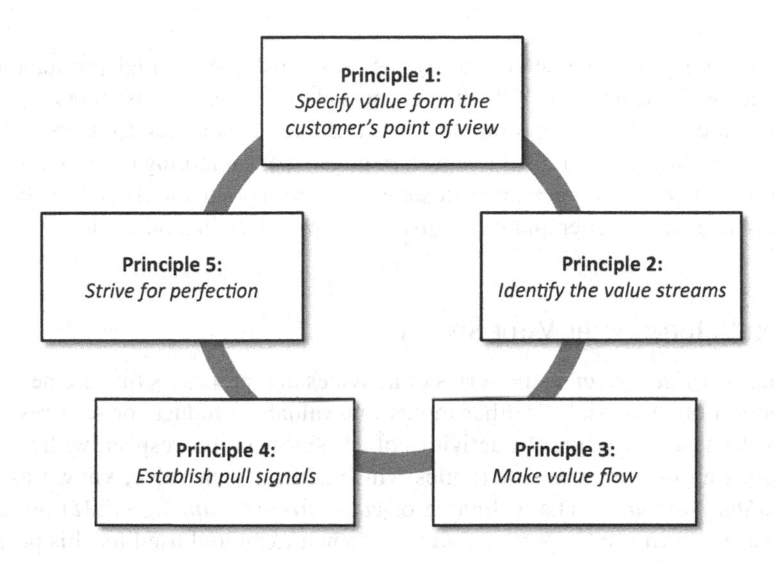

FIGURE 9.6 Principles of Lean

Figure 9.6 summarizes the five principles underpinning Lean. It should be noted that, due to the extremely wide adoption of Lean, different wordings of these principles are accepted. The following briefly highlights these principles, outlining important methods and concepts closely associated with each.

Principle 1: Specify Value from the Customer's Point of View

Lean is based on the conviction that the purpose of a business is to deliver value to the customer. This understanding of value includes the overall performance and features of a product or service in addition to offering it at an attractive price. Taking this view, value is defined from the customer's perspective. This implies that anything that is not valuable to the customer and for which the customer is not prepared to pay is not considered valuable in Lean. This thought extends beyond the operations in the business, influencing design decisions, research and development activities, as well as the overall business strategy.

As introduced in Chapters 2 and 3, it is asserted in this book that the main purpose of a business is to return profits to its shareholders. This suggests that creating customer value is a means to the end of achieving profit objectives. In other words, managers often see customer value as a purely *instrumental* criterion in achieving business success.

As stressed by the first principle of Lean, the idea of specifying value from the customer's point of view may appear disconnected from, or even run against, the goal of maximizing profit. This possible conflict indeed creates tension around the implementation of Lean. This is normally resolved by assuming or demonstrating that commercial success and sustained levels of high profitability coincide with the delivery of the highest possible value for the customers.

In this context, it is important to stress that Lean is not intended to be a rigorous and abstract theoretical framework targeted at explaining profit maximization. Rather, it is a practical philosophy containing many tools and methods aimed at creating better and more sustainable practices in a business.

Principle 2: Identify the Value Stream

In Lean, the *value stream* is the series of activities and processes that are necessary to transform raw materials or other inputs into valuable products or services. Lean stresses the need to analyze the activities of a business in this respect, with the goal of minimizing or eliminating activities which add no such value, known as *non-value-adding activities.* The technique of *value stream mapping* (*VSM*) presented in this section forms an important and well-known Lean tool used for this purpose. Value stream mapping allows the identification of problems and opportunities for the implementation of solutions. An important additional function of value stream mapping is to uncover previously unknown links between operations functions and supporting departments (such as sales and procurement) and outside partners (such as suppliers and customers).

Lean places a great emphasis on using data directly obtained from the processes and operations in the business. For this reason, it is essential that value stream maps utilize such data. By doing so, managers can develop a sense of what is actually happening in operations and how these activities and processes can be quantified and analyzed.

The completion of value stream maps for manufacturing operations is typically aimed at the calculation of the *value-added ratio*, which expresses the time in which an operation adds value to a product unit, known as the *value-added time*, as a share of the overall duration of the process, which is referred to as the *total lead time.* Expressed as a percentage, the value-added ratio is defined as follows:

$$\text{Value added ratio} = \frac{\text{Value-added time}}{\text{Total lead time}} \times 100 \tag{9.1}$$

To further introduce value stream mapping, this section presents an illustrative example of making tea in a kiosk. Of course, most real operations systems are far more complex than this and may involve many more steps. While useful in illustrating the scope of a complex operations system, an extensive value stream map depicting the whole system can be too complex to provide any real insight. For more manageable results, the overall value stream map can be broken down into smaller, more digestible elements – not unlike a cup of tea!

Figure 9.7 presents the value stream map for the tea-making process using symbols and notation accepted in Lean. As can be seen, the upper part of the figure depicts the information flows occurring between different business functions

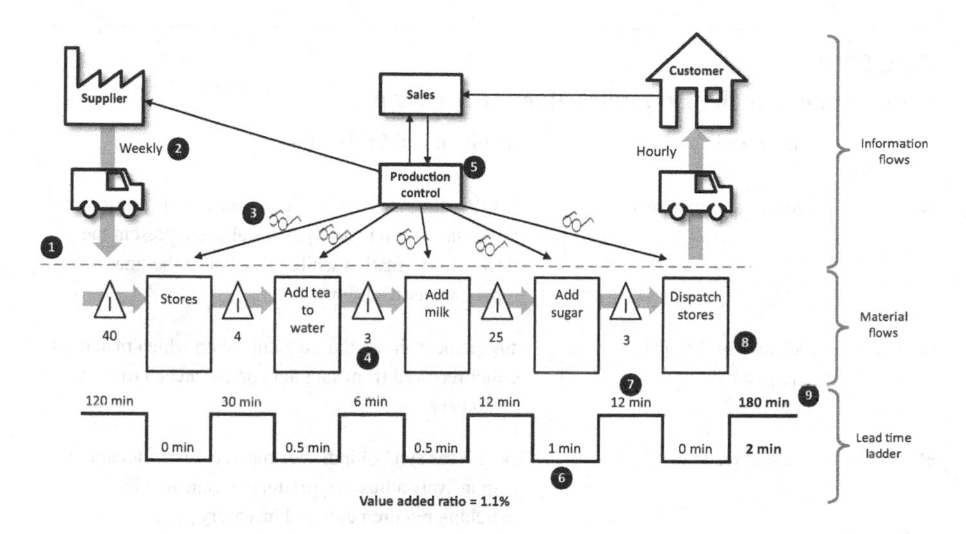

FIGURE 9.7 Value stream map of the tea-making process

(thin arrows) and flows of materials from suppliers and products to customers (thick arrows). The middle part of Figure 9.7 shows the material flows in the operations system and the five process steps used in tea-making. More generally, this component of the value stream map illustrates how work-in-progress passes through the operations system. It also shows the level of inventory tied up in queues waiting for each process. The bottom part of Figure 9.7, known as the *lead time ladder*, summarizes the value-added time and non-value-added time associated with each process step, showing the data required for the calculation of the value-added ratio. Supporting explanations are provided in Table 9.3 for the elements marked by circled numbers in Figure 9.7.

Once the value stream has been characterized in this manner, the next step in a Lean implementation is typically to initiate a program of improvement to better align the operation with Lean principles and methods. This requires the development of additional value stream maps showing desired future states, allowing managers and workers to understand how the operation can be improved.

Principle 3: Make Value Flow

Lean follows the idea that any work should be carried out in a way that produces an even and level flow of activity in a business. This means that the rate with which products and services flow through operations systems should be as constant as possible. In the ideal case, the speed with which this work flows will be determined by the customers, avoiding the build-up of any inventory. Again, an important objective is to minimize processes and activities during which no value is added. To achieve this, it will usually be necessary to restructure operations and reduce batch sizes to reduce queuing. One key technique to achieve an even flow is the adoption of *just-in-time* (*JIT*) operations. In this method, the components or raw materials required

TABLE 9.3
Explanations for the Example Value Stream Map

Element Number	Name and Symbol	Definition and Explanation
❶	Process and information line	This line separates the upper and the middle portion of the value stream map. Symbols above represent the flows of information while those below the line represent material flows
❷	Delivery or dispatch frequency	This element shows the frequency with which material is either received from suppliers or dispatched to customers
❸	See and respond	The glasses symbol indicates that there is a function to responsively adjust the production schedule by matching requirements and inventory
❹	Level of inventory 4	The triangle symbol containing the "I" denotes inventory and the number below shows the level of inventory held between production operations
❺	Supporting functions or departments	The boxes in the upper part of the value stream map show departments that are involved in the manufacturing process and the information flows between them
❻	Value-added time	The numbers on the lower segments of the line, referred to as the *value ladder*, show the time during which a product unit is undergoing a value-adding process
❼	Non-value-added time or waiting time	The numbers on the upper segments of the line show the time during which a product unit is waiting, typically in a queue at a machine
❽	Databox or process box	Although not shown in the example value stream map, each process box lists key information about a process. This often includes scrap rate, cycle time, and overall equipment effectiveness (OEE) as presented in this chapter
❾	Sums of non-value-added time and value-added time	These sums are used for the calculation of the value-added ratio, as defined in Equation 9.1

by a process are delivered to the operation just before they are used. This allows significant reductions in queuing time and inventory. Just-in-time can thus be defined as follows:

Just-in-time operations are processes in which the required inputs, materials, or components are delivered or made available immediately before they are used, minimizing the need to hold inventory and the resulting costs.

Apart from eliminating most of the inventory and the need for storage, just-in-time yields significant additional flexibility since different variants of materials, components, or products can be delivered when and where they are needed. An implication of this approach is that the sizes of deliveries are normally significantly smaller. Additionally, just-in-time operations often encourage important suppliers to locate their operations near their customers, in a process known as *co-location*. Co-location minimizes the costs of frequent deliveries and the risk of disruption in case of supply problems.

A further major concept used in Lean to make value flow is the notion of *one-piece flow*. One-piece flow takes the minimization of batch sizes in the production of integral products to its logical end point by moving only one product unit through the operation at a time. Where one-piece flow is achieved, batches are thus eliminated altogether. While minimizing the quantity of work-in-progress present in the operation at any point of time, adopting one-piece flow often requires significant restructuring of operational processes and facility layout, so that machines and processes are placed next to each other and the work can progress smoothly one piece at a time. Apart from eliminating queuing, one-piece flow also aids the detection of quality issues very early since there are no batches to complete before inspection processes can take place.

One piece flow, or the minimization of batch sizes, often requires frequent changeover of tooling. To allow this to be done rapidly and efficiently, Lean offers the *single-minute exchange of dies* (*SMED*) approach. While the name suggests that this applies to the exchange of dies used in hydraulic presses, SMED is now considered a general approach aiming to dramatically speed up all changeover activities in which equipment or processes are adapted or reconfigured so they can process a different variety. SMED carries many benefits, including a reduction in the number of machines required, smaller batch sizes, less need to hold inventory, and more flexible patterns of operation. Together, these steps result in a higher level of capacity utilization. In reality, single-minute changeover processes are rarely achieved – nevertheless, changing processes in less than ten minutes is often realistic and dramatically faster than using traditional approaches. To implement SMED, an organization must define rigorous procedures and standardize processes and equipment. It must achieve a team culture that enables a highly coordinated "pit stop" style of working and ensure that there are suitable setting aids and jigs to eliminate guesswork. It is also important that the used equipment conforms to exact specifications so that units or tools are perfectly interchangeable with minimum scope for misalignment or other errors.

Principle 4: Establish Pull Signals – Only Operate as Needed

Once demand forecasts have been made for batch operations systems and line operations systems, as defined in this chapter, the production of a defined quantity of products is usually initiated. As the products are manufactured, they progress through the operations system with one process being completed after another. Because this style of operation is "pushed" by plans and demand forecasts, operations systems that work in this way are referred to as *push systems*. A significant disadvantage of push systems is that they are prone to work-in-progress being held up in queues caused by the capacity constraints in some systems. Such constraints are commonly referred to by operations managers as *bottlenecks*.

Lean avoids this situation by structuring operations as *pull systems*. In a pull system, the pace of operations is determined by customer demand, arising as orders, for the products made. In consequence, this form of operation does not require demand forecasts or plans. Instead, such systems rely on *pull signals* between processes, so that one stage can request a unit of work-in-progress from the preceding stage. The goal of this approach is to eliminate costly queues at each process interface. If implemented successfully, a pull system allows the timely completion of the work while removing most inventory.

This means that, as work-in-progress flows downstream through the operations system, a flow of pull signals moves upstream in the opposite direction. By organizing an operation in this way, each stage will be aware of the requirements of the next. In extensive Lean adoptions across multiple businesses, pull signals can travel between businesses, obviating the need for traditional purchase orders.

The pull signals passed between the stages or processes in a Lean operation are usually referred to using the Japanese term *Kanban*. The main requirement of Kanbans is that they are highly visible to the workers running the process to avoid confusion and to give an instant indication of the status of the manufacturing cycle. For this reason, Kanbans are often physical objects or tokens, such as cards or colored balls.* Where necessary, Kanbans will also hold additional data specifying the part or material required and other relevant parameters.

Principle 5: Strive for Perfection – Delivering Exactly What the Customer Needs

The self-proclaimed goal of Lean is to achieve perfection. This ambition is not limited to the quality and delivery of the product or service to the customer but includes every process and internal activity in the business. Naturally, this state of perfection is elusive and will never be realized. Nevertheless, striving toward this goal is seen as worthwhile by Lean proponents since there is always an opportunity for improvement. This thought also shapes the idea that for a business to defend itself against other businesses in a competitive environment, there is no room for complacency. This ongoing struggle for perfection in Lean is known as continuous improvement,

* While Kanbans are frequently physical objects, they are increasingly being replaced by electronic methods.

or *Kaizen* in Japanese. This principle also implies that a Lean implementation can never be truly complete. For this reason, Lean practitioners often refer to Lean adoptions as "journeys".

The following aspects are often included in programs aiming to establish a culture of striving for perfection through continuous improvement:

- acknowledgment of the workers as holders of expertise and empowering them to proactively make suggestions for improvements;
- adopting team-based organizational forms with team leaders rather than formal, hierarchical leadership arrangements;
- allocation of time for team meetings, often held daily, to discuss and review the current processes and to discuss any potential suggestions for changes; and
- facilitation of improvements using a formal improvement cycle such as the *plan-do-check-act* (*PDCA*) cycle presented in Chapter 12.

The idea of involving all employees in improvement activities stands in stark contrast to traditional operations systems. In systems such as mass-production, workers will generally be assigned responsibility for a single task, with the improvement of the processes being the responsibility of specialists such as production engineers. Moreover, in traditional, non-Lean operations, there is often the tendency to attempt the implementation of large-scale improvement projects that address tough problems once and for all using the right combination of human motivation and technical innovation. Such *solutionism*[*] is alien to Lean and the idea of Kaizen.

The Special Role of Waste in Lean

One of the major findings of the study of the Western motor industry by Taiichi Ohno, as described in Chapter 1, was the high amount of waste created in the traditional mass-production process. This led to a sophisticated understanding of waste and a classification of different kinds of waste. Importantly, in Lean, waste is identified through the eyes of the customer. This implies that anything that does not contribute to value for the customer amounts to waste and should be eliminated. Current Lean approaches consider eight different types of wastes.

Waste of Scrap and Rework

Lean takes the position that scrap and rework are wastes that are generally avoidable by reducing the number of defects. The aversion against processes creating defects in Lean is due to two reasons: first, if a defect escapes the operations system undetected and reaches the customer, it may cause severe problems as well as dissatisfaction. Second, if a defect is detected, the ensuing rework increases the operations workload and disrupts the normal production flow. If the frequency with which defects occur in the operations system cannot be reduced substantially, it is a common response in

[*] Many technology-oriented managers think that "solutionism" is a good thing. In reality, the term has strong negative connotations. A brief discussion of this subtle yet important point is provided by McAfee and Brynjolfsson (2017).

Lean to question the structure and design of the entire operations system to eliminate defects in the future.

Waste of Overproduction

In Lean, overproduction is seen as a waste that not only generates inventory but also disrupts production since more work-in-progress and product units must be handled and stored. To eliminate overproduction, it is important to understand why it has occurred. Three reasons are generally seen to cause overproduction. First, an operations system can overproduce by accident, for example, if one process unintentionally generates more output than the overall system can absorb. Second, overproduction can occur by intent, for example, through an operation manager's misguided belief that increasing batch sizes leads to lower unit costs. Third, overproduction can occur through carelessness, for example, if insufficient effort is put into generating accurate demand forecasts. Once the cause for overproduction has been established, it can be curbed.

Waste of Inventory

Any component or unit of work-in-progress that is sitting idly in an operations system and not undergoing a value-adding process constitutes a form of waste. This includes all forms of inventory, as highlighted throughout this chapter. Holding inventory and work-in-progress is frequently the result of a lack of coordination between the operations function in a business and other functions such as purchasing and marketing. Inventory levels can normally be reduced by restructuring the operations processes and through better coordination of supporting functions. It is important to note that despite best efforts to minimize inventory it cannot normally be eliminated fully. A residual level of inventory and work-in-progress will be present in most businesses even where Lean has been adopted successfully.

Waste of Motion

The waste of motion occurs if workers or machines transport items unnecessarily before, during, or after a process. The waste of motion represents a loss of time, floor space, and effort. Operations managers implementing Lean will reduce this form of waste through better planning of processes and improved facility layout.

Waste of Processing

Processing products in ways that are not required or valued by the customers does not add value. Instead, it consumes time and incurs costs and is, therefore, a waste. This frequently occurs if processes are needlessly repeated, for example, polishing before and after another process, or if tolerances are specified unnecessarily tightly. This highlights the idea that quality is not an end in itself but a necessary requirement for the delivery of customer value, as discussed in detail in Chapter 12.

Waste of Transportation

The unnecessary transportation of work-in-progress is a waste. Considering global supply chains, it is not uncommon for work-in-progress to travel repeatedly between continents to complete a manufacturing process. This should be avoided. In addition to producing additional costs and an environmental burden, excessive transportation occupies management resources and leads to further avoidable inventory.

Waste of Waiting

Machine availability and worker time are frequently wasted by waiting for work-in-progress to arrive from preceding processes or due to rework. Moreover, waiting often occurs as inventory waits in batches between operations, for example, as a result of overproduction. The waste of waiting can be reduced by achieving a more level flow of work through the operations system.

Waste of Underutilized People

The eighth and final waste in Lean* reflects the unused talents and skills residing in individual workers. As stressed in this chapter, Lean is a humanist approach that assumes that all employees can offer useful expertise, knowledge, abilities, and experience beyond the tasks they may be assigned to. To avoid wasting valuable opportunities, Lean aims to further develop and empower employees to the mutual advantage of the business and its workers.

Common Lean Methods

Lean is frequently described as a toolkit that an adopter must learn to use over time. Due to the wide remit and scope of Lean, the challenge of transforming a business toward Lean operations can be daunting. This is best addressed by breaking down the Lean transformation into a series of achievable steps. It is important to understand, however, that most Lean tools cannot be deployed in isolation. For example, for one-piece flow to be viable it must usually be combined with a pull system and a just-in-time method of component delivery. Without the support of these complementary elements, the benefits of one-piece flow would be severely curtailed.

Building on the introduction of Lean principles and the focus on waste, the remainder of this chapter very briefly introduces a range of additional and widely known Lean methods and tools.

The 5S

The *Sort-Set-Sweep-Standardize-Sustain* (*5S*) approach is the basic workplace organization method in Lean. Its goal is to control the working environment throughout the business to promote efficiency and effectiveness. 5S is usually part of a deliberate effort of standardization that aims to create a shared understanding among workers of how the workplace should look and how the work should be undertaken. The following briefly characterizes each element in 5S:

* Taiichi Ohno, whom the reader has encountered in Chapter 1, originally identified seven types of waste. The waste of underutilized people was added later as a useful extension.

- *sort* refers to the identification of what is really necessary in the working area to perform the job at hand. Unnecessary objects should be removed swiftly and regularly;
- *set in order* refers to allocating a designated space to all required items. The items should be easily and ergonomically accessible without unnecessary movements;
- *sweep* or *shine* refers to making a proactive* effort to keep the working area clean and orderly. This includes performing routine maintenance on machinery and equipment. This aspect applies to the storage of digital information as much as to physical items;
- *standardize* refers to the creation of documented standards for processes so that they become embedded in the working practice. This is supported through the use of schedules, charts, posters, and lists; and
- *sustain* refers to ensuring that the standards are adhered to and conduct does not roll back into an undesirable state. This should be checked through regular audits by operations managers.

Visual Tools and Control Boards

A clear understanding of the way that an operations system works greatly enhances the sense of team membership and responsibility among workers. Lean supports this by employing many different forms of visual tools, ranging from simple visual aids, such as floor markings indicating where equipment or bins should be placed, to more complex displays, such as charts indicating the training status of staff. Good visual tools should be kept as simple as possible so workers can understand them at a glance. Table 9.4 summarizes different kinds of visual tools and devices used in Lean.

Root Cause Analysis

In the event of non-conformance of processes to agreed standards, unexpected failure, or accidents, the Lean response is to investigate the event thoroughly to eliminate its underlying reason, known in Lean as the *root cause*. As a rule of thumb, Lean teaches that such investigations should trace back causes over five levels to identify underlying reasons that can be considered root causes. This technique is known as the "*five whys*". The following simple example illustrates this approach.

An important shipment of parts is delivered late to a customer. The operations manager has performed a root cause analysis.

- An initial inquiry ("1st Why") indicates that the late delivery was caused by the following reason: the computer numerically controlled (CNC) machining center used to process the product was out of tolerance.

* Please refer to the end of Chapter 10 for a deeper explanation of the term "proactive".

- Further investigation ("2nd Why") reveals the following cause for this: an incorrectly sized cutting tool was used in the machining process.
- Questioning of the machine operator ("3rd Why") identifies the reason for this: an approved supplier has shipped an incorrect cutting tool.
- Further research by the operations manager ("4th Why") reveals the following cause for this: the incorrect part number was selected from the supplier's new catalog.
- An additional investigation identifies the following underlying cause for this ("5th Why"): the procurement team was not informed of the new catalog.

Through this investigation, the confusion about new supplier catalogs is established as the root cause of the problem by the operations manager. The process for updating supplier catalogs in the procurement department is amended immediately by the procurement team.

TABLE 9.4
Visual Tools and Devices in Lean

Visual tool	Description
People and staffing boards	Simple charts indicating what team members should be doing and what their responsibilities are
Displays showing training status	Training matrices showing what training each team member is expected to receive and what level has been achieved. In the ethos of Lean, this will not be understood as micromanagement but will be recognized as a means of developing staff skills
Inventory markings	In a Lean environment, there will be clearly marked areas for inventory as it is received, processed, and completed at a workstation. Any work-in-progress outside these areas will be noticed immediately and its status will be investigated
Location indications for tools and storage	Storage areas for tools and jigs will have markings on the floors and walls, showing immediately if a tool is missing. Shadow boards for hand tools fall in this category. Lean normally avoids storage units with doors since their contents are not immediately visible
Results boards and displays	Important performance measures are shown in prominent places so that every worker is able to see how the operation is performing. Typical parameters include lead time, right-first-time rate, average cost, and the total output
Process and equipment signage	Each process and machine will have information such as maintenance status and performance prominently displayed. Access to documents such as working procedures, maintenance status, and technical manuals should be easy and clearly marked

In the provided example, the correct remedial response is to change the processes so that an incorrect cutting tool selection cannot occur in the future. Moreover, it will be necessary for the operations manager to locate faulty parts that have been processed or shipped. An incorrect response would be to try to produce more faulty products at a higher rate so the next delivery can meet the customer's schedule.

An alternative way of identifying the root cause of a problem is to draw an *Ishikawa* diagram (named after its inventor Kaoru Ishikawa), which is also known as a *fishbone* diagram. Ishikawa diagrams can be characterized as a graphical method of relating a possible outcome to its causes. As illustrated in Figure 9.8, the process of drawing an Ishikawa diagram begins with the identification of a particular – normally problematic – outcome. A set of categories of causes contributing to this outcome forms the main limbs of the diagram. Figure 9.8 summarizes six important categories of causes. More detailed specific causes are added to complete the diagram. In this way, an Ishikawa diagram provides an insight into how the detailed reasons contribute to the identified causal categories leading to an observed outcome. As with the five whys, the goal of the analysis is to systematically distinguish a symptom, which may be recurring, from its underlying cause, which can then be eliminated.

Overall Equipment Effectiveness

In even the most successful Lean organizations, there will be times when equipment is unavailable due to breakdown or when non-conforming parts are produced because of technical issues. To assess the frequency of such adverse events, *overall equipment effectiveness* (*OEE*) has emerged as an important and versatile metric. One particular strength of OEE is that it allows the comparison of the performance

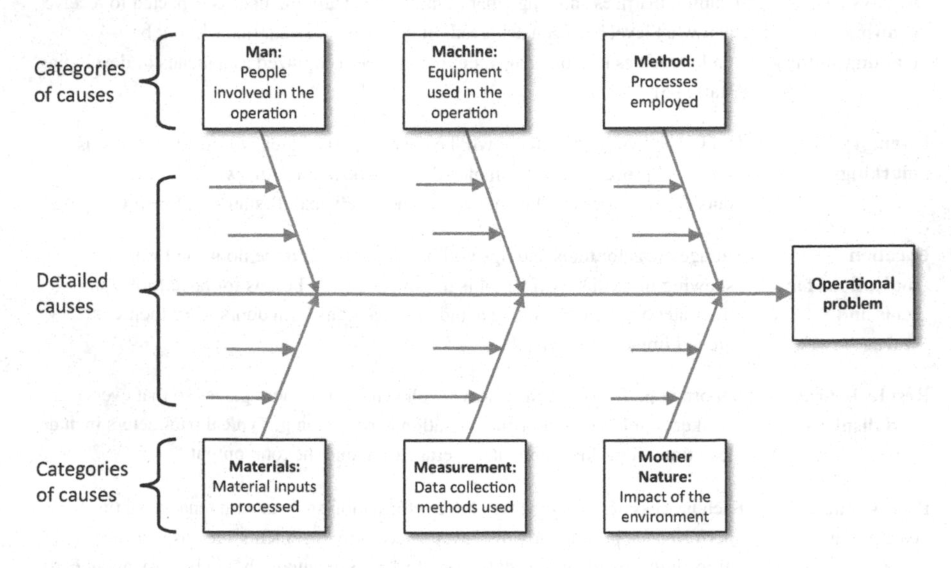

FIGURE 9.8 Establishing the root cause of an operational problem through an Ishikawa diagram, adapted from Holweg et al. (2018)

of dissimilar types of equipment, such as milling machines and heat treatment furnaces. OEE is a numerical value associated with a machine or piece of equipment expressed in percentage terms. It is defined as:

$$\text{OEE} = (\text{Availabilty} \times \text{Performance} \times \text{Quality}) \times 100 \quad (9.2)$$

The elements *availability*, *performance*, and *quality* usually enter OEE in percentage terms. Availability denotes the share of the overall working time that the machine is active on average, excluding any time during which it is in a broken-down state, undergoing changeover, or otherwise unavailable. Performance is the ratio of the number of units that are, on average, processed during the availability over the number of units that could have been processed during this time. This is affected by aspects such as loading, clamping, and unloading. Finally, in OEE, quality is the ratio of the average number of non-rejected parts over the average total quantity of processed parts. This implies that there is an inspection process that detects faulty products. The following illustrative example shows how OEE can be calculated.

A shift in a plant has a duration of 9 hours. Due to planned maintenance and team meetings, the working time of the machine under investigation is 8 hours (480 minutes). During a typical day, the machine breaks down once, resulting in a delay of 20 minutes. Changeovers during the operation consume a total of 40 minutes. Thus, availability is:

$$(480 - 20 - 40) / 480 = 88\% \quad (9.3)$$

The average cycle time of the machine is one minute, meaning that the machine can manufacture 420 parts during the available operating time. However, due to loading and unloading the machine the team only processes 350 parts on average. Therefore, performance is:

$$350/420 = 83\% \quad (9.4)$$

Finally, of the 350 manufactured parts, 50 are typically rejected due to noncompliance with the specified tolerances. On this basis, quality is calculated as:

$$300/350 = 86\% \quad (9.5)$$

Applying the OEE formula, the result is an OEE metric of 63%.

Due to its nature, OEE is mainly used in manufacturing operations systems. For non-continuous operations systems, OEE levels between 85% and 92% would be considered world-class. An OEE level of 63%, as estimated in the example, indicates some room for improvement. It is important to note that very high levels of OEE may also be the result of large batch sizes, which are likely to create wastes in other areas of an operations system.

CONCLUSION

After defining operations management and presenting the core idea of viewing operations as transformation systems, this chapter briefly outlined the main functions of operations management. It introduced a group of operations systems that can be applied to characterize the operations in most businesses and highlighted the core idea that the operations system employed by a business is constrained by the availability of technology. After providing brief overviews of supply chain management, demand forecasting, and ERP systems, this chapter gave an introduction of Lean as the dominant philosophy shaping operations in the present. The chapter pointed out the core idea that in Lean, the delivery of value to the customer is the most important goal of operations processes, rather than profit maximization.

The chapter on operations management in *Management – An Introduction* by David Boddy (2017) usefully presents operations management in the context of other management topics. Dedicated and authoritative introductions to operations management are provided in *Operations Management* by Nigel Slack and colleagues (2013) and in *Introduction to Operations Management* by John Naylor (2002). The textbook *Process theory: The principles of operations management* by Matthias Holweg and colleagues (2018) provides an up-to-date overview of operations management with and emphasis on processes. The book *The Fourth Industrial Revolution* by Klaus Schwab (2017) accessibly characterizes currently emerging operational methods around digital manufacturing and their likely effects. A very useful introduction to practical Lean methods is provided in *The Lean Toolbox* by John Bicheno and Matthias Holweg (2016).

REVIEW QUESTIONS

1. Complete the below characterization of operations.
   (Question type: Fill in the blanks)
   "The transformation system operated by a business is one that takes in ________, such as ________, processes these into ________, such as products and services for customers. Note, however, that any such transformation system will also generate ________."

2. Which of the following are general functions of operations management?
   (Question type: Multiple response)
   ☐ Marketing strategy
   ☐ Procurement logistics
   ☐ Scheduling software implementation
   ☐ Controlling operations
   ☐ Strategic focus
   ☐ Inventory management

3. To be truly efficient, businesses should not hold inventory under any circumstances.

   (Question type: True/false)

   Is this statement true or false?

   O True O False

4. Match the provided scheduling activities to the relevant time horizons.

   (Question type: Matrix)

	Dispatching	Aggregate scheduling	Master scheduling	ERP forecasting
A few months into the future	O	O	O	O
Immediate	O	O	O	O
Up to five years into the future	O	O	O	O
This is not an accepted scheduling activity and has no time horizon	O	O	O	O

5. Order the provided production operations systems according to their ability to process products in high volumes in descending order, with the first being the highest, etc.

   (Question type: Ranking)

   - Line and continuous operations systems
   - Job-shop operations systems
   - Project operations systems
   - Batch operations systems

6. Complete the following figure showing two supply chain structures by inserting the correct labels.
   (Question type: Labeling)

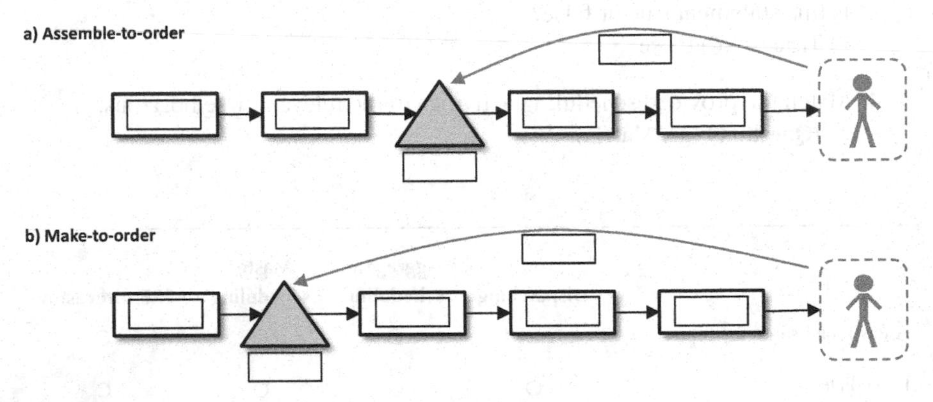

FIGURE 9.9 Insert the appropriate labels

7. Which of the following describes the shown pattern of demand appropriately?
   (Question type: Multiple choice)

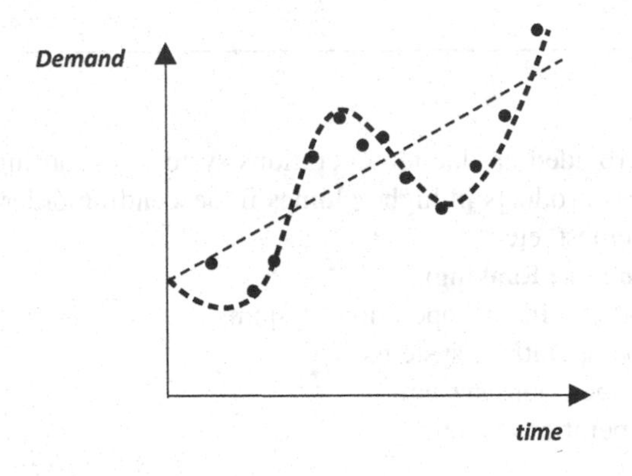

FIGURE 9.10 Observed demand level over time

O Random variation around a trend
O Nonlinear trend
O Declining trend
O Seasonality and trend
O Only seasonality

8. Which of the following is not a waste in Lean?
   (Question type: Multiple choice)
   O Waste of overproduction
   O Waste of inventory

- O Waste of opportunity
- O Waste of waiting
- O Waste of underutilized people

9. Which of the following statements about Kanbans are true and which are false? **(Question type: Dichotomous)**

True	False	
O	O	A Kanban is a push signal
O	O	Kanbans are always machine-readable
O	O	A Kanban travels through operations against the flow of the work
O	O	Kanbans facilitate demand planning
O	O	Kanbans are an essential part of pull systems
O	O	A Kanban can be a physical object
O	O	A Kanban is purely a signal and never contains additional information

10. As part of a Lean analysis, you are tasked with the investigation of the effectiveness of a CNC-machining process. Data collection has yielded the following information:
    - The available time during a shift in the factory is 480 minutes.
    - During each shift, the machine has been inactive for 25 minutes due to breakdown or changeover on average.
    - The mean cycle time for the machine is one minute.
    - Due to loading and unloading, only 320 parts have been processed on average.
    - Of the parts processed during a shift, 14 parts are rejected due to manufacturing faults on average.

    (Question type: Calculation)
    Calculate the OEE for the CNC machine.

REFERENCES AND FURTHER READING

Bicheno, J. and Holweg, M., 2016. *The lean toolbox: A handbook for lean transformation*. 5th ed. Buckingham: PICSIE books.

Boddy, D., 2017. *Management: An introduction*. 7th ed. New York: Pearson Education.

Hayes, R.H. and Wheelwright, S.C., 1979. Link manufacturing process and product life cycles. *Harvard Business Review*, 57(1), pp.133–140.

Holweg, M., Davies, J., De Meyer, A., Lawson, B. and Schmenner, R.W., 2018. *Process theory: The principles of operations management*. Oxford: Oxford University Press.

McAfee, A. and Brynjolfsson, E., 2017. *Machine, platform, crowd: Harnessing our digital future*. New YorK: WW Norton & Company.

Naylor, J., 2002. *Introduction to operations management*. 2nd ed. Pearson Education.

Schmenner, R.W., 1986. *How can service businesses survive and prosper? Sloan Management Review (1986-1998)*, 27(3), p.21.

Schwab, K., 2017. *The fourth industrial revolution*. New York: Currency.

Slack, M., Brandon-Jones, A., and Johnston, R., 2013. *Operations management*. 7th ed. London: Pearson.

10 Managing People

OBJECTIVES AND LEARNING OUTCOMES

The objective of this chapter is to introduce the area of management concerned with the management of people, also referred to as human resource management. After defining this field and setting out its scope, this chapter briefly introduces the activities of human resources management as a dedicated function in a business. This is followed by an introduction of a range of broader topics relating to the management of people, including influencing people, leadership, and motivating people. This chapter additionally provides an overview of current trends in human resource management and, for completeness, briefly presents a perspective on the important topic of managing the self. This chapter aims to achieve the following learning outcomes:

- an appreciation of the general field of human resource management;
- the ability to define human resource management and human capital, including an understanding of the activities that make up human resource management;
- knowledge of a range of models of influencing people, including personality traits models, behavioral models, situational models, models based on power and tactics, and networking models;
- an understanding of traditional content theories of motivation, including Maslow's hierarchy of needs, Herzberg's two-factor theory, and McGregor's theory X and theory Y;
- knowledge of current topics in human resource management;
- understanding Covey's seven habits model of managing the self;
- an understanding of key terms, synonyms, and accepted acronyms; and
- an appreciation of important thinkers in the field of human resources management.

THE SCOPE OF HUMAN RESOURCES MANAGEMENT

The management of activities and processes in a business aimed at influencing the attitudes and conduct of employees is referred to as human resource management. Human resource management is generally seen to involve four different aspects: influencing employees, motivating employees, organizing the flow of human resources, and managing the instruments available to reward good employee performance. Generally, there is a consensus that effective human resource management improves the performance of a business. For this reason, it usually forms an area of priority for managers. In particular, it is thought that investing in the *human capital*

DOI: 10.1201/9781003222903-12

embodied by staff leads to greater business success. Human capital can be defined as follows:

IMPORTANT DEFINITION

Human capital is the overall explicit and implicit knowledge, habits, skills, abilities, behaviors, social attributes, and personality traits embodied by people that can be used in the attainment of business goals.

As stressed in Chapter 1, managers rely on the direction of the effort of other people to reach organizational goals. Hence, it is unsurprising that human resource management and investing in human capital forms an important part of most managers' activities. In larger organizations, there are usually dedicated *human resources managers* facilitating or supporting such activities.

The approach such human resource managers take in businesses are generally thought to fall into distinct categories which are characterized by the degree to which they tend to intervene and by whether they display a more long-term strategic approach or a short-term tactical approach.* Table 10.1 summarizes the framework resulting from this categorization, identifying the approaches of *change agents*, *advisors*, *regulators*, and *service providers*. The majority of human resource managers adopt a strategic and non-interventional attitude.

In practice, the management of human resources and human resources processes normally involves a set of specific tasks, irrespective of the approach adopted. These include *planning human resources*, *analyzing jobs*, *recruiting new staff*, *managing rewards*, and *maintaining equal opportunities and diversity*. To introduce the field of human resource management, it is instructive to briefly characterize each of these broad activities.

TABLE 10.1
Typical Approaches to Human Resource Management as a 2-by-2 Matrix

	Interventional attitude	Non-interventional attitude
Strategic focus	• Actively pursuing a change of the business and its culture to increase profitability. Human resources managers following this approach are described as *change agents*	• Facilitating rather than taking action, offering expertise and advice to other managers in the business. Human resources managers following this approach are described as *advisors*
Tactical focus	• Engaging in the formulation and implementation of human resources policy and making sure that this is observed. Human resources managers following this approach are described as *regulators*	• Responding to requests for support by managers and providing support in the administration of human resources processes. Human resources managers following this approach are described as *service providers*

* Boddy (2017).

Planning Human Resources

A significant role of human resource management is the planning of staffing so that the requirements of the business can be met while ensuring that enough people with the right qualifications can be hired. In many larger businesses, there are forecasting processes to anticipate how staff requirements will change in the future. In smaller businesses, human resource planning will normally be carried out in a more reactive way.

Human resource planning will need to consider the number of staff joining, the number of staff leaving, and any transfers of staff between roles within the business. Moreover, aspects related to employee attitudes need to be taken into account, including the productivity of employees. This aspect can be assessed by managers with special measures of productivity such as the employee productivity index, the frequency with which staff tends to resign, known as *staff turnover*, and the available data on staff absence due to illness, holidays, and other reasons. Table 10.2 briefly defines and summarizes four such metrics involved in human resource planning.

Businesses will want employee productivity to be as high as possible and staff turnover and absence rates to be as low as possible.

Analyzing Jobs

An important role in human resource management is the analysis and specification of *jobs** within the business. Jobs are also known as *roles*, *posts*, or *positions*. The specific requirements and characteristics of a job, including the necessary skills and competencies, training level, responsibilities and duties, level of responsibility, and other important parameters are usually captured in a document called the *job description,* also known as the *job specification* or *role profile*. The specification of a job will additionally focus on the team working abilities and soft skills required from a person in a particular role. Once the job description is agreed upon within the business, the position can be filled by recruiting a suitable candidate.

Recruiting New Staff

The objectives of recruitment in a business are twofold. Initially, managers will want to have the largest pool of suitable applicants to recruit from. This is achieved by advertising an available job widely to solicit as many applications from qualified candidates as possible. This often involves managers' personal contacts, conferences and conventions, careers fairs, education institutions, recruitment advertisements, employment agencies, social media, and, increasingly, internet recruitment platforms. It is important to realize that the candidate's and the employer's interests are often opposed in this situation. While the employer will want the number of suitable applicants to be as large as possible, candidates will want to have as few competitors for the available job as possible.

* Risking a statement of the obvious, the term "job" has a different meaning in this context from the one introduced in Chapter 7.

TABLE 10.2
Metrics of Worker Performance and Attitudes

	Description	Explanation
Employee productivity index	$\frac{\text{Productive activity}}{\text{Unit of time}}$	An example for an employee productivity index could be the number of product units processed by an employee over a period of time, such as the number of cars sold by a car salesperson per month
Staff turnover	$\frac{\text{Number of staff resigned}}{\text{Unit of time}}$	This metric expresses the number of staff leaving a business over a period of time. It is seen as an important indicator of worker satisfaction. Staff turnover is affected by relative pay levels in other organizations
Absence rate	$\frac{\text{Total absence per unit of time}}{\text{Number of staff}}$	This metric shows the mean total absences, normally measured in days, per member of staff over a period of time. It forms an indicator of worker satisfaction that is independent of relative pay
Absence rate per manager	$\frac{\text{Total absence per unit of time under a manager}}{\text{Number of staff under a manger}}$	This metric shows the mean total absences per member of staff over a period of time in a team under a manager. This metric also reflects manager performance

The next stage in the recruitment process is the selection of the best applicant from the pool of available candidates. The selection process can be implemented in many different ways but usually follows the logic of progressively eliminating candidates from the pool of applicants until the most promising candidate has been identified. This will usually begin with a review of each candidate's *curriculum vitae* (*CV*), or *resume*, to determine if the candidate matches the job description. Successful candidates will then be *shortlisted* and invited for interviews, which may or may not take place in person and may or may not follow a formal structure. Frequently, there are multiple rounds of interviews with different managers in the business. Large businesses often ask applicants to complete personality tests and to attend assessment centers at which they are subjected to various tests by assessors, generally with the goal of better identifying the applicants' competencies and their potential for future development. It is important to realize that there is

TABLE 10.3
Elements of Reward Management

	Description
Fixed salary	The fixed monthly pay an employee is entitled to before additions or deductions. Also known as the *base salary* or *basic salary*
Time rate pay	Pay based on the time worked
Payment by results	Pay based on the amount of work performed or completed
Skills-based pay	Pay based on the level of knowledge, skill, or qualifications held by the employee
Performance-related pay	Pay based on the employee or the business reaching or exceeding agreed objectives. Where this form of pay supplements fixed pay, it is called a *bonus*. Bonuses paid immediately to reward exceptional performance are called *spot bonuses*
Non-monetary benefits packages	Additional rewards, usually not in the form of money, that are tailored to employees' preferences. These include health care, generous maternity or paternity arrangements, and a luxurious company car
Employee stock options	This form of reward grants employees the right to buy company shares at an agreed, and usually favorable, price for a period of time

significant subjectivity and inaccuracy in all forms of selection processes* so it is by no means guaranteed that the most suitable or deserving applicant is offered the job. This again highlights why rational and highly qualified candidates may be put off if there are large numbers of applicants for a job.

Managing Reward

In human resources management, *rewards* are used to align the objectives of the business with the objectives of the employees. Employees can be rewarded for their work in a business in a range of different ways, most of which involve some form of payment to the employee. The basic and most important form of reward is an employee's *fixed salary*. Table 10.3 summarizes common elements of pay in a reward management system. A typical pay structure contains a fixed salary and may contain a combination of other elements. Also known as a *pay package*, it is normally agreed upon toward the end of the selection stage in the recruitment process. While there is often scope for negotiation, it is common for businesses to exclude applicants with excessive demands.

The composition of the pay structure is specified in an employee's contract of employment. Generally, the more senior a position on a business is, the higher the share of performance-related pay and employee stock options. Additionally, the contract of

* Beardwell and Thompson (2014).

employment may entitle the employee to a *severance package*, also referred to as *redundancy pay*, which may include a combination of additional benefits of the types listed in Table 10.3 if the employee leaves the company under certain conditions.

Pay structure is often subject to various laws, regulations, and agreements with the government and organizations such as trade unions. Depending on these rules, businesses are often required to make various kinds of employer contributions to the state. Importantly, the employees themselves are likely to be tax liable for most forms of pay, including flexible benefits such as the use of a company car, normally requiring the payment of income tax.

If the performance of an employee is poor, performance-related pay will be reduced or not paid at all. If the performance of the employee is unacceptably poor over a longer period of time, the employment may be terminated. In most businesses, this is a rare occurrence. Fixed salaries are not normally reduced due to poor performance.

Managing Equal Opportunities and Diversity

As briefly touched upon in Chapter 1, there is a current emphasis on improving gender balance and diversity in many roles. These developments generally reflect that the composition of the workforce in many countries has become more diverse in the past and is continuing to do so. Diversity can be defined as follows:

Diversity encompasses the different attributes of people and any characteristic that leads to one person being identified as different from another person.

In human resources management, many activities are unavoidably about distinguishing between different individuals. Such distinctions are normally made on the basis of employees' potential in the recruitment process and in terms of performance in the management of rewards. For this to be done fairly it is vitally important that there is no room for discrimination by gender, ethnicity, nationality, disability, age, or any other group attribute. The *equal opportunities* approach has been developed to prevent unfair discrimination; it involves the setting up of formal procedures that aim to eliminate the influence of such unfair behavior. *Diversity management* is an alternative approach based on the conscious inclusion of employees from different backgrounds in various business functions through targeted policies and programs.

Extracting benefits from diversity in pursuit of business objectives requires a coordinated, systematic approach and commitment from managers. However, there is strong evidence that businesses benefit from diversity in certain circumstances*. This is not only because a diverse sales workforce will be able to relate better to a diverse customer audience but also because greater diversity among employees across all business functions promotes new, creative problem solving and a more harmonious and inclusive working style.

* Gaudiano (2020).

INFLUENCING AND LEADING

Due to the social nature of their occupation, managers rely on influencing others to achieve their objectives. They may need to influence not only colleagues in their own business but also other outside stakeholders such as customers, suppliers, investors, decision-makers in governmental agencies, and trade union representatives. In this process of influencing, managers will frequently have to reach beyond functional and organizational boundaries and address individuals that are indifferent or perhaps hostile toward their objectives. There is a wide range of well-known theories proposing explanations of how the process of influencing and leading others works. This section briefly outlines a range of such theories.

MODELS OF MANAGERS' PERSONALITY TRAITS

There are several competing frameworks that explain the process of influencing others through the personality traits of managers. One such theory is known as the *big five* model, also referred to as the *five-factor* model or the *openness, conscientiousness, extraversion, agreeableness, and neuroticism* (*OCEAN*) model. In its basic version, this model suggests that there are five broad dimensions in a manager's personality determining the manner in which the manager influences others. These dimensions are associated with a range of personal attributes. Figure 10.1 summarizes the basic model and offers very brief characterizations of the behaviors matching each trait. The message of this model is that the degree of fit between a manager's work situation and their personality traits is decisive for successful influencing and leadership. A particular trait that is useful in one situation might not be useful in another. This theory, of course, implies that managers should be aware of their own personality traits to effectively manage.

Another well-known trait-based model is the distinction between *transactional* and *transformational* managers. Transactional managers prioritize bargaining with other people to achieve their objectives. This means that they influence others to pursue goals, such as obtaining certain rewards, while simultaneously achieving their own goals. Transformational managers, on the other hand, influence others by encouraging them to realize higher ideals and aspirations. This can be done, for example, by formulating a desirable and unambiguously beneficial future vision for the business in which everyone benefits. Transformational managers are also known as *charismatic managers*. As with the big five model, this model challenges human resource managers to match the characteristics of a manager to the circumstances present in a particular managerial situation.

BEHAVIORAL MODELS

An alternative set of models explains the process of influencing others through the behavior exhibited by managers. One particular group of *behavioral models* expresses the conduct of managers as a combination of two dimensions. The first dimension is the *concern for results* (also known as *initiating structure*). This refers to a manager's focus on the achievement of goals and objectives, clarity in communication with

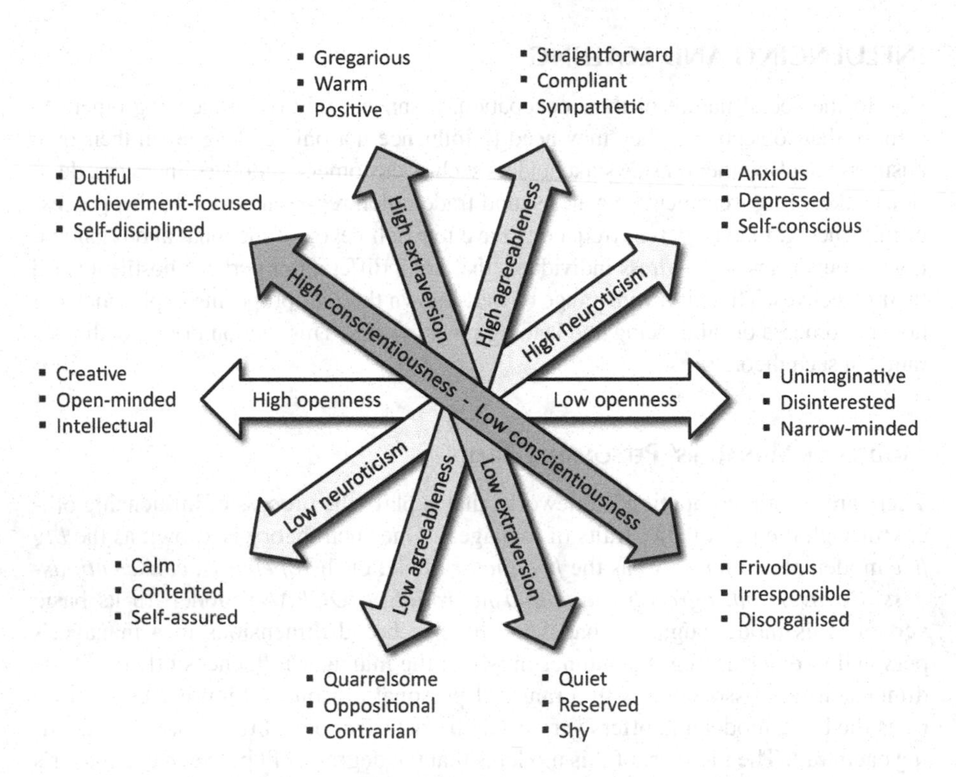

FIGURE 10.1 Big five model personality traits, adapted from Boddy (2017)

subordinates, and an emphasis on the following of rules and directions. Behaviors associated with concern for results include delegation of closely defined tasks to subordinate employees, clear specification of levels of expected performance, stating job requirements, detailed planning of work, and enforcing adherence to procedures.

The second dimension is the *concern for people* (also known as *consideration*). This reflects an emphasis on the needs of the employees and an appreciation of them as individuals. Managers of this kind normally do not rely on their formal position and seniority to influence people. The following behaviors are associated with concern for people: showing appreciation of above-average levels of performance, setting realistic expectations for employees, supporting employees if they have personal problems, demonstrating availability and approachability, and a readiness to reward high performance.

Forming a coordinate system using these two dimensions, a classification of common combinations of behaviors among managers can be established, as shown in Figure 10.2. This model is known as the *leadership grid*.

The lower-left corner of the leadership grid combines low concern for people with low concern for results. Managers in this behavioral category are passive and choose inaction wherever possible. They show *indifference* toward their staff and their objectives. Since this results in a management style that is likely to lead to poor results and low innovativeness, this combination is also known as *impoverished style*. The lower-right corner of Figure 10.2 reflects a combination of high concern for results and low concern for people. Managers in this category show *controlling* forms of

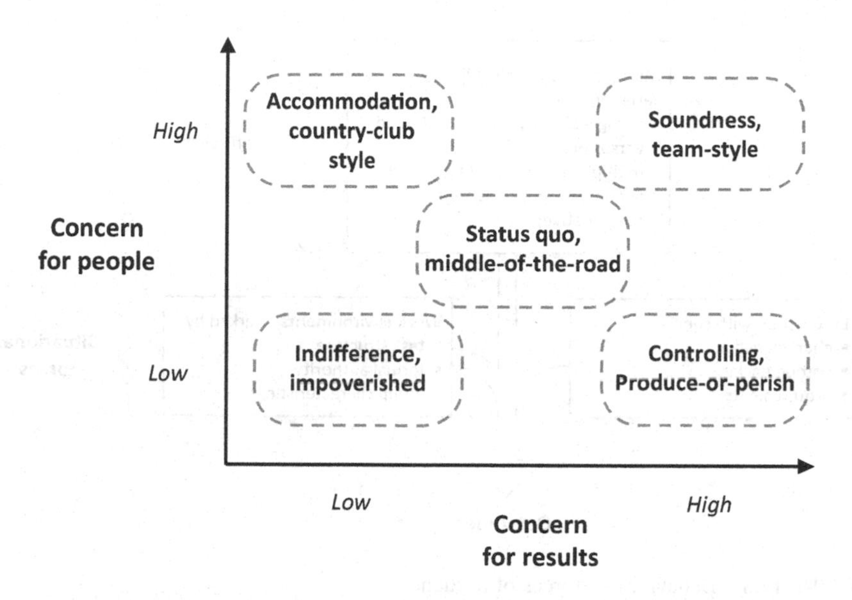

FIGURE 10.2 Leadership grid with descriptors, adapted from Blake and Mouton (1964) and McKee and Carlson (1999)

behavior, exert pressure on employees, and usually set out the work to be done in a systematic way. This management style is frequently found in situations of crisis; it is also known as *produce-or-perish*. The center of the figure reflects behavior known as *preserving the status quo* or the *middle-of-the-road* style. Managers showing this kind of behavior are seen to balance and moderate the needs of the business against the needs of the employees. Since this approach is based on compromises is it likely that some performance objectives will not be met. The top-left in Figure 10.2 shows the *accommodating* style combining a high concern for people with a low concern for results. This attitude is also known as the *country-club style* of management. Managers using this approach tend to pay great attention to the needs and feelings of their employees. Despite creating a friendly work environment, the result will likely be low performance and thus detrimental to the business. Finally, managers falling in the category in the top-right of the leadership grid show the *sound*, or *team-style*, approach to management. This combines a high concern for people with a high concern for results. Relying on committed employees and an atmosphere of teamwork, this approach is marked by trust and respect. Since it avoids trade-offs between the concerns for people and results, the performance of managers adopting this behavior is likely to be high. The leadership grid suggests that this is the ideal form of manager behavior and should be aspired to.

Situational Models

Situational models of influencing others are based on behavioral models of management but add the idea that managers adapt their style to the circumstances they face. Such models are also known as *contingency models* and generally propose that managers adapt their leadership along a specified range of dimensions and under the influence

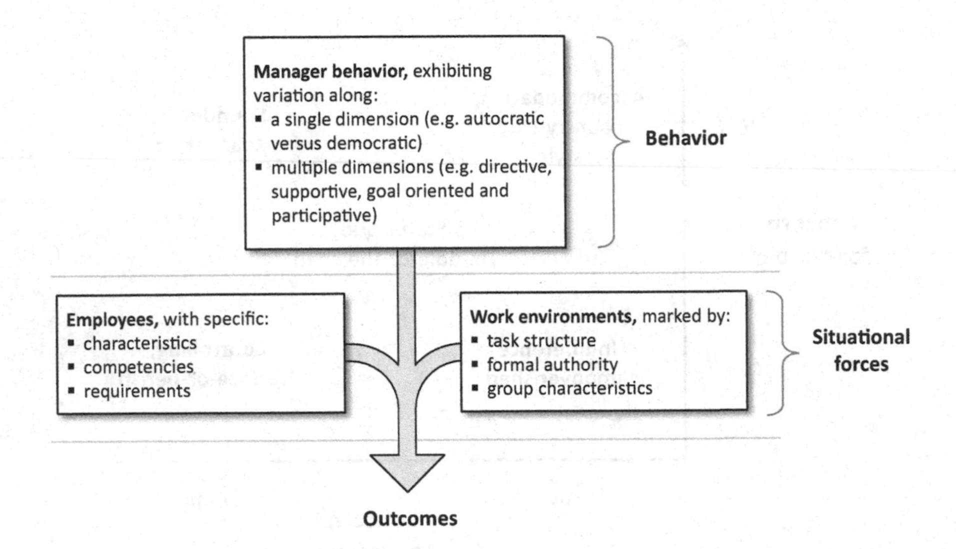

FIGURE 10.3 Behavior and forces of influence

of a set of forces, including the characteristics of the employees and the features of the work environment. Figure 10.3 summarizes such models of manager behavior, showing how the outcomes of management result from manager behavior and a set of situational forces flowing from the employees and the work environment itself.

This perspective suggests that multiple behaviors are appropriate at different times and the correct behavior depends on the situation. *Directive behavior* or *autocratic behavior* tends to be beneficial if there is considerable uncertainty and ambiguity; in such situations, managers can establish control by setting clear expectations, processes, and rules to follow. *Supportive behavior* delivers good results in situations that are characterized by unpleasant, dull, or frustrating tasks. In such situations, managers should show concern for employees' welfare and set a good example by themselves participating in the work. *Achievement-oriented behavior* is beneficial in demanding situations that are non-recurring and ambiguous. By articulating challenging goals and setting the expectation that workers will succeed, managers are likely to deliver positive outcomes. Finally, *participative and democratic behavior* tends to deliver good results if the work is non-repetitive and if there is a good chance that the employees can successfully complete their tasks.

Models Based on Power and Tactics

Based on the hierarchical nature of most businesses, as outlined in Chapter 2, managers exert power over their subordinates in some form.* In the context of management, power can be thought of as the capacity to produce an intended change

* In some cases managers will also try to exert power over their peers or even over more senior colleagues, especially in the form of coercive power, referent power and expert power, as outlined in this section.

or effect. There are multiple sources of power over other people and managers are likely to draw their power from a combination of these sources. According to a traditional model,* there are five main sources of power:

- *Legitimate power*, referring to the power that flows from a manager's formal position of seniority. Their position gives managers legitimate authority to make certain decisions, such as discretion on how to spend an allocated budget or which supplier to buy raw materials from.
- *Reward power*, referring to a manager's ability to issue rewards and compensation to incentivize others. In practice, a manager might reward an employee through payment of a spot bonus or through promotion.
- *Coercive power*, referring to the ability to make others comply by instilling fear of negative consequences, punishment, or harm. Coercive power is expressed through practices such as issuing threats, reprimanding others publicly, cultivating an imposing physical presence, and belittling or demeaning others. The use of coercive power is always problematic among adults in a professional setting and its exercise often amounts to bullying. Managers resorting to coercive power show that they are incapable of resolving situations in another way. In this sense, relying on coercive power is very often a sign of weakness.
- *Referent power*, referring to the power that is derived from the personal attractiveness and charisma of an individual. To have referent power, managers will attempt to make others identify with them or like them. Stemming from a manager's personality, this source of power strongly depends on the specific situation and the individuals involved.
- *Expert power*, referring to an individual's mastery of indispensable abilities, knowledge, or expertise. In the business environment, expert power often takes the form of deep administrative knowledge or special technical expertise in a particular matter. Expert power is also frequently used by individuals to shield themselves from criticism by non-experts since it is not normally possible for non-experts to fully understand or judge a situation involving experts. From this viewpoint, the exercise of expert power can be a sign of weakness, especially if the expert is unable to provide sufficient explanations that are understandable by non-experts.

As implied in the concept of legitimate power, a manager's position in the business often determines the power they wield. A manager's legitimate power is likely to be large if they do not require approval from more senior managers for non-routine decisions, their activities are central to the business's success, and if they interact frequently with external stakeholders or more senior colleagues. Some managers systematically seek to increase their power through behavior that expands their influence in the business. Such activities are known as *political behavior* or *office politics* and are usually seen as counter-productive by colleagues.

* French, Raven and Cartwright (1959).

TABLE 10.4
Influencing Tactics and Brief Explanations

	Description	Source(s) of power
Applying pressure	When a manager applies pressure, demands or threats are made or persistent reminders are issued	Legitimate power, coercive power
Claiming legitimacy	This tactic legitimizes demands by referring to formal authority or to alignment with agreed policy	Legitimate power
Coalition building	Seeking the support of others through persuasion	Referent power, legitimate power, expert power
Personal appeals	Appealing to loyalty or personal integrity in order to influence someone	Referent power
Inspirational appeals	Arousing confidence and enthusiasm through an appeal to values	Referent power
Exchange or transaction	Promising an exchange of favors or reciprocity, highlighting mutual benefit	Legitimate power, reward power
Ingratiation	Making others feel favorable about the influencer or putting them in a good mood	Referent power
Consultation and involvement	Involving others in decision-making or soliciting their participation	Legitimate power
Rational persuasion	Use of rationality, logic, and evidence to persuade others to take a position or action	Expert power

Source: Adapted from Yukl and Falbe (1990)

An alternative and more direct way to explain the process of influencing is through the application of *tactics*. Tactics, which can be employed consciously or unconsciously, are ways of securing goals set out in a strategy. Influencing tactics can be used when interacting with many kinds of stakeholders, including subordinate employees, peers, senior managers, shareholders, and outside collaborators. Tactics employed by managers draw on specific sources of power, as described in the previous section. Table 10.4 presents a range of influencing tactics and identifies the sources of power they draw on.

As a general pattern, rational persuasion techniques are likely to work best when influencing those in more senior positions. Tactics based on exchange, personal appeal, and claiming legitimacy are frequently adopted when exerting influence over peers and colleagues at the same level of seniority. Tactics making use of inspirational appeal and applying pressure are frequently used when influencing and leading employees at subordinate levels.

Networking Models

The term *networking* refers to a person's activities directed at building relationships with others to support their work goals or career ambitions. Networks consist of valuable contacts and form an important source of information. In this way, networks allow managers to increase their influence. Suitable approaches to networking are determined by the level of seniority of a manager. Junior managers may prioritize networking with junior and senior colleagues in their own business, whereas senior managers in large businesses are likely to focus on powerful outside stakeholders such as governmental administrators, politicians and trade union representatives. Moreover, with the continued trend toward flatter hierarchies, as discussed in Chapter 2, the importance of networks is increasing. A range of different types of networks have been defined,* each with its own characteristics:

- *practitioner networks,* which are normally joined by individuals sharing a particular career path or area of specialization;
- *privileged power networks,* which are normally open to people in senior positions and only accessible by invitation;
- *people-focused networks,* which are joined by people in search of emotional support or friendship;
- *ideological networks,* which are joined by people sharing cultural, religious, or political values, often with the objective of promoting these; and
- *strategic networks,* which are available for the purpose of creating and strengthening links between organizations and businesses.

As can be seen from this list, some networks are based on personal preferences. Other networks are based on characteristics that are of no professional relevance, such as ideological networks. In general, managers are expected to exercise care and create transparency when their network membership influences decisions, such as awarding contracts or hiring new staff. Managers must avoid unduly favoring fellow network members, which would constitute *nepotism* and may amount to criminal conduct punishable by law.

MOTIVATING PEOPLE

In Chapter 1, management was characterized as the active direction of human effort. Naturally, the level of motivation exhibited by managers and employees plays a decisive role for the business. In this context, motivation can be defined as the drive within an individual that evokes and supports their commitment to an activity or undertaking. In a quote widely attributed to the US President Dwight D. Eisenhower, motivation is characterized as "the art of getting people to do what you want them to do because they want to do it"†. All businesses benefit from motivated employees and managers.

* Boddy (2017).

† Despite the source of this quote being unclear, it is included in this book because it pithily captures an important aspect of motivation.

This section summarizes traditional and established theories of motivation known as *content theories of motivation*. However, there are many other theories of motivation that exceed the scope of this book, including process theories of motivation, behavioral modification, and deliberately designing work for motivation. It is important to note that theories of motivation are often woven into other management theories such as Taylor's approach of scientific management (presented in Chapter 2) and, perhaps most importantly, into the framework of economics in the form of the axiom of non-satiety (introduced in Chapter 3).

Most scholars of motivation have focused on the identification of human desires and needs to find opportunities to intervene, for example, to increase the motivation of staff in a business. This section presents three such theories that are very well-known: Abraham Maslow's *hierarchy of needs*, Frederick Herzberg's *two-factor theory*, and Douglas McGregor's *theory X and theory Y*.

Abraham Maslow's Hierarchy of Needs

FAMOUS THINKER

Abraham Maslow (1908–1970)

Abraham Maslow was a clinical psychologist who developed an extremely influential theory of motivation known as the hierarchy of needs.* Maslow's theory is based on the idea that people are subject to a range of needs and will try to fulfill whichever need is most pressing in a given situation. Characterized by Maslow as *states of felt deprivation*, needs collectively form a hierarchy so there are lower-order needs and higher-order needs. People are generally seen to prioritize lower-order needs until they are at least partially fulfilled. Maslow's theory is often shown in the form of a pyramid model, as summarized in Figure 10.4.

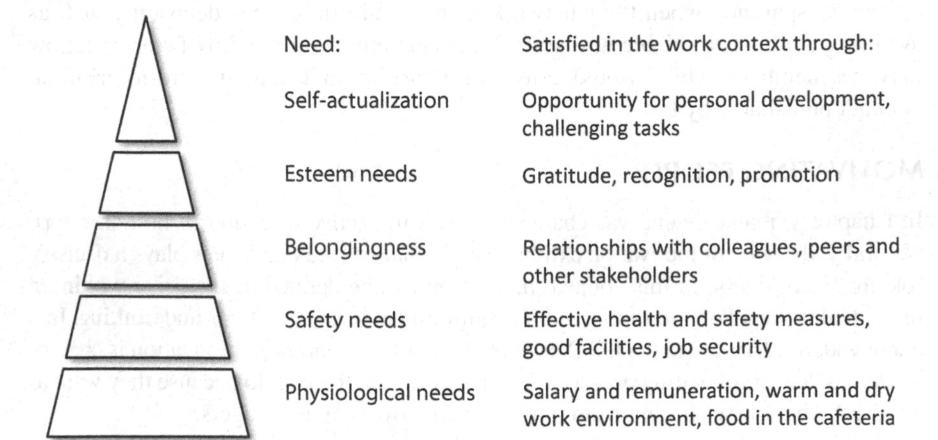

Need:
Satisfied in the work context through:
Self-actualization
Opportunity for personal development, challenging tasks
Esteem needs
Gratitude, recognition, promotion
Belongingness
Relationships with colleagues, peers and other stakeholders
Safety needs
Effective health and safety measures, good facilities, job security
Physiological needs
Salary and remuneration, warm and dry work environment, food in the cafeteria

FIGURE 10.4 Maslow's hierarchy of needs

* Maslow (1970).

In this model, the most basic needs are *physiological needs* that must be met for people to survive, including access to air, water, and food. In the context of work, these needs are addressed through a salary and a warm and dry workplace. When these needs are met, a different set of *safety needs* are typically triggered. These needs include protection from violence, excessive disturbance, and harmful levels of pollution. In the work environment, safety needs are addressed through effective health and safety policies and the reassurance provided by job security. Once safety needs are satisfied, the next set of needs addressed relates to *belongingness* and *emotional needs*. These are fulfilled through human contact, friendship, and intimacy with others. At work, these needs are met mainly through personal relationships with colleagues, co-workers, and peers. The subsequent group of *esteem needs* relates to approval and recognition; at the workplace, this is usually expressed in the form of commendations, gratitude, praise, and promotion. The final and highest set of needs relates to *self-actualization*. These needs are fulfilled through self-improvement, education, and betterment. At the workplace, this takes the form of opportunities for personal growth and the ability to successfully confront challenging tasks.

It should be stressed that, as a clinical psychologist, Maslow knew that this ordering of needs does not apply to all people in the same way and in every situation. Additionally, the hierarchy of needs should not be interpreted in a mechanistic way such that one need must be fulfilled completely before the next higher need is pursued. Like many theories of this kind, the hierarchy of needs is instead meant to reflect a common pattern found in the world.

Frederick Herzberg's Two-Factor Theory

Frederick Herzberg (1923–2000)

FAMOUS THINKER

Where Maslow's theory focused on differences between different kinds of motivation, the work of fellow psychologist Frederick Herzberg, whom the reader has already encountered in Chapter 7, focused on the importance of the nature of work for motivation. Herzberg proposed that the determinants of job satisfaction can be split up into *motivator factors* that motivate people to invest effort in their work and factors surrounding work causing dissatisfaction, labeled *hygiene factors* by Herzberg.

Table 10.5 provides a range of examples for motivator factors and hygiene factors, in approximate descending order of importance.

Herzberg's groundbreaking thesis was that job satisfaction and positive feelings toward one's work come from the work itself, from performing tasks that create a sense of achievement and accomplishment. This implies that managers cannot order or demand motivation from employees. Through careless actions and bad

CORE IDEA

leadership, managers can, however, destroy motivation in employees by creating adverse hygiene factors. This led Herzberg to conclude that to motivate employees, work should be interesting and challenging. Managers should therefore ensure that hygiene factors, in particular those involving company policy, administration, and supervision processes, do not create significant dissatisfaction. In a significant departure from what is thought by most people, even in the present time, Herzberg concluded that the level of pay is a hygiene factor of relatively low importance. This directly contradicts the idea that high pay can be used to motivate employees in a sustained way.

Douglas McGregor's Theory X and Theory Y

FAMOUS THINKER

Management scholar Douglas McGregor proposed a dual view of motivation containing two theories, *theory X and theory Y*, which describe opposing models of employee motivation. Managers adhering to theory X stress the necessity of close supervision, formal reward systems, and penalties through managers. In contrast, those adhering to theory Y focus on the role of intrinsic job satisfaction and management without the close supervision of employees. In practice, managers are likely to base their practices on theory X in some areas and on theory Y in others. Moreover, the prevalence and appropriateness of theory X and theory Y will depend on the operations processes and technology employed by a business, as discussed in Chapter 9.

Douglas McGregor (1906–1964)

Theory X draws on the belief that the typical worker has little ambition and motivation to do their work. It assumes that workers tend to shirk responsibility and are motivated by their own self-serving goals. Theory X further assumes that workers see their employment as an exchange in which commitment, effort, or time is exchanged against pay, which echoes the *transactional management style* presented in this chapter. In McGregor's theory X, it is the manager's role to direct, regiment, and coerce employees to achieve an acceptable level of motivation. This makes theory X prevalent in businesses that are characterized by an unskilled or low-skilled labor force and high degrees of repetition and specialization.

In theory Y, managers assume that employees are motivated internally, which is an idea shared with Herzberg's two-factor theory. Employees are assumed to see work as a natural activity and to take responsibility for successfully executing their work. This means that in theory Y workers do not need close supervision and instruction by their managers. For this reason, theory Y aligns closely to the *soundness* or *team-style* approach in the leadership grid presented earlier in this chapter, combining a high concern for people with high concern for results. Theory Y-based approaches to management are prevalent in businesses that rely heavily on highly skilled workers that frequently need to exercise their judgment to do their job.

TABLE 10.5
Determinants of Worker Satisfaction in the Motivator-Hygiene Theory

Motivator factors	• Sense of achievement • Recognition for achievement • The nature of the work itself • Being in a state of *flow*, in which the challenge of the work balances with the workers' abilities • Responsibility • Personal advancement • Growth and development
Hygiene factors	• Company policy and administration • Quality of supervision and control • Relationships with more senior colleagues • Working conditions • Salary • Relationship with peers • Compatibility with personal life, *work-life balance* • Relationship with junior colleagues • Status • Security

CURRENT TOPICS IN HUMAN RESOURCE MANAGEMENT

Recent topics in human resource management often revolve around the idea that competitive success and strategic advantage can be realized by improving the human capital available to the business. This section introduces two aspects that are currently in the focus of many human resource managers, *flexible patterns of work* and *high-performance work* practices.

Flexible Patterns of Work and Working from Home

The trend toward offering employees more flexible patterns of working has existed for a significant time. Jobs can be made more flexible by allowing employees to adjust the time spent working through offering flexible working hours, job sharing, and part-time work. Flexibility can also be offered by hiring workers on short-term contracts or using *zero-hours* contracts that do not require the business to state minimum working hours. Importantly, however, zero-hours contracts should not oblige the contractor to take on the work offered. Especially in the so-called gig economy, as discussed in Chapter 7, many platform businesses rely on such flexible forms of work. This development is enabled by the low transaction costs associated with information technology allowing the management of such arrangements.

A further form of flexibility at the workplace is *functional flexibility*, which allows managers and employees to adapt the tasks required by a job to suit the needs or preferences of the employees. This can be achieved by promoting teamwork and worker autonomy as well as by providing additional training where required. Reward management practices in a business can be structured to promote flexibility, for example, by offering a base salary and additional payments upon successful completion of tasks.

An additional long-standing trend, which has become hugely relevant during the global coronavirus pandemic in the early 2020s, is *working from home*. Also known as *telecommuting*, *teleworking*, *remote working*, or *flexible workplace*, working from home is an arrangement in which employees are not obliged to be physically present at their place of work, such as an office building or a warehouse. This is enabled by information technologies such as videoconferencing, instant messaging, remote desktop technology, virtual private networks, file exchange, email, and telecommunications. The available empirical evidence suggests that working from home improves the employees' ability to limit their working hours to an acceptable level, known as *work-life balance*, and increases job satisfaction for many employees.* Additional advantages of working from home are the elimination of commuting which can free up significant amounts of time, reduce cost and eliminate a significant environmental burden.

However, working from home also has problematic sides. Some workers suffer from reduced motivation and feel isolated due to the absence of face-to-face interaction with colleagues, customers, and suppliers. This may make it more difficult to build or maintain positive working relationships over time. The broad adoption of working from home has also been criticized on the grounds of equality since not all jobs are suitable for working from home. Jobs requiring manual work, which are often lower-paid, cannot normally be done from home.

High-Performance Work

High-performance work is a relatively recent human resource management approach that can be defined as stimulating "more effective employee involvement and commitment to achieve high levels of performance".† High-performance work aims to achieve this by coaxing workers to increase the effort they put into their work and by promoting a higher utilization and development of the available skills. An important role in this is played by increasing the levels of identification and attachment employees feel toward their business. High-performance work addresses the following specific areas:

- *new approaches in human resource management*, including new forms of reward management, incentive structures, staff appraisal, and training programs;
- *work organization*, including an emphasis on teamwork and worker autonomy, participation of employees in decision-making, and careful job design;

* Gajendran and Harrison (2007).

† Belt and Giles (2009, p. 5).

- *employment relations*, including enhanced job security and more selective recruitment processes;
- *management and leadership*, including changes to strategic management, line management, and business development; and
- *improved organizational development,* referring to the ability to implement ongoing change to how the business operates.

Research on the impact of high-performance work suggests that its adoption allows businesses to become more effective and competitive by increasing the competence, motivation, and commitment of employees.[*]

COVEY'S HABITS OF MANAGING THE SELF

FAMOUS THINKER

Stephen R. Covey (1932–2012)

As Mahatma Gandhi famously argued, if "a man changes his own nature, so does the attitude of the world towards him".[†] In this sense, the most important person any manager must learn to manage is the self. A vast and often highly questionable body of work known as *self-help literature* is available on this topic. One particularly successful book of this sort that has found acclaim in the management community – and deserves to be read – is Stephen R. Covey's book *The 7 Habits of Highly Effective People* (1989). This section briefly presents Covey's ideas.

Covey's understanding of personal effectiveness focused on the ability to achieve results while caring about how those results have been achieved. According to Covey, this can be achieved by aligning oneself to a set of supposedly timeless and universal virtues as a guide for effective living. These include "integrity, humility, fidelity, temperance, courage, justice, patience, industry, simplicity modesty, and the Golden Rule".[‡] On the basis of these virtues, which Covey calls *principles*, it is possible to formulate a set of habits that, when practiced, should give individuals a high level of personal effectiveness. Ultimately, Covey argues that this enables individuals to collaborate deeply and meaningfully with other mature individuals to reach a beneficial and deeply cooperative state of *interdependence*.

Covey proposes an ordering of seven such habits, as summarized graphically in Figure 10.5. The first three habits concern attitudes toward the own private conduct, independent of others. The second three habits are about attitudes toward social behavior and about how to form meaningful cooperation with others. The seventh habit concerns renewing and maintaining one's resources, health, and energy. The seven habits are briefly characterized in the following:

---

[*] Zhang et al. (2018).

[†] Government of India (1964, p. 158).

[‡] Covey (1989, p. 18).

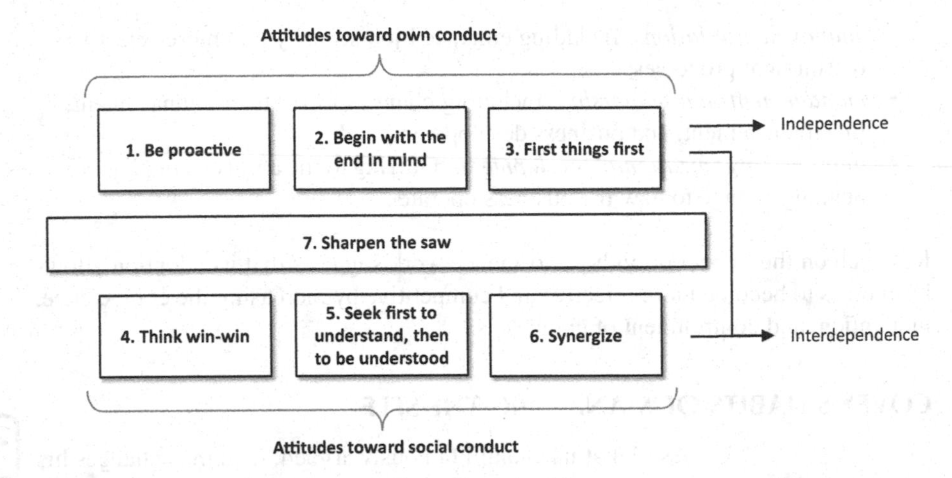

FIGURE 10.5 Covey's seven habits model

1. *Be proactive.* The first and most fundamental habit extolled by Covey requires an individual to take full responsibility for their own reactions to all experiences. This is driven by the insight that all people must constantly choose how to respond to situations, even in the worst conditions imaginable. The implication of this imperative is that, as human beings, individuals are compelled to strive to be better people than they are. Note how this corresponds to Maslow's need for self-actualization. This notion appears in many different guises throughout human culture. For example, this is what is meant when the (fictional!) character of Captain Jean-Luc Picard, portrayed as an extremely talented human resources manager in the Star Trek franchise, declares, "and that is what it is to be human. To make yourself more than you are".*

2. *Begin with the end in mind.* The second habit makes the point that effective and planned action must be guided by a concrete vision for a desirable future state. In this sense, all things created by human action are in fact created twice; they are first designed mentally and then created in the exterior reality.
3. *Put first things first.* The third habit is about prioritization and contains the counterintuitive insight that the majority of one's time should be spent with important but not urgent tasks. This stresses the importance of organizing activities in a way that allows individuals to bring to bear their strengths, talents, and planning in an optimal way. This line of thought is strongly associated with the US President Eisenhower, who was encountered earlier in this chapter already, and forms the basis of a

* Star Trek: Nemesis (2002).

strategic prioritization tool known as the *Eisenhower grid* or *Eisenhower matrix*.

4. *Think win-win.* The fourth habit is about seeking mutually beneficial solutions with other people wherever possible. This is grounded on the belief that situations in which all parties can perceive themselves as winners ultimately produce the best and most stable outcomes.
5. *Seek first to understand, then to be understood.* The fifth habit is about communication and stresses that the content of an individual's communication is subordinate to more fundamental aspects such as personal credibility and genuine empathy with the other side.
6. *Synergize.* The sixth habit is to seek solutions combining the strengths of people in teamwork. This approach enables individuals to achieve goals that no one could achieve in isolation. This habit corresponds to the soundness behavior in the leadership grid presented in this chapter.
7. *Sharpen the saw.* The final habit reflects the need to balance effort with the renewal of one's own personal resources by choosing a good lifestyle. This includes the balancing of health, energy, mental, spiritual, and emotional needs.

CONCLUSION

After defining human capital and the scope of human resource management, this chapter has introduced a range of widely recognized models of influencing and leading other people. The chapter has also introduced the topic of motivating people, focusing on what are known as content theories of motivation. This included a summary of Herzberg's two-factor theory of motivation, which makes the somewhat counterintuitive point that a high level of pay should not be expected to produce a high level of motivation in workers. Presented in this chapter as a core idea of managing people, this thought has the uplifting implication that the power of money over people has limits, and that this needs to be acknowledged by managers. The chapter continued with an outlook on current topics of human resource management, including the emphasis on flexibility and working from home and the concept of high-performance work. The chapter closed with a brief summary of Covey's seven habits approach to managing the self, in which the imperative to be proactive was singled out as another core idea of managing people.

The textbook *Human Resource Management: A Contemporary Approach by* Beardwell and Thompson (2014) is an authoritative textbook devoted to human resource management. The textbook *Management – An Introduction* by David Boddy (2017) contains a useful and extensive section on human resources management. The final section of this chapter is based on the book *The 7 Habits of Highly Effective People* by Stephen R. Covey (1989), which is a very worthwhile book at the intersection of business literature and self-help. The book *Getting Things Done* by David Allen (2001) is recommended for those aiming to increase their own productivity.

REVIEW QUESTIONS

1. Which of the following are elements of human resources management?
   (Question type: Multiple response)
   - ☐ Motivating employees
   - ☐ Development of customers
   - ☐ Making sure products and services meet human needs
   - ☐ Making sure the business conducts itself in an ethical and human way
   - ☐ Recruiting employees
   - ☐ Devising reward structures

2. Complete the below definition of human capital.
   (Question type: Fill in the blanks)
   "Human capital is the overall ________ knowledge, habits, skills abilities, behaviors, social attributes and personality traits embodied by ________ used for the attainment of ________".

3. Which of the following labels appropriately describes a human resources manager who takes a non-interventional and strategic approach?
   (Question type: Multiple choice)
   - ○ Human capitalist
   - ○ Service provider
   - ○ Change agent
   - ○ Regulator
   - ○ Advisor

4. You are analyzing a company that produces car parts. You are concerned that there might be a problem with the attitudes of factory floor workers. Unfortunately, the human resources department has lost all records of absence. The following data for the last year of operation are available:
   - The measured absence rate among factory floor workers is 20.6.
   - The company has three factories, employing 400 factory floor workers in total.
   - Average output for each factory is 12,000 car seats per month.

   (Question type: Calculation)
   Reconstruct the total number of days the factory floor workers have not been at work.

5. Match the provided examples of rewards to the correct reward types.
   (Question type: Matrix)

	Skills-based pay	Non-monetary benefits packages	Employee stock options	Time rate pay
Granting the right to buy shares	○	○	○	○
Pay determined by a time clock	○	○	○	○
Payments based on the level of knowledge	○	○	○	○
Provision of a company car	○	○	○	○

6. The single benefit of diversity is that it allows the workforce to relate better to a diverse audience of customers.
   (Question type: True/false)
   Is this statement true or false?
   ○ True ○ False

7. Identify the appropriate location for sound style, controlling style, accommodation, indifference, and status quo in the following image:
   (Question type: Image hotspot)

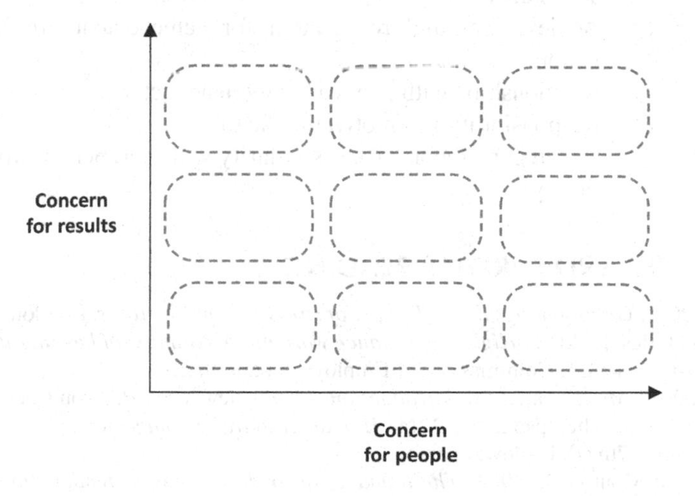

FIGURE 10.6 Identify the appropriate locations in this image

8. Which of the following statements is true of coercive power?
   (Question type: Multiple choice)
   - ○ It is the only real form of power
   - ○ It can be based on the threat of violence
   - ○ It is ineffective
   - ○ It is illegal
   - ○ It is expressed through the practice of bullying

9. Order the provided human needs according to Maslow's hierarchy of needs, with the first being the highest, etc.
   (Question type: Ranking)
   - Esteem needs
   - Safety needs
   - Belongingness
   - Physiological needs
   - Self-actualization

10. Which of the following statements reflect Herzberg's two-factor theory appropriately?
   (Question type: Dichotomous)

Yes	No	
○	○	Managers cannot order employees to be motivated
○	○	High salaries are an excellent motivator factor
○	○	It is based on factors of motivation, dissatisfaction, and acceptance
○	○	Achievement and recognition for achievement are hygiene factors
○	○	Relationships with peers are a hygiene factor
○	○	Responsibility is a motivating factor
○	○	McGregor's theory Y has similarity with Herzberg's two-factor theory

REFERENCES AND FURTHER READING

Allen, D., 2001. *Getting things done: The art of stress-free productivity.* London: Penguin.

Belt, V. and Giles, L., 2009. *High performance working: A synthesis of key literature.* Wath-upon-Dearne: UK Commission for Employment and Skills.

Boddy, D., 2017. *Management: An introduction.* 7th ed. New York: Pearson Education.

Beardwell, J. and Thompson, A., 2014. *Human resource management: A contemporary approach.* 7th ed. Harlow: Pearson.

Blake, R. and Mouton, J., 1964. *The managerial grid: The key to leadership excellence.* Houston: Gulf Publishing Co, 350.

Covey, S.R., 1989. *The 7 habits of highly effective people.* New York: Simon & Schuster.

French, J.R., Raven, B., and Cartwright, D., 1959. The bases of social power. *Classics of Organization Theory*, 7, pp.311–320.

Gajendran, R.S. and Harrison, D.A., 2007. The good, the bad, and the unknown about telecommuting: Meta-analysis of psychological mediators and individual consequences. *Journal of Applied Psychology*, 92(6), p.1524.

Government of India, 1964. *The collected works of Mahatma Gandhi XII*. The Publications Division, Ministry of Information and Broadcasting.

Maslow, A.H., 1970. *Motivation and personality*. 2nd ed. New York: Harper and Row.

McKee, R.K. and Carlson, B., 1999. *The power to change*. Austin: Grid International.

Star Trek: Nemesis, 2002. *Film directed by Stuart Baird*. United States: Paramount Pictures.

Yukl, G. and Falbe, C.M., 1990. Influence tactics and objectives in upward, downward, and lateral influence attempts. *Journal of Applied Psychology*, 75(2), p.132.

Zhang, J., Akhtar, M.N., Bal, P.M., Zhang, Y., and Talat, U., 2018. How do high-performance work systems affect individual outcomes: A multilevel perspective. *Frontiers in Psychology*, 9, p.586.

11 Marketing

OBJECTIVES AND LEARNING OUTCOMES

This chapter introduces marketing as a major field in management. After defining marketing and outlining its scope, this chapter begins by introducing the theory of service-dominant logic as a main foundation of modern marketing practice. It proceeds with a summary of digital marketing and a range of contemporary marketing concepts. This is followed by brief introductions to established marketing topics including customer engagement, segmenting the market, and the marketing mix. The final section in the chapter draws connections between organizational culture and marketing activities. This chapter aims to achieve the following learning outcomes:

- an appreciation of the general field of marketing;
- the ability to define marketing and the ability to explain the theory of service-dominant logic;
- understanding fundamental aspects of digital marketing and the ability to construct simple customer lifetime value models;
- understanding the relationship between customer engagement and business models;
- understanding the basics of customer relationship management;
- knowledge of the concept of market segmentation;
- the ability to name and explain the elements of the marketing mix, including how the topic of branding sits within this framework;
- understanding the relationship between organizational culture and marketing;
- an understanding of key terms, synonyms, and accepted acronyms; and
- an appreciation of important thinkers in marketing.

DEFINING THE SCOPE OF MARKETING

Marketing is traditionally understood to be a set of activities undertaken in the commercial context involving the conception, pricing, advertising, and distribution of products or services to address customer requirements. Because marketing activities are so commonplace, most people across the globe are exposed to advertising and sales activity daily. This includes online advertising, sales promotions and discount offers, television and radio advertisements, print media, and celebrity and influencer endorsements.

As this chapter will show, however, advertising activities form only a small part of the wider field of marketing. Reflecting this broad scope, a modern general definition of marketing is as follows: *

* Chartered Institute of Marketing (2015, p. 2).

DOI: 10.1201/9781003222903-13

"Marketing is the management process responsible for identifying, anticipating, and satisfying customer requirements profitably."

As part of this definition, the notion of *customer satisfaction* is traditionally used in marketing to describe the degree to which a customer perceives that a product or service meets their expectations. An important thought underpinning this thought is that if managers in a business understand the requirements of current and prospective customers, they will be able to generate products or service offerings that will satisfy these requirements. As introduced in Chapter 7, customers may be private individuals buying products or services for themselves or their family members, referred to by managers as consumers, but also businesses, institutions, and governments.

While many organizations are likely to have a dedicated marketing team or department, other functions within businesses also relate to marketing activities. As a consequence of this interrelatedness, many marketing specialists in businesses cooperate closely with colleagues in other roles. For example, many marketers collaborate routinely with the operations department to ensure that production processes and product characteristics are such that customer requirements can be satisfied effectively. Marketers may also cooperate closely with financial planners or technical staff in a business to determine whether a particular price for a product or service offered by the business is adequate.

Marketing usually distinguishes between three different kinds of customer requirements:

- *Customer needs*, which are states of felt deprivation that customers are trying to address. Customer needs include physical, social, emotional, and material needs. It is important to note that the needs exist prior to the marketing activity, in a sense that they are not created by it. Customer needs can be identified by techniques such as customer needs analysis, as introduced in Chapter 7.
- *Customer wants*, which are customer needs shaped by customer traits such as their personality (if the customer is an individual) or their organizational culture (if the customer is an organization or a business).
- *Customer demands*, which are needs and wants that coincide with customers' willingness to pay. Customer demands reflect the general concept of demand introduced in Chapter 3.

The identification of customer wants allows businesses to collaborate with customers to generate an attractive *market offer*, which embodies the combined products or services the business intends to supply to the customers to satisfy the identified wants. Note that there is a significant overlap between the concept of the market offer in marketing and the value proposition as part of business models, as discussed in Chapter 7. While the market offer focuses on the discovery of needs and wants in customers, the value proposition is geared toward creating different kinds of fit between the changeable characteristics of a product or service and the needs of the customers.

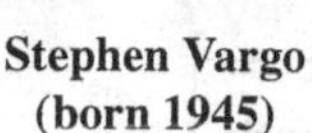

Stephen Vargo (born 1945)

Robert Lusch (1949–2017)

An exciting development that has changed the field of marketing since the early 2010s is the emergence of *service-dominant logic*.* In a series of landmark publications, marketing scholars Stephen Vargo and Robert Lusch established the idea that organizations, businesses, markets, and society as a whole are based on the exchange of services, even where physical products are exchanged. This is possible because physical products can be interpreted as providing a valuable stream of services to their users over time. Moreover, this theory argues that the universal purpose of businesses is to apply their competencies to serve customers.

Service-dominant logic has many implications within and beyond marketing that exceed the scope of this book. One major and relevant aspect emerging from service-dominant logic is that it recognizes that value does not flow from suppliers to customers, as expressed, for example, in Porter's value chain presented in Chapter 2 or in the traditional upstream-downstream supply chain view summarized in Chapter 7. Instead, service-dominant logic stresses that the value of a product or service is co-created by both the supplier and the customer. This reflects the simple fact that businesses can never know with certainty how exactly customers derive value from products or services and what this value actually is. This suggests that the co-creation of value for any given product or service is a complex and unique process involving many material, social, and cultural elements.

Service-dominant logic has greatly enriched the field of marketing and has replaced older theories of marketing. Up to the 1950s, marketing was seen mainly as a process of communicating and selling to customers. From the 1950s to the 2010s, the attention of marketing shifted to the idea of delivering value to customers. As described in this section, the contribution of service-dominant logic is a shift toward the idea that value is in reality always created through the partnership between businesses and customers. Figure 11.1 illustrates this development.

DIGITAL MARKETING

The internet has become the central venue for commercial success in most industries. Management scholar Michael Porter, who is encountered throughout this book, observed presciently that the "key question is not whether to deploy internet

* Vargo and Lusch (2004).

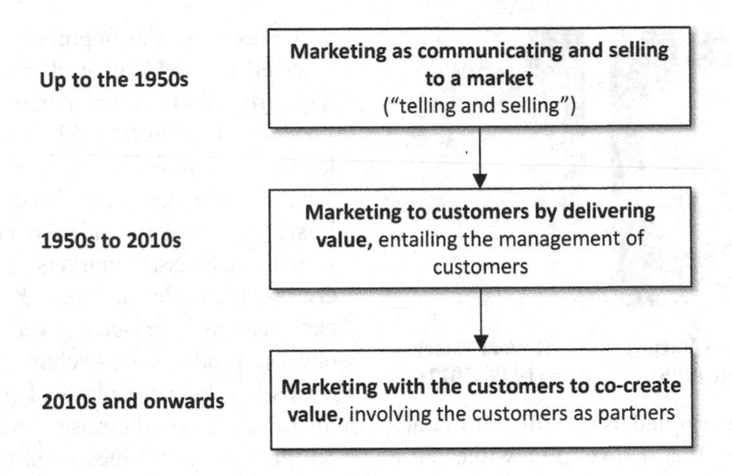

FIGURE 11.1 The changing understanding of marketing, adapted from Lusch et al. (2007)

technology—companies have no choice if they want to stay competitive—but how to deploy it".* Correspondingly, the topic of digital marketing has taken center stage in marketing. It can be defined as follows:

Digital marketing is the pursuit of marketing goals through the use of digital media, digital data, and information technology.

Stated broadly, digital marketing supports the implementation of various types of business models introduced in Chapter 7 and establishes the many different aspects of what is known as a business's *online presence*. Digital marketing activities include the development of a business website, proprietary applications, as well as engaging in social media activities, search engine optimization, online advertising, email marketing, and building relationships with the online presences of other organizations. The objective of such activities is to acquire as many new customers as possible and to maximize sales to the existing customer base.

In businesses embracing digital marketing, relationships with customers are often managed with the help of databases of customer information, by approaching customers with personalized messages and advertisements, through placing online advertisements, and by offering online systems for customer services. Collectively, the management of customer relationships in this way is known as *customer relationship management* (*CRM*).

One key element in customer relationship management is the concept of *customer lifetime value* (*CLV*). Customer lifetime value is the total value of a customer or a customer group to a business resulting from transactions throughout the relationship between the business and the customer. It can be defined as follows:

Customer lifetime value is the total net benefit arising from a customer or a group of customers to a business over their relationship with the business.

* Porter (2001, p. 64).

Customer lifetime value can be established for past interactions with customers or projected for future customers. There are many different ways to estimate customer lifetime value and some of these methods are technically demanding. Nevertheless, accurate estimates of customer lifetime value are important for the planning of marketing activities since they provide an answer to the question of how much a business can invest in acquiring a new customer and still make a profit. The following worked example provides a simple model for the estimation of customer lifetime value.

A Worked Example of a Customer Lifetime Value Model

Consider a small brewery for craft ales that supplies to a regional market. The brewery is re-evaluating its marketing strategy to increase sales in its market. Its current marketing activity is to send out a monthly newsletter featuring events and new beer varieties received by 20,000 email contacts and social media followers.

The senior managers require a model of the profit obtained from the typical customer over a period of five years. They ask this estimate to be generated on the basis of an initial cohort of 10,000 customers. The following assumptions are made:

- Not all customers in the initial cohort can be retained at the end of each year. It is assumed that annual retention improves over the duration of the customer lifetime, starting at 50% at EOY 1 and increasing in 5% increments to 70% at EOY 5.
- The average annual revenue attributable to each customer increases over the duration of the customer relationship, starting at $200 at EOY 1, increasing in $40 increments to $360 at EOY 5. Moreover, it is assumed that annual customer revenue can be established by multiplying the remaining size of the cohort at EOY with the anticipated annual customer revenue at that time.
- The net profit margin is estimated at 10%.
- The value of future flows of money is not modified to reflect their current value.

On the basis of the data available, a model is estimated, as shown in Table 11.1. In this model, the number of customers at each EOY is obtained by multiplying the number of customers at the end of the previous year with the retention level in the current year. The total annual revenue in a given year is obtained by multiplying the estimated annual customer revenue and the number of customers in that year. Net annual profit is obtained by applying the profit margin of 10%. Finally, the customer lifetime metric is obtained by dividing the cumulative net profit by the size of the initial customer cohort ($273,547/10,000=$27).

TABLE 11.1
Customer Lifetime Value Model (Worked Example)

End of Year (EOY)	Retention at EOY (%)	Customers	Annual customer revenue ($)	Total annual revenue ($)	Net annual profit margin (10%) ($)	Cumulative net profit ($)	Customer lifetime value ($)
0	-	10,000	-	-	-	-	-
1	50	5,000	200	1,000,000	100,000	100,000	10
2	55	2,750	240	660,000	66,000	166,000	17
3	60	1,650	280	462,000	46,200	212,200	21
4	65	1,073	320	343,200	34,320	246,520	25
5	70	751	360	270,270	27,027	273,547	27

Digital marketers typically distinguish between three different kinds of digital *channels* through which customers can be reached. These can be characterized as follows:

- *Paid media*: analogous to non-digital forms of marketing such as TV or print advertisements, some online marketing channels require the business to pay for visitors or to display advertisements. This approach can involve third parties known as *affiliates*, which are paid a commission for the promotion of the business's product and services.
- *Owned media*: another channel through which a business can reach a target audience is through its own websites, blogs, social media accounts, and email. Non-digital owned media include product brochures and the advertising that occurs within stores and retail outlets.
- *Earned media*: Forms of earned media include the exposure created through social marketing, including any communications, ratings, and reviews published by customers. Earned media also include the attention resulting from engaging with important organizations, people, and prominent users of social media known as *influencers*.

It is important to note that digital marketing activities will frequently be complemented or supported by traditional marketing activities such as TV advertisements or in-person sales. The coordination and integration of digital and non-digital marketing activities into a seamless whole is known as *multichannel marketing* and forms a main challenge in modern marketing.

CUSTOMER ENGAGEMENT AND BUSINESS MODEL KINDS

The objective of building *customer engagement* has emerged as highly important for all businesses and forms a central objective in marketing. Customer engagement reflects the idea that there is a two-way interaction between a customer and a business and that this transaction takes place through different channels. Customer engagement can be defined as follows:

Customer engagement is the sustained interaction between a customer and a business through online and offline communication channels.

The objective is to strengthen the relationship a customer has with a business through mental, emotional, and social factors as well as through the objects the customer owns. Successful customer engagement will allow a business to capture its customers' attention and interest on an ongoing basis. It has been suggested that customer engagement has four dimensions* as illustrated in Figure 11.2:

* Haven (2007).

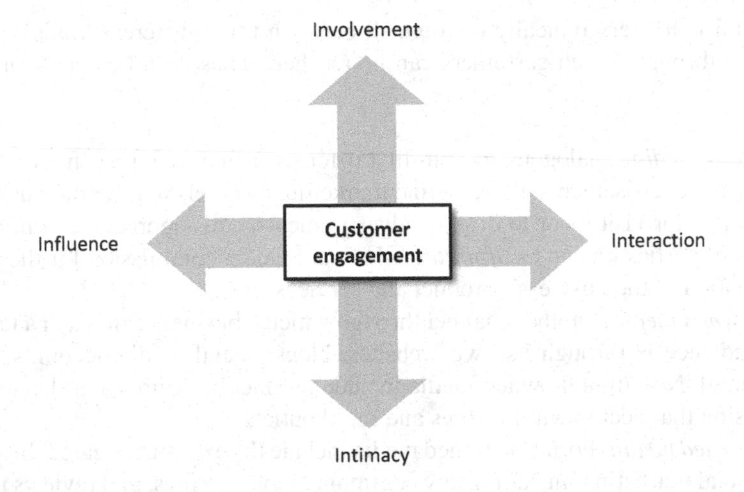

FIGURE 11.2 Dimensions of customer engagement

- *Involvement*: this dimension refers to the measurable aspects of the relationship between a business and its customers. This can be occurrences such as visits to a store or the time spent viewing a website.
- *Interaction*: this dimension measures actions through which customers make contributions to the business. This includes actual purchases, requests for additional information, product reviews or ratings, social media contributions, and the visible use of a product or service in public.
- *Intimacy*: this dimension reflects the thoughts and feelings of customers toward the product or service offered by the business. This can be measured by capturing customers' opinions, their perspectives, and their level of emotional involvement.
- *Influence*: this dimension reflects the likelihood that customers will encourage other customers to buy a product or service. This includes individual actions of customer advocacy to promote awareness, customer loyalty, and the likelihood of making repeat purchases. Influence can be measured using surveys, questionnaires, and phone interviews.

Marketing in B2B versus B2C

The appropriate approach to building customer engagement depends heavily on the adopted business model. In businesses pursuing a B2B model, as explored in Chapter 7, transactions occur between businesses, for example, in the form of sales from a manufacturer to a wholesaler. In this setting, the value of individual orders tends to be high and transactions often result from a high degree of involvement and interaction. B2B sales frequently involve a bidding process in response to customers' requests for proposals. Moreover, the decision-making process leading up to transaction in this setting is likely to take a prolonged period of time. This can range from days to months, depending on the processes within the customer business and the volume and nature of the order. Businesses are often risk-averse so the sale will frequently involve product samples or prototypes. Because such purchasing decisions are often made collectively by groups of managers or committees, each member needs to be persuaded. For this reason, B2B marketing often requires salespeople to establish personal relationships with customers. Conferences and industry trade shows are common venues to build such relationships.

Owned media such as business websites are the primary marketing channel in the B2B setting, allowing businesses to share information on their products or services to a competent audience. Traditional mass advertising, such as TV advertisements, is not common or effective in the B2B setting. Due to the goal-oriented and transactional nature of B2B, the importance of intimacy and influencing tactics is likely to be low in B2B.

Businesses engaged in B2C, such as manufacturers of fast-moving consumer goods, tend to take a different marketing approach that more evenly involves the cited dimensions of customer engagement. The focus is normally on the creation of a strong relationship by eliciting an emotional response, therefore requiring intimacy to build customer engagement. Generating demand in a mass customer audience in a crowded market is usually a very expensive process that requires a significant marketing budget.

In situations in which businesses are not able to sell directly to the consumer, B2C marketers may also face the task of convincing other businesses interfacing with consumers, such as retailers, to offer their products or services. For this reason, effective B2C marketing frequently also includes an element of B2B marketing. Another situation in which B2C may rely on B2B activities is when a business engages the services of an online platform to gain access to relevant target audiences. As explored in Chapter 7, there is often ambiguity about who is a customer and who is a supplier in transactions involving platforms.

DEVELOPING CUSTOMER ENGAGEMENT THROUGH CUSTOMER RELATIONSHIP MANAGEMENT

Customer relationship management encompasses a multitude of activities by which customer engagement can be improved. As discussed, the goal of such activities is to maximize sales by creating a targeted and personalized relationship with the customers. Specific activities include the following:*

- *Cost-effective targeting*: by collecting data on customers, it is possible to target customers with specific marketing communications. This information, or access based on such information, can be obtained cost-effectively through social media platforms.
- *Permission marketing* (or *inbound marketing*): some actions by customers, such as visiting a certain website, signing up to a mailing list, or registering their interests, allow marketers to identify highly relevant groups of customers for further marketing activities.

* For more information, see Chaffey and Ellis-Chadwick (2019).

- *Tailored marketing communications*: coupled with sophisticated software systems, information on customers allows the tailoring of marketing communications to small groups of customers or to individuals.
- *Sense-and-respond communications*: by closely monitoring customer behavior during interactions, for example, during interactions with a company website, or when a customer fills in online feedback forms, timely and relevant marketing communications can be designed and delivered to the customer.
- *Automated communications*: once the necessary systems are in place, targeted communications can be sent to customers in an automated way, incurring little additional cost.
- *Loyalty programs*: the data required for customer relationship management can be obtained by customer loyalty schemes allowing the identification of individual patterns of interaction and habits. This information can also be used to incentivize or reinforce certain customer behaviors, for example, by awarding additional reward points for specific transactions.

SEGMENTING THE MARKET

The general basis for the commercial success of any business relies on its ability to meet the needs of customers, which may be diverse. The process of splitting an overall market into smaller groups that exhibit more similar needs is referred to as *market segmentation*. It can be defined as follows:

IMPORTANT DEFINITION

Market segmentation is the practice of dividing a market comprised of diverse customers into more homogenous groups.

As indicated in this definition, market segmentation relies on a set of factors along which the population of potential customers can be separated into sub-groups*. This is normally done on the basis of the following factors:†

- *Demographic factors*: in many cases, the most straightforward and effective method to segment markets is by using demographic factors, which relate to the structure of the population measured by statistical means. Such factors often include age, gender, family status, education, nationality, religion, and ethnicity. This form of segmentation is used frequently in B2C marketing.
- *Geographic factors*: an intuitive way to segment a market is by geographic factors, especially the location of customers. This allows businesses to vary products, services or marketing activities according to local preferences or to target geographical locations. This form of segmentation is important for B2C and B2B marketing.

---

* Market segmentation is different from customer classification, introduced in Chapter 7. While customer classification is focused on the needs and attributes of groups of customers in relation to a product or service, market segmentation aims to divide customers into groups on the basis of broader factors.

† Boddy (2017).

- *Socioeconomic factors*: factors relating to social and economic features can be used to segment a market. Commonly used socioeconomic factors include income, membership of social groups, cultural variables, and different lifestyles. This form of market segmentation is usually directed at identifying groups that share aspirations and have similar ideas about how they want to live. It applies to the B2C context in particular.
- *Behavioral and psychological factors*: it is common for marketers to distinguish between different customers according to their attitude toward a product or service or according to how it will be used. Additionally, there may be psychological and cognitive factors to consider, such as the tendency toward loyalty or the ability to process complex information. This form of segmentation is relevant for both B2C and B2B marketing.

Following market segmentation, managers face the task of selecting a target market from the segments they have identified. This selection is normally informed by criteria such as whether there are customer demands that the business can satisfy, whether the projected profit from the target market is large enough to justify the effort, and what growth prospects the market segment exhibits.

Managers will also want to obtain accurate information about the customers in a target market. This involves the development of *customer profiles* focusing on the characteristics of customers. Customer profiles include aspects such as location, language abilities, available income, what kinds of other products or services are bought, how products are purchased, and how the products or services are delivered. As discussed at the start of this chapter, modern theories of marketing assume that value is co-created between suppliers and their customers. This implies that different customers will engage in different forms of value-co-creation. For example, a supplier of flat pack furniture such as IKEA may identify that some groups of customers are prepared to assemble their furniture themselves while others are more likely to hire external help to do this.

THE MARKETING MIX

E. Jerome McCarthy (1928–2015)

FAMOUS THINKER

The instruments available to marketers to create as much demand for the market offer as possible are traditionally summarized in what is known as the *marketing mix*. Originally proposed as the four 'P's of *product, price, promotion, and place*, by marketing scholar Jerome McCarthy in 1960, this was later extended by the 'P's *people, process*, and *physical evidence* to better accommodate the characteristics of services. The purpose of the instruments contained in the marketing mix is to *position* products or services in a way that makes them as attractive as possible to the potential customers in the market. More specifically, the goal of positioning is to convince potential customers that a product or service is more likely to satisfy their customer needs than any competing offering. Figure 11.3

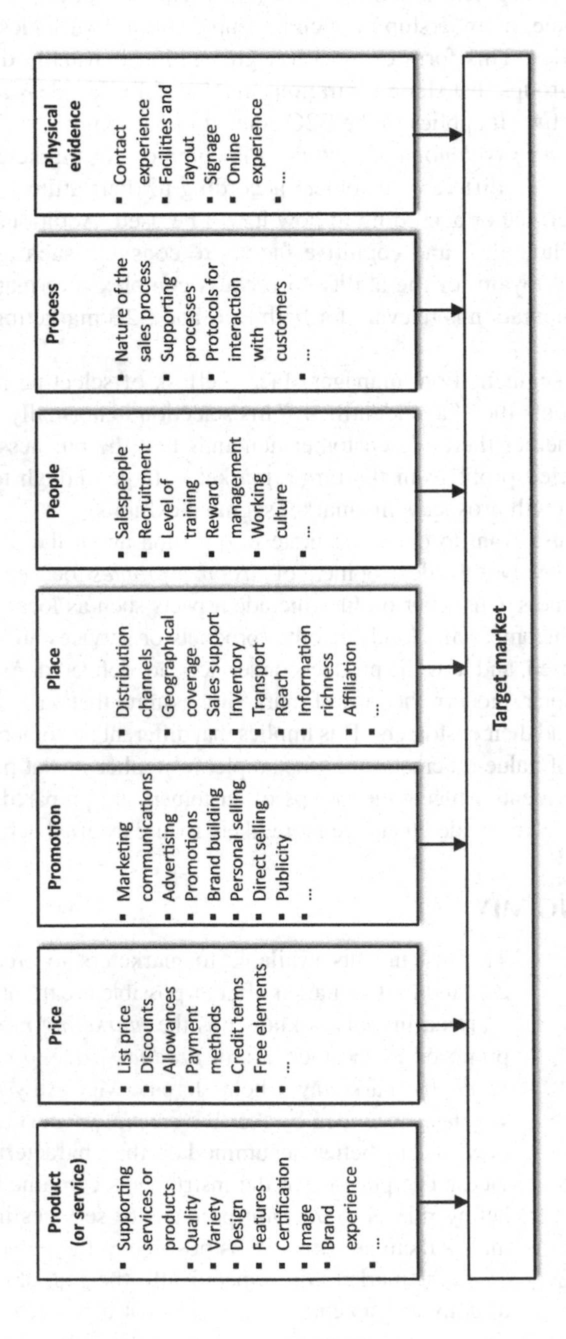

FIGURE 11.3 The marketing mix with seven elements

summarizes the components of the marketing mix and provides examples of important factors. This section presents brief summaries of each element in the marketing mix.

Product (or Service)

Product refers to the products or services that make up the market offer to the customer audience in the target market. This is often a combination of aspects since a physical product is usually supported through services such as delivery or support, and likewise, a service may include some physical objects that facilitate or enhance the service, such as in-flight meals on board a commercial flight. Products and services can be varied in many different ways, including quality, design, and features. Important ancillary features that can be adapted include packaging, insurance, technical and customer support, availability, and certification. Moreover, some products are only available through specific distributions channels. For example, a retailer might offer a smaller selection of products on their website than in their brick-and-mortar outlets. Moreover, some businesses pursue *bundling* strategies in which some products or services are available only in combination with other products or services.

A highly important aspect of products or services as part of the marketing mix is their *branding*. A branded product or service is one which is differentiated from other products or services that satisfy the same need. This helps businesses establish a relationship with customers by creating a promise that only the brand can deliver. It can be defined as follows:

Branding is the practice of distinguishing a product or service offering from competing offerings through its overall features and characteristics as perceived by the customer.

IMPORTANT DEFINITION

Different customers relate to branding in different ways. Some customers prefer branded products or services due to functional or physical characteristics associated with a brand. Others are attracted to branded items for emotional reasons or because of the image conveyed. Importantly, successful branding allows sellers to charge substantially higher prices, thereby increasing profitability.

The effectiveness of branding is often taken to underline the idea that the value of a product or service, including the value of the brand experience, is co-created by both the business and the customer. In this sense, branding carries multiple benefits: it allows the co-creation of greater perceived value, it acts as a deterrent and protection against competitors, and it is an important tool to build and reinforce trust among customers.

Price

Businesses face the challenge of setting the appropriate price for their products or services. As discussed in Chapter 3, the general goal of pricing, from the perspective of the business, is to maximize profit. While economics teaches that the profit-maximizing price is determined by the market, at least in the long term, in some circumstances marketers will want to vary this price as part of their activities, for example, to promote sales, to capture market share, or to signal that a product is of high value. For the use of pricing as a marketing instrument, the concept of own-price elasticity of demand introduced in Chapter 3 is of central relevance.

A product's *list price*, which is the final price recommended by a business selling through intermediaries such as retailers, gives customers important signals about its quality. It may well be the case that an increase in the list price leads to higher sales volumes in some situations, which contradicts the economic law of demand (as presented in Chapter 3). As part of marketing activities, some businesses may also want to offer products or services at discounted prices or offer customers trade-in allowances in exchange for old products. Businesses may also decide to make their offerings more attractive by providing customers a wide variety of payment methods or by showing lenient payment options. Such arrangements are known as *credit terms*. To increase the attractiveness of a product or service, some businesses find it beneficial to include free additional elements to their product free of charge for a period of time, such as maintenance or customer support.

An increasingly common marketing approach enabled by information technology is *dynamic pricing*, also known as *surge pricing*, in which businesses systematically change prices in an ongoing process. This allows businesses to exploit fluctuations in market demand, normally with the objective of maximizing profits. The digital systems needed to run dynamic pricing employ *pricing rules*, or *pricing algorithms*, that often consider competitor prices, supply and demand in the market, and many other factors of relevance. Dynamic pricing is increasingly used in B2C and B2B settings, including transportation, entertainment, retail, and utilities. Such approaches often form an important element of the mediating technology at the heart of platform business models, as discussed in Chapter 7.

Promotion

Promotion refers to the *marketing communications* employed to create awareness of the products or services among potential customers. Successful promotion will create interest and persuade potential customers to take a closer look at what is offered. The most important form of promotion is *advertising*, which is used to transmit pre-defined, possibly customized, messages to the target audience of a product or service. Before the emergence of the internet, this process was largely impersonal since it was not normally feasible for B2C businesses to communicate directly with members of their target audience. The emergence of social media marketing and consumer-facing internet platforms has changed this dramatically, allowing platform businesses to offer highly targeted advertising access using information obtained from individual consumers. This makes it possible to bring customized advertising to tightly defined groups of people through digital devices, including mobile phones, tablets, and computers.

CORE IDEA

A relatively recent form of promotion utilizing social media is *viral marketing*, also known as *buzz marketing*. Applied to the online context, viral marketing utilizes word-of-mouth, in which awareness of brands and specific marketing communications spreads from one user to other using electronic systems, for example, in the form of social media posts. If implemented successfully, social media will amplify the reach of the marketing communication. The main advantage of viral marketing is that it forms a very cost-effective way to reach large, even global, audiences. By exploiting word-of-mouth from peers and social media contacts, viral marketing communications are normally rated highly by consumers. The main disadvantage of viral marketing efforts is that it is difficult to craft the right kind of content to engage customer audiences. This means that there is no guarantee that the viral effect will be achieved.

Other forms of promotion include *sales promotions*, *personal selling*, and *publicity*. Sales promotions play an important role in creating the necessary traction in a market to develop and build the brand owned by a business. The goal of sales promotions is usually to persuade customers who are already considering a purchase. The purpose of personal selling is to provide customers with the information needed to buy the product in person. It is a highly interactive process that gives customers an opportunity to ask questions about a product or service, which is not possible in other forms of promotion. If a personal selling approach is used outside of traditional commercial venues, such as shops for B2C business models, the term *direct selling* is used. Publicity, also known as *public relations*, is an umbrella term for activities aimed at establishing a positive public image of a business. Such activities normally rely on successful engagement with media businesses and increasingly use social media platforms. Note that other activities within the business, such as corporate social responsibility, as outlined briefly in Chapter 2, can be used to generate positive publicity for the business.

Place

The place element of the marketing mix refers to the distribution of products or services to the customers. This can be organized by the business directly or indirectly through intermediaries, such as retailers in the case of B2C. The goal of marketers is to place products or services prominently in the minds of potential buyers. As with the other elements of the marketing mix, it is important that the place is aligned with the other aspects to form a coherent overall approach. For example, a product that is marketed as a luxury item and supported by the required branding should be made available only at suitable retail outlets. In contrast, goods aimed at more price-sensitive customers can be made available more widely in as many different outlets as possible. For physical products, traditional differentiators of place include distribution channels and networks, geographical coverage of outlets, sales support locations, the level of inventory at specific locations, and the ability to transport stock.

Place plays a very important role in digital commerce, where it refers to the quality and accessibility of the information presented online by businesses. Here it has been argued that three general forms of *navigational advantage* are crucial in ensuring successful placement in digital marketing.* These are as follows:

- *Reach*: this form of navigational advantage is about the customers' ability to access and connect to a product or service. For product or service offerings directed at consumers online, this focuses on the ability to capture the attention of potential customers, frequently referred to by digital marketers as *eyeballs*.
- *Richness*: this form refers to the depth and detail of the information transmitted to customers and well as the level of detail in the information received about the customers. Obtaining rich information is decisive in building strong customer engagement.
- *Affiliation*: this form of navigational advantage refers to the various interests represented by the seller of a product or service. Often it is desirable to show to customers that a seller acts in their interest, for example, when acting as an intermediary for another business. This aspect is particularly important for marketing strategies involving platform business models in which more than two parties are involved in transactions.

People

In the marketing mix, the people element refers to the way in which the business interacts with customers and other stakeholders. In many businesses, the emphasis will be on salespeople or employees interacting directly with customers. People are especially important in many service industries, such as professional services, financial services, and hospitality, where the service is generated directly by workers interacting with customers.

It is important to note that workers have an impact on the public perception of a business, for example, in the way they respond to queries or complaints or how they conduct themselves in public. As with the other elements of the marketing mix, employees should represent the business in alignment with other marketing activities. The people element of the marketing mix, therefore, involves coordination with the functions of recruitment, training, and reward management. Well-executed human resource management, as discussed in Chapter 10, along with cultivating a suitable working culture, is likely to contribute to effective marketing efforts in the people dimension.

Process

The process element of the marketing mix refers to the group of activities in a business that results in the delivery of products or services. While normally focusing on the sales process, other marketing-relevant processes include the development of new products, promotion activities, and customer service. Most marketing-relevant

* Evans and Wurster (1999).

processes draw on ideas or strategies of how a business interacts with customers. For example, a restaurant may have processes that define how customers are promptly greeted, seated, served, and led out of the premises after dining so that other customers can be served.

When devising processes, businesses often face the tradeoff of delivering a high-quality sales experience and minimizing the cost incurred through the sales process, for example, in terms of staff costs. General objectives when designing processes, therefore, include the following:*

- *Minimizing response time*: aiming to respond to customers as fast as possible and with as little variance in response time as possible.
- *Minimizing clear-up time*: minimizing the average staff time spent on each interaction and minimizing the number of interactions.
- *Maximizing customer satisfaction*: measuring customer satisfaction and adapting processes to maximize satisfaction, especially if this does not run against the other listed objectives.

Physical Evidence

The last element of the marketing mix is physical evidence. This refers to the expression of a product or service in a physical, or tangible, form. It involves any physical aspects of an encounter between the business and the customer, including equipment, furniture, and facilities. In this, physical evidence may also refer to less obvious aspects, including interior design, color schemes, and facility layout. Some marketing approaches are directed at creating lasting proof that the transaction or encounter has occurred, in the form of souvenirs, mementos, or commemorative documents featuring business signage. In digital commerce, physical evidence focuses on the customers' experience of using the business's digital resources and website.

ORGANIZATIONAL CULTURE AND MARKETING

Different businesses exhibit different *organizational cultures*. Organizational cultures are the various norms, values, assumptions, and beliefs held by the members of an organization that guide how an environment is perceived and responded to. In the context of marketing, distinct cultures are often referred to as *business orientations*. Four major business orientation distinguished in marketing are as follows:†

- *Product orientation*: businesses with a product orientation prioritize the development and perfection of a product or service. This may entail significant research and development efforts to generate novel products or services, or the provision of highly controlled and consistent products or

* Chaffey and Ellis-Chadwick (2019).
† Adapted from Boddy (2017).

services designed to meet specific customer needs. Since this approach is likely to be influenced by a technology push mindset, as discussed in Chapter 5, a product orientation may fail to closely match customer needs.

- *Production orientation*: businesses with a production orientation emphasize the ability to generate large volumes of products or services as efficiently and economically as possible. This orientation may yield significant advantages over less-efficient competitors but is associated with the risk of being unable to enter new markets and is seen as incompatible with variety or frequent change.
- *Selling orientation*: a selling orientation implies that a business is guided by the objective of selling products and services as effectively and rapidly as possible. This orientation is normally associated with aggressive sales techniques and focused on customers who have urgent needs and want to buy products or services with the least possible hassle, known as *distress purchases*. This orientation is also found in businesses in which perishable excess stock has to be sold rapidly, such as in a fish market. Businesses may also temporarily adopt a selling orientation if they require cash urgently, for example, to make urgent debt repayments.
- *Marketing orientation*: in businesses with a marketing orientation, marketing activities permeate the entire business, and a strong emphasis is placed on satisfying customer needs and co-creating value with the customer. In businesses with such an orientation, almost all employees, even those not directly engaging with the customers, will have a strong awareness of customer needs. While fostering a marketing orientation is seen to be an effective approach in competitive markets, it is also seen as a challenge to create and maintain, due to the need for extensive coordination between various business functions.

CONCLUSION

After defining the scope of marketing, this chapter has briefly outlined the theory of service-dominant logic as a core idea and a currently dominant view of marketing. Following this, various other major topics in marketing were introduced, including digital marketing, customer engagement, and segmenting the market. The marketing mix was presented as a framework reflecting the main tools and instruments available to marketers. As part of the promotion element of the marketing mix, this chapter briefly introduced the second core idea of this chapter, viral marketing. The chapter ended with a short summary of the relationship between organizational culture and the orientations businesses can take toward marketing.

Many useful marketing textbooks are available to the interested reader. A relevant and up-to-date textbook focusing specifically on digital marketing is *Digital Marketing* by Dave Chaffey and Fiona Ellis-Chadwick (2019). A more advanced version of the marketing mix with additional dimensions is presented in the textbook *Marketing Management* by Philip Kotler and Kevin Keller (2014). As discussed, the theory of service-dominant logic has revolutionized the scholarly literature on marketing over the last ten years. The original research articles referenced in this chapter such as *Evolving to a New Dominant Logic for Marketing* by Stephen Vargo and Robert Lusch (2004) are interesting and accessible. Perhaps more so than other fields in management, marketing has been criticized on the grounds of its impact on contemporary culture. A particularly influential and valid criticism of the practice of branding was contributed by Naomi Klein in her 1999 book *No Logo*. A pointed criticism of the extensive use of customer data has been made by Shoshana Zuboff in her 2019 book *The Age of Surveillance Capitalism*.

REVIEW QUESTIONS

1. Marketing is best defined as:

   (Question type: Multiple choice)

   O Persuading potential customers to buy a particular product
   O Generating awareness of a product or service
   O Identifying, anticipating, and satisfying customer requirements
   O Placing a particular product before potential customers

2. Which of the following is not normally considered to be part of marketing activity?

   (Question type: Multiple choice)

   O Sales and selling activity
   O Advertising
   O Product development
   O Developing marketing campaigns
   O Market research

3. Which of the following statements are true or false according to the theory of service-dominant logic?

   (Question type: Dichotomous)

True	False	
O	O	Value is sometimes co-created
O	O	Value is always co-created
O	O	Value is created for the customer by the manufacturer
O	O	The transaction of physical goods can be reduced to a service
O	O	The transaction of a service can be reduced to a physical good
O	O	The business cannot know with certainty what the value of its offering is
O	O	The process that creates value is complex and influenced by culture

4. Complete the below paragraph on the concept of a market offer.

   (Question type: Fill in the blanks)

   "The accurate identification of customer ________ allows businesses to collaborate with ________ to generate an attractive market offer, which embodies the combined ________."

5. You are analyzing the marketing activities of a fast food outlet. You are trying to measure the profit obtained from the average customer over the course of the relationship between the business and the customer, based on a cohort of 1,000 customers. You are making the following assumptions:
   - Not all customers in the initial cohort can be retained each year.
   - You assume that some customers will remain customers for six years.
   - The retention rate starts at 80% after the first year but decreases by 5% in each following year.
   - The annual revenue attributable to each customer is estimated at $150.
   - The net profit margin is estimated at 5%.
   - The value of flows of money is not discounted over time.

   (Question type: Calculation)

   Calculate customer lifetime value using the method presented in this chapter.

6. Which of the following are accepted dimensions of customer engagement in the framework presented in this chapter?

   (Question type: Multiple response)

   ☐ Involvement
   ☐ Intermediation
   ☐ Independence
   ☐ Interaction
   ☐ Intimacy
   ☐ Influence

7. Which of the following are characteristics of customer engagement in B2C?

   (Question type: Multiple response)

   ☐ Normally focused on creating an emotional response in customers
   ☐ Development of customers' competencies
   ☐ B2C marketing is designed to address human problems
   ☐ Successful B2C campaigns tend to be very expensive
   ☐ B2C marketing frequently involves an element of B2B marketing
   ☐ B2C marketing relies heavily on commissions
   ☐ B2C marketing frequently involves bidding processes

8. Match the provided business model orientations to the applicable factors used in market segmentation.
   (Question type: Matrix)

	B2B	B2C	B2B and B2C	This is not a factor in market segmentation
Socioeconomic factors	O	O	O	O
Behavioral factors, concentrating on how the product is used by a business	O	O	O	O
Political factors	O	O	O	O
Geographic factors	O	O	O	O

9. Identify the appropriate header for the following aspects in the marketing mix: product packaging, location of a shop, attitude of sales staff, payment within 60 days, and rapid response to customer queries.

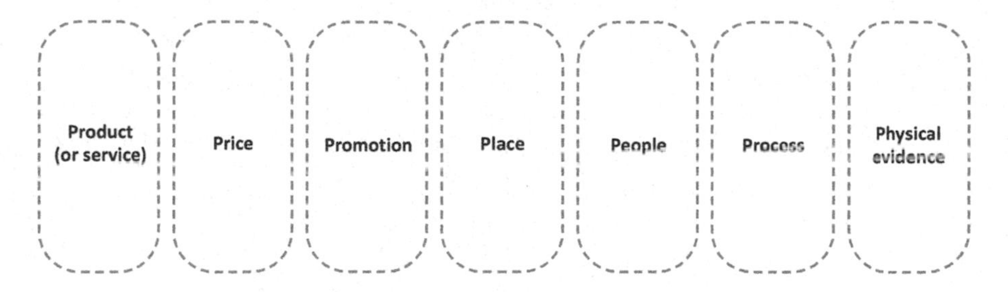

FIGURE 11.4 Identify the appropriate locations in this image

(Question type: Image hotspot)

10. In a business with a marketing orientation, marketing activities are performed by highly trained, best-in-class marketers in a special department with significant authority.
    (Question type: True/false)
    Is this statement true or false?
    O True O False

REFERENCES AND FURTHER READING

Boddy, D., 2017. *Management: An introduction*. 7th ed. New York: Pearson Education.

Chaffey, D. and Ellis-Chadwick, F., 2019. *Digital marketing: Strategy, implementation and practice*. 7th ed. Harlow: Pearson.

Chartered Institute of Marketing, 2015. *Marketing and the 7Ps. A brief summary of marketing and how it works*. Maidenhead, United Kingdom: The Chartered Institute of Marketing (CIM).

Evans, P. and Wurster, T.S., 1999. Getting real about virtual commerce. *Harvard Business Review*, 77, pp.84–98.

Haven, B., 2007. Marketing's new key metric: Engagement. Forrester Research Inc.– For Marketing Leadership Professionals, August 8.

Klein, N., 1999. *No logo: No space, no jobs, no choice*. New York: Picador.

Kotler, P. and Keller K.L., 2014. *Marketing management*. 14th ed. Harlow: Pearson Education.

Lusch, R.F., Vargo, S.L., and O'Brien, M., 2007. Competing through service: Insights from service-dominant logic. *Journal of Retailing*, 83(1), pp.5–18.

Porter, M.E., 2001. Strategy and the Internet. *Harvard Business Review*, 79(3), pp.62–78.

Vargo, S.L. and Lusch, R.F., 2004. Evolving to a new dominant logic for marketing. *Journal of Marketing*, 68(1), pp.1–17.

Zuboff, S., 2019. *The age of surveillance capitalism: The fight for a human future at the new frontier of power*. London: Profile books.

12 Managing Quality and Risk

OBJECTIVES AND LEARNING OUTCOMES

The objective of this chapter is to introduce the areas of quality management and risk management. The chapter begins with definitions of both topics and outlines how they are connected. It proceeds with an introduction to quality management, including brief presentations of quality management systems, graphical methods used to analyze quality, the setting of quality standards, robust quality methods, statistical process control, and Six Sigma. This is followed by a short introduction to the risk management process, introducing the three main elements of risk identification and assessment, risk treatment, and risk review and management. This chapter aims to achieve the following specific learning outcomes:

- an appreciation of the general areas of quality management and risk management, and an understanding of how these fields are connected;
- an ability to define quality and risk in the commercial context;
- an understanding of basic methods of quality management, including the plan-do-check-act cycle and graphical methods of analysis;
- knowledge of methods of setting quality standards, including benchmarking, the quality loss function, and failure modes and effects analysis;
- an understanding of robust quality and statistical process control and of the role of sampling-based approaches in quality management;
- understanding the basics of the Six-Sigma philosophy;
- knowledge of the fundamentals of risk management, including risk identification and assessment, risk treatment, and risk review;
- understanding key terms, synonyms, and accepted acronyms; and
- appreciation of key thinkers in quality management.

THE CONTEXT OF MANAGING QUALITY AND RISK IN A BUSINESS

The operational perspective on the activities of a business presented in Chapter 9 stresses that in any business there will be some form of output, be it a physical product, a service, an experience, or some form of information. *Quality* can be characterized generally as the attribute inherent to such an output that determines if the requirements of stakeholders, most importantly those of the customers, are satisfied. Managers, therefore, face the challenge of understanding and capturing such requirements and expectations and setting up the activities of the businesses so the outputs can meet these. Focusing on the customer as the most relevant stakeholder, quality can be therefore be defined in a customer-centric way:

DOI: 10.1201/9781003222903-14

Quality is the satisfaction enjoyed by customers as the outcome of the short-term and long-term performance of a product or service.

As implied by this definition, any factor that can have an effect on customer satisfaction forms a determinant of the quality of a product or service. It is generally assumed that the overall quality of a product or service is the result of a mix of six dimensions:*

- *Functionality:* this dimension of quality refers to the capability of a product or service to function in the intended way. This aspect also refers to the scope of functionality, reflecting the fact that some products or services cover more functions than others.
- *Performance*: this dimension of quality refers to how well or to what extent the product or service delivers its function. An offering that performs highly in some relevant way, such as a car that achieves a high top speed, is considered to be of high quality in this sense.
- *Reliability*: this dimension refers to the consistency with which the performance of the product or service is delivered over time. High levels of reliability normally correspond to perceptions of high product or service quality.
- *Durability*: this dimension refers to the level of robustness exhibited by a physical product. A highly durable product is often seen as a high-quality product. This aspect is likely to be more important in products that will be used heavily, such as professional-grade equipment.
- *Customization (and customizability)*: this dimension of quality refers to the extent to which a product or service fits, or can be made to fit, a specific or personal need. This aspect is particularly relevant in offerings that can be combined with additional complementary elements, such as in the case of hard- and software. Examples for quality through customization and customizability can be found in consumer electronics such as smartphones but also in services, such as banking. Correspondingly, customized products or services are often seen as high quality.
- *Appearance*: this dimension refers to the aesthetic value of a physical product, mostly reflecting how good a product looks and feels. This aspect is normally important for customers who want or need to convey a certain image. It may also have practical implications, such as clear and easy-to-understand user interfaces. Unsurprisingly, beautiful products are normally associated with higher quality than ugly products.

There is a close connection between quality management and operations management. This is often based on the assumption that the operations systems employed by a business determine the quality of products or services. In consequence, the operations in a business must be managed to ensure or maximize the quality of the outputs

* Boddy (2017).

of the business. Collectively, such activities form the field of *quality management*, which can be defined as follows:

Quality management is the activity of overseeing and controlling all processes and tasks needed to maintain the desired level of quality.

Failure to meet expectations by failing to deliver acceptable levels of quality is likely to have negative consequences for a business. In minor cases, such failure will result in the need to undertake rework at additional cost or delay the delivery of a product or completion of a service. In more serious cases, the customer will reject the product or service on the grounds of poor quality, seeking an alternative supplier. This is more problematic than it may first appear since the customer is likely to be vocal about the poor quality of the product or service, damaging the painstakingly built reputation of the business in the marketplace. In the worst cases, quality problems will result in actual physical harm to customers or even loss of life. A current example of an extremely poorly executed quality management process is provided by the aircraft manufacturer Boeing. In the late 2010s, Boeing delivered the latest version of its 737 aircraft with a catastrophic design flaw that led to two plane crashes, killing 346 people in total.

This example highlights the close connection between the issues of quality and risk. As evident throughout this book, every activity in a business carries a degree of uncertainty, especially if it is unique or new. For the general context of business, risk can be defined as follows:

Risk is the possibility of an event that, should it occur, will have an effect on the achievement of objectives. The level of risk is determined by the likelihood that the event will occur and the magnitude of its impact on objectives.

An activity or a project will generally be considered risky if either the likelihood of an event or the magnitude of its impact is large. In particular, it will be seen as risky if its potential impact results in harm to people, including loss of life, or in a significant or unrecoverable financial loss. It is important to note that the definition of risk is worded in a way that also allows for opportunities that may have positive impacts on the objectives of the business, for example in the form of *windfall* profits. Normally, however, the focus of risk management in business is on negative impacts resulting from various forms of failures and their costs.

Managing risks is the objective of the field of *risk management*. Generally stated, the methods of risk management aim to provide managers with a systematic approach to dealing with risk and uncertainty. This usually requires a range of activities toward the identification, assessment, and control of risks in order to protect or advance the interests of the business. Risk management can thus be defined as follows:

Risk management is the identification, assessment, and control of financial and non-financial risks to stakeholders, together with the identification of procedures to avoid or minimize their impact.

There is a significant overlap between the activities of risk management and quality management. This is because quality problems pose a significant *operational risk* in most businesses. As this term suggests, operational risks relate to the operations in the business and can thus be controlled by managers. For this reason, quality management can be interpreted as addressing operational risks in a business, especially if the quality-related risks are predictable risks that the business knows it faces, or if they are unpredictable risks the business knows it may face*.

QUALITY MANAGEMENT

Quality Management Systems

In a business, the set of processes aimed at consistently meeting quality standards is known as a *quality management system*. While early quality management systems developed in the first half of the 20th century focused on reducing undesirable variation occurring on industrial production lines, modern quality management systems emphasize the relationships and cooperation between stakeholders and are aimed at detecting problems proactively to achieve continuous improvement over time.

Quality management systems must be aligned with business goals and the nature of the product or service offered. Moreover, the emphasis of modern quality management systems is on preventive actions to eliminate problems before they materialize while at the same time maximizing the chances that mistakes can be detected and corrected as soon as they occur. Because quality management systems aim to achieve continuous improvement of time, they are often structured as improvement cycles. A well-known improvement cycle is the so-called *plan-do-check-act* (*PDCA*) cycle, consisting of four activities:

- *Planning:* in order to reach satisfactory performance, all aspects of a business that can affect the quality of products or services require active planning. This includes a clear statement of objectives, formulation of standards, and the development of *standard operating procedures* (*SOPs*).
- *Execution* (*"Doing"*): execution refers to the carrying out of activities with the necessary expertise and using the appropriate resources. An important part of execution is the collection of quality-relevant information from the various processes employed by the business.
- *Checking*: in the checking activity, the data collected during the execution stage are evaluated. Specifically, the data are compared to expected values to detect variation, non-conformities, inefficiencies, and opportunities for improvement. Moreover, the data collection processes themselves are also evaluated.
- *Action*: in the action step, defective products or workpieces are removed and the processes in the operation are improved. This stage includes the re-evaluation of operational risks and the updating of standards and SOPs.

* Williams et al. (2006).

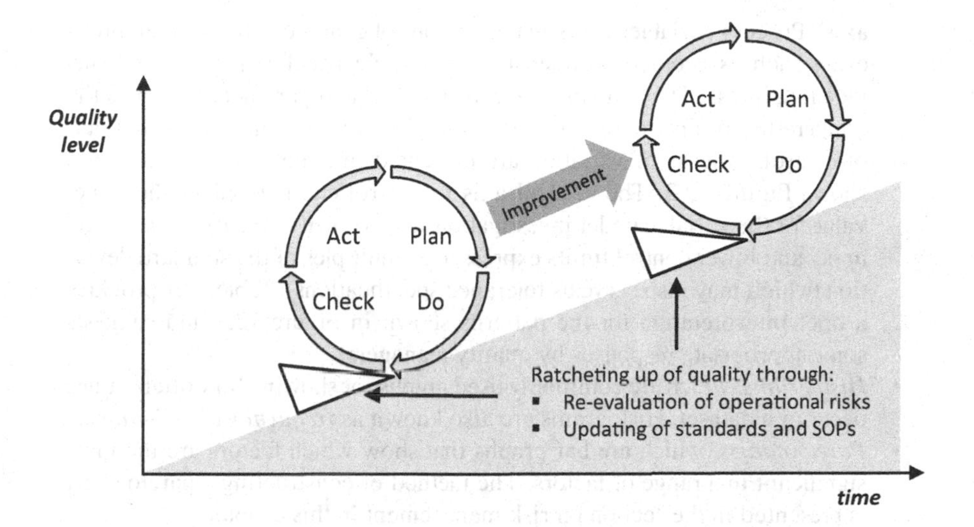

FIGURE 12.1 Illustration of gradual quality improvement through the PDCA cycle

The PDCA cycle is also known as the *Deming cycle, Shewhart cycle,* and under many other names. It is important to note that many variations of this cycle exist and are accepted. Figure 12.1 illustrates the process of continuous improvement through the repeated application of the PDCA cycle. This image expresses the gradual *ratcheting up* of quality levels through the iterative re-evaluation of operational risks and the updating of standards. This idea is strongly associated with Lean, as presented in Chapter 9.

Graphical Methods to Investigate Quality

Quality managers frequently rely on graphical tools to build awareness of problems and to record quality information. Many different graphical formats are used specifically for this task. A set of basic and indispensable tools of this kind are known as the *basic seven tools of quality.* These tools can be outlined briefly as follows:

- *Ishikawa diagrams,* also known as *fishbone diagrams,* identifying and categorizing multiple potential causes for a problem, as presented in Chapter 9.
- *Check sheets,* also known as *defect concentration diagrams,* which are simple spreadsheets prepared for the collection and analysis of quality-relevant data, such as occurrences of defects.
- *Control charts.* Accepting that it is not realistically possible to produce to specification with perfect consistency, the aim of control charts is to show how quality-relevant process variables fluctuate over time and to indicate that a process is *in control,* which is defined as a state in which only random variation occurs. To achieve this, time is presented as the horizontal axis and the variation of a process variable is presented on the vertical

axis[1]. Process variables investigated in control charts can be physical properties such as geometric dimensions, mass, electrical resistance, and surface roughness. They can also be processes-related parameters such as the temperature or flow rate of a fluid. Examples of typical in-control and out-of-control patterns of variation are presented in simplified control charts shown Figure 12.2. The vertical axis in control charts displays the target value for the variable under investigation, also known as the *center line*, and upper and lower control limits expressed as multiples of the standard deviation (which may also serve as tolerance specifications). Table 12.1 provides a brief interpretation for the patterns shown in Figure 12.2 and suggests some appropriate responses by quality managers.

- *Histograms*, which are commonly used graphs for showing how often values occur in a dataset. Histograms are also known as *frequency distributions.*
- *Pareto charts*, which are bar graphs that show which factors are the most significant in a range of factors. The method of constructing a pareto chart is presented in the section on risk management in this chapter.
- *Scatter plots*, which are graphs showing pairs of numerical data with one variable on each axis, to visually identify relationships such as correlation between variables.
- *Stratification*, which is a technique used to separate data into distinct groups or layers. When quality data from different sources have been mixed, the information contained in the data can be difficult to understand. Stratification techniques are used in combination with other data analysis methods to sort the data so that patterns can be identified.

Setting Quality Standards

As evident in the control charts introduced in the previous section, quality management requires the definition of targets and metrics that help managers decide when to take action, such as control limits for product and process variables. Three methods to generate such standards are outlined in this section, *benchmarking*, the *quality loss function*, and *failure modes and effects analysis (FMEA).*

Benchmarking

Benchmarking identifies and examines best practices or current standards of performance. The scope of a benchmarking exercise can be limited to single or multiple departments in a business, across several businesses in an industry, or include activities across the entire globe. Broadly speaking, the goal of benchmarking is to capitalize on the success of others in order to produce an improvement in the own processes. A typical benchmarking project in quality management involves the following six steps:*

* Cook (1997).

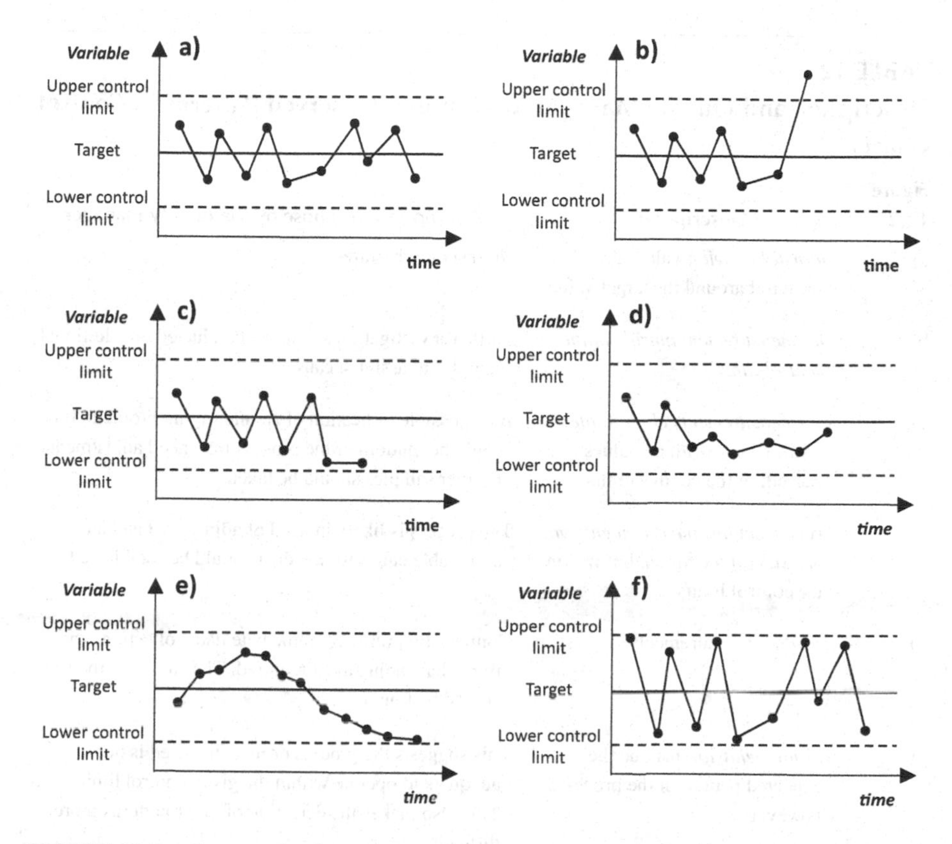

FIGURE 12.2 Commonly observed patterns shown in simplified control charts

1. *Identify and develop an understanding of the targeted processes.* The benchmarking process begins by collecting accurate and detailed information on a particular aspect under investigation.
2. *Selection of factors for benchmarking.* The second step involves establishing an understanding of which factors are of relevance in the analysis and then selecting a set of factors for inclusion.
3. *Data collection.* In this step, the required data for the selected factors are collected. This can be challenging, as the means to collect this information may not yet be established. The required data may come from existing internal data pools, from specialist publications, or from the reports produced by developmental groups or research organizations, both academic or industrial.
4. *Analysis of data and identification of gaps.* In this step, the collected dataset is subjected to analysis using appropriate methods to generate the required insight. As the outcome of the benchmarking process, a report is generated that highlights any gaps or limitations in the used data.

TABLE 12.1
Description and Quality Manager Response to Observed Patterns in Control Charts

Figure 12.2	Description	Appropriate response by the quality manager
a)	*Normal behavior,* values are scattered around the target value	No response required
b)	*One measurement outside of the control limits*	Further investigation should be conducted on additional samples to assign a cause
c)	*Two measurements close to the control limit,* with no values exceeding the control limits	As a possible indication of an underlying problem, this could be random or the process may need adjustment. Further samples should be taken
d)	*Six consecutive measurements on one side of the target,* but within the control limits	The process is likely in need of adjustment and an assignable cause for the drift should be established
e)	*Trending* measurements	Confirmation through immediate data collection, with immediate adjustment required. The cause for the trend should be found
f)	*Erratic behavior,* may be the expected pattern of the process, however	This suggests the process or machine used is not adequate to operate within the given control limits. This also makes the detection of other patterns more difficult

Source: Adapted from Naylor (2002).

5. *Plan and implement improvements.* This step involves the formulation of actionable steps and plans to implement changes on the basis of the benchmarking report.
6. *Review and adaptation of the benchmarking approach.* The final stage refers to the modification of the benchmarking process and the factors used as criteria in the analysis. The main aim of the review stage is to improve future repetitions of the benchmarking activity.

Benchmarking techniques are best applied to the evaluation of existing products or services and to compare processes among similar businesses. It cannot normally be applied to competitor's products or services that are not yet commercially available in the marketplace, due to the lack of available data.

The Quality Loss Function

Genichi Taguchi (1924–2012)

The concept of the *quality loss function* (*QLF*) emerged out of the realization that conformance-oriented quality management processes, which focus on establishing if control limits are exceeded, often ignore quality problems that do not result in actual rejection because the deviation from the target levels is not large enough. As famously recognized by engineer and statistician Genichi Taguchi, such occurrences are likely to result in significant additional costs and other problems in a product's life cycle.

To overcome this problem, thc quality loss function (QLF) aims to estimate the total cost resulting from a deviation in a specific target value in the long run. To form a plausible cost estimate, this analysis must include a diverse set of costs, such as service and warranty costs, inspection and scrappage costs, and the loss of customer confidence. In basic forms of the quality loss function, these costs are captured in a constant cost parameter C, which is highly process- and application-dependent. The following basic model proposes that the total quality loss L, measured in money terms, resulting from the production of an item is a function of the cost term C multiplied with the square of the deviation of an engineering characteristic y from its target value m:

$$L = C \times (y - m)^2 \tag{12.1}$$

Despite its simplicity, the quality loss function allows quality managers to gauge the cost impact of a deviation from a process target. It also expresses the intuition that this impact does not behave in a linear way, implying that progressively larger deviations will quickly lead to extremely high costs to the business.

Failure Modes and Effects Analysis

Products or services that fail will not be considered to be of high quality by customers. One of the most common and versatile tools to assess the failure of products and processes, as well as consequences and risk, is *failure modes and effects analysis (FMEA).* FMEA can be applied in a reactive way to understand how a failure has happened and in a more proactive way to prevent failure. As the name suggests, FMEA focuses on *failure modes*, which are descriptions of the results of failure. FMEA supports this with explanations of the *failure mechanisms* that have led to these occurrences.

The goal of FMEA is to identify possible failure modes and mechanisms to illuminate the consequences that the failure modes have on the performance of products, services, or processes. It also seeks to identify methods of detecting failure modes and ways of preventing them. In the quality management context, FMEA is a valuable tool for the setting of quality standards since it alerts quality managers to how failures can occur and which technical factors are involved. FMEA has important applications in risk management, outlined later in this chapter, as well as in engineering design, reliability engineering, and analyses of maintainability, safety, and survivability.

The traditional way of constructing an FMEA is by tabulating the functions or elements of a product or process and then identifying the relevant failure modes for each. The next step is to add a description of the failure mechanism for each failure mode, taking into account connections to the application, the environment, and the operating method. This is followed by a statement of how the failures will be detected, which may be by a human or through an automated sensor, and how the failures may be compensated. In turn, this is followed by a description of the expected or immediate effects of the failure on people, objects, or the wider environment. The final element is a description of preventive measures that can be taken to stop the failure mode from occurring, constituting an opportunity to improve products or processes and to eliminate the scope for mistakes.

As is clear from this very brief description, successful FMEA requires detailed technical knowledge of products and processes and must be performed in an iterative way throughout the product life cycle, as outlined in Chapter 6. Moreover, there are different sub-types of FMEA such as *functional FMEA*, *design FMEA*, and *process FMEA*, each with different methods and objectives. FMEA may also include criticality analysis by taking into account the severity of the consequences of failure modes and their likelihood of occurring, closely related to the idea of risk maps presented in the treatment of risk management in this chapter. Extended in this way, FMEA is also known as *failure mode effects and criticality analysis (FMECA).*

Robust Quality

Earlier approaches to quality management focused on measuring whether products conform to specifications, also known as *goal post philosophies*. As a response to the perceived shortcomings of this way of thinking about quality, Taguchi developed the concept of *robust quality*. Instead of viewing quality as embodied by conformance to specifications, robust quality refers to the performance of the products in use, including unexpected environments, atypical patterns of use, mild misuse, deviation from maintenance schedules, and so on.

Taguchi recognized that the best opportunity to tackle undesirable variation in manufacturing processes and in the useful lives of products is during the design phase. Taguchi also understood that many quality problems stem not from issues in individual components but from the interaction of small variations in multiple components. The implementation of robust quality, known as the process of *robustification*, is commonly associated with three stages:

1. *System design*. This stage refers to the conceptual development and design of the products, and manufacturing processes that will be used. This includes the specification of appropriate subsystems, components,

materials, production technologies, and ensuring maintainability of the overall system.
2. *Parameter design.* Once the design for the overall system has been created, the parameters and values of the system must be set. Robust design methodology accepts that there is a tradeoff between robustness and cost.
3. *Tolerance design.* Once the parameter design is completed and the relationships between specific parameters and performance are understood, the final stage focuses on using this information to specify product and process tolerances. Since achieving tight tolerances is positively related to cost, it is important to avoid the overspecification of tolerances. Moreover, the knowledge gained through the previous two stages of the robust quality process allows quality managers to concentrate on controlling variation in the few components and dimensions that are critical for performance.

Taguchi's methods resulted in a leap forward in the field of quality management, which was focused on conformance-oriented methods at the time. Taguchi recognized that the choice of suitable parameters is not determined by the requirements of the manufacturing system. To unlock higher levels of quality, it is instead necessary to select parameters that minimize the impact on product performance originating from variation in a much wider range of interacting factors. These include manufacturing-related factors, the use-phase environment of the product, and accumulating damage over a product's life span.

Statistical Process Control

Modern quality management makes heavy use of statistical techniques to analyze the available data and uses graphical techniques to extract meaning from such information. An important family of methods going beyond basic graphical techniques is known as *statistical process control (SPC)*. A major application of statistical process control is to assess the quality of products after they have been made. However, statistical process control can also be embedded within manufacturing processes to preventively address quality problems.

Generally stated, the objective of statistical process control is to identify variation in transformation processes, such as those used to generate a product or service, in order to achieve consistency and stability. To do this, statistical process control is based on a *sampling* approach in which a reduced number of product units (or instances of service provision) are picked out for quality inspection, rather than analyzing the entire output, which is referred to in quality management as *full inspection*.

Sampling carries a number of advantages over full inspection. It tends to be faster and less costly and is, in some cases, the only way of collecting data, for example, if full inspection would disrupt processes or inconvenience customers. Moreover, a sampling-based approach permits *destructive testing*, in which a small number of produced units are intensively evaluated, leading to their destruction. Additionally, it has been proposed that sampling-based quality inspection may in fact be more accurate than full inspection since a greater effort can be expended on testing each unit.*

Sampling-based statistical process control is conceptually based on two kinds of errors. *Type I* errors occur if the assessment of the sample falsely suggests that something is wrong with a process when it is, in reality, working correctly. Such errors can occur, for example, if measurements in the sample are close to the control limit and this is not representative of the whole production. If such deviations are detected in a sample, the appropriate response will be for the quality manager to order immediate additional inspections. If a Type I error has occurred, then additional inspections should reveal this at an additional cost.

In contrast, *Type II* errors occur if the assessment of the sample falsely suggests that everything is in order when in fact the process is not working correctly. This situation can arise in three ways:

- a problem occurs in the process and a faulty observation is in the sample but it is not detected as such;
- a problem occurs in the process leading to some fault but the fault is not contained in the sample; or
- the fault is detected in the sample but incorrectly interpreted as random variation when in reality there is a systemic cause.

Quality managers consider Type II errors significantly more dangerous than Type I errors. The consequence of such errors is that problems remain undetected, which may lead to severely negative consequences for a business, including complaints from customers and legal action. The possibility of the occurrence of Type II errors can be reduced through the adoption of random sampling techniques in which product units or instances of a service are randomly selected for inspection.

When using statistical process control, the focus is usually on two characteristics in the collected quality-relevant data: their absolute values, for example, in terms of physical dimensions, temperatures and durations, and the dispersion around their means. As seen in Figure 12.2, control charts require the identification of a process target, also known as *set standard*, which is shown as a mean line. There are multiple ways to specify appropriate targets. They can be set based on technical judgment by a designer or technical expert, form the result of a study into the capability of the used processes, or be inferred from large amounts of historical data.

The notation used in the statistics of quality management is somewhat complicated by the fact that most methods are aimed at the calculation of statistics relating to multiple samples, with each sample containing a number of measurements, or *observations*. To handle this, let X_{ij} represent the measurement of the *i*th unit in the *j*th sample. Further, suppose that each sample is of size n_j. In this case, the mean for the *j*th sample $\bar{X}_j$, also referred to as *X-bar*, is calculated as follows:

* Naylor (2002).

$$\bar{X}_j = \frac{\sum_{i=1}^{n} X_{ij}}{n_j} \tag{12.2}$$

If measurements from a correctly running process are used to set the process target, the process target is the grand mean $\bar{\bar{X}}$ of all measurements across all k samples:

$$\bar{\bar{X}} = \frac{\sum_{j=1}^{k} \sum_{i=1}^{n_j} X_{ij}}{\sum_{j=1}^{k} n_j} \tag{12.3}$$

However, to be able to judge whether a sample lies within specified tolerances, it is necessary to define control limits. In cases in which the analyzed data have sample sizes that are greater than 2 and smaller than 10, which is the most common case in practice, the specification of control limits requires the definition of the range R_j of the data in the jth sample:

$$R_j = X_{ij_{max}} - X_{ij_{min}} \tag{12.4}$$

This value is used to estimate the average range across all k samples, $\bar{R}$:

$$\bar{R} = \frac{\sum_{j=1}^{k} R_j}{k} \tag{12.5}$$

Once $\bar{R}$ is established, is possible to specify the upper control limit and the lower control limit. This is done by applying the sample size-specific anti-biasing constant A_2, which can be obtained from an appropriate source, so that:

$$\text{Upper control limit} = \bar{\bar{X}} + A_2\bar{R} \tag{12.6}$$

$$\text{Lower control limit} = \bar{\bar{X}} - A_2\bar{R} \tag{12.7}$$

The next step in the evaluation of the data is the calculation of the upper and lower control limits for an additional chart, known as the *R chart*, which is used to indicate whether the ranges R_j found in the samples indicate any disturbance. Together, the use of the $\bar{X}$ and R charts, referred to by quality managers simply as *X-bar and R*, can be used to judge whether or not the sample statistics, sample mean $\bar{X}_j$ and sample range R_j, indicate the presence of a disturbance with a special cause that must be further investigated. As evident in the equations 12.6 and 12.7, the control limits are determined using the process target $\bar{\bar{X}}$, the average range $\bar{R}$, and the constant A_2.

It should be noted that X-bar and R forms only one approach among many in statistical process control. Many other configurations may be appropriate for quality managers to answer a wide variety of questions about process stability.

In statistical process control, the use of a sampling approach has subtle but profound implications. This is because of a statistical concept known as the *central limit theorem (CLT)*. According to the central limit theorem, the distribution of sample means, such as $\bar{X}$ in the notation used in this chapter, will approach a normal distribution* for samples of a reasonably large size. Rather remarkably, this is the case even if the underlying variable, X, is not normally distributed. This allows quality managers to engage in methodologically sound statistical techniques on the basis of the data they collect.

Six Sigma

A popular and very widely adopted set of techniques based on statistical process control is the Six Sigma methodology. Originating from the electronics industry, Six Sigma focuses on eliminating defects and reducing variation. Somewhat similarly to the Lean philosophy presented in Chapter 9, Six Sigma techniques are influenced by Japanese culture and methods. In recent quality management practice, there is a trend to combine both methods into a Lean Six Sigma approach.

In processes with one parameter of importance, as discussed in the basic method of statistical process control presented in this chapter, the required level of this parameter is usually written as a *specification width* around a *central value*. For example, a dimensional parameter of this sort could be specified as 700 ± 2 mm. Assuming that there is random variation in this parameter and it follows the normal distribution with the population mean μ, the specification width can be expressed as a multiple of the population standard deviation σ of this parameter. If a defect can be defined as any production unit exhibiting a value outside of the specification width, the defect rate of a process can be stated.

Using a specification width of $\pm 1\sigma$ results in a defect rate of approximately 31.73%, which equates to 317,300 defective units per million units produced. This is obviously excessively high for any industry. Applying a specification width of $\pm 3\sigma$ dramatically reduces the defect rate to 0.27%, which equates to 2,700 defective units per million produced, which would still be considered unacceptably high in many businesses. Adopting a specification of $\pm 6\sigma$ almost eliminates defects, resulting in a defect rate of 0.0000002%, which equates to 0.002 defective units per million produced. This level of defects is considered excellent by quality managers.

To reflect the fact that processes performance tends to deteriorate in the long run and that an increase in variation is likely, the defect rate normally used in Six Sigma is based on a shift in the mean μ by 1.5σ, resulting in the long-term performance of a specification width of only $\pm 4.5\sigma$. For this reason, Six Sigma targets a defect rate of 0.00034%, corresponding to 3.4 defective units per million units produced.

* A normal distribution, also known as Gaussian distribution, is a continuous probability distribution for a random variable. Normal distributions are common in statistics and are frequently used in the natural and social sciences to represent random variables if the real distribution is not known.

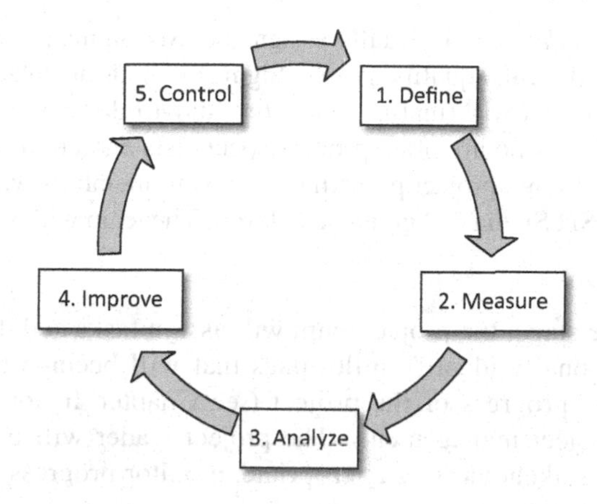

FIGURE 12.3 The define-measure-analyze-improve-control (DMAIC) cycle

As with Lean, the successful application of Six Sigma depends upon a sympathetic, quality-minded culture throughout the whole business. In fact, there is significant conceptual overlap between both philosophies. Most importantly, both stress the use of real-world data and place heavy emphasis on improving quality over time.

The methodological core of Six Sigma is the *define-measure-analyze-improve-control (DMAIC)* cycle of process improvement, illustrated in Figure 12.3. As an iterative process aiming to create a gradual improvement in processes, this cycle is similar to the plan-do-check-act cycle presented in Figure 12.1. The following briefly characterizes each element of the DMAIC cycle.

Define

The *define* stage is initiated when a quality problem has been identified. Its purpose is to clearly establish that there is a problem before resources are committed to addressing it. While some problems may be obvious to managers, for example, if products are returned from the customers, other problems may be harder to detect, for example, if they manifest themselves in queues building up between processes. Additionally, it must be clear that the problem is specified in a way that a solution can be obtained and that measurement is possible. Further, the resolution of the problem should yield an improvement of some sort to the business and it should be possible to state a point in time at which the solution must be achieved.

During the definition stage, a project team with a clear structure and responsibilities is formed. A typical Six Sigma project team will comprise:

- a *Six Sigma champion*, who is a seasoned Six Sigma expert possessing knowledge of the overall Six Sigma strategy in the business;
- *the project sponsor,* who carries the responsibility for the project. The project sponsor is likely to be a senior manager in the business who has responsibility for the process as well as the staff involved;

- *the project leader*, who is usually assigned to Six Sigma projects on a full-time basis and is fully qualified in Six Sigma methods (to "black belt" level). The project leader will run the project on a day-to-day basis; and
- *team members*, who are often process specialists and are usually involved with the Six Sigma project part-time. The team members will be partially qualified in Six Sigma (to "green belt" level). The team will be kept as small as possible.

During the define stage, the project team will assign tasks and dates to the project. It will additionally identify milestones that will become the basis for the assessment of the progress of the project (see Chapter 16 for an introduction to methods in project management). The project leader will organize reviews, involve essential stakeholders as appropriate, monitor progress, and clarify any changes that are necessary.

Measure

The *measure* stage establishes the performance of the process, both as a baseline and as a means of evaluating changes as they are implemented. The resulting data will also be used to determine whether a significant improvement has been made upon project completion. The measure stage typically involves the following activities:

- *developing process metrics*, which refers to finding appropriate quantitative measures of capturing the issue at hand, often referred to as *key performance indicators* (*KPIs*);
- *collecting process data*, which refers to compiling the data from the available sources most likely using a sampling-based process as introduced in this chapter;
- *checking the data quality*, which refers to ensuring that the data contain the required information;
- *understanding process behavior*, which refers to creating models of the process to facilitate an understanding of process behavior among stakeholders;
- *establishing the capability of the process and its potential*, which refers to the identification of the current performance of the process and its potential performance levels.

Analyze

The objective of the *analyze* stage is to develop a theoretical understanding of the process to apprehend the causes of failure. A wide range of methods and tools are available to assess the collected data to inform such analyses. These methods include:

- process mapping and root cause analyses, as outlined in Chapter 9;
- FMEA methods presented earlier in this chapter; and
- additional methods briefly summarized in the risk management part of this chapter, including *Pareto analysis*.

Improve

The *improve* stage involves the identification, selection, testing, and deployment of a solution to the problem at an acceptable level of risk. The resulting improvements are then assessed against the key performance indicators developed. The improve stage typically involves the following activities:

- generating and documenting a range of potential solutions from a range of options for the purpose of downselection (as discussed in Chapter 6), focusing on the least complex and lowest risk solutions;
- selecting the best solution – some solutions will be easier to apply or be more effective than others;
- testing the solution systematically, for example using the plan-do-check-act cycle or the benchmarking approach presented in this chapter;
- anticipating the risks associated with the improvement, for example, using the risk management techniques presented in this chapter; and
- creating an implementation plan, possibly involving an initial pilot study, and deploying the improvement.

Control

As the final stage in the cycle, the *control* stage embeds the improvements aiming to produce enduring change. This requires the implementation of control systems, also involving statistical process control, changes to the SOPs and process documentation, and putting in place mechanisms to make sure that the workers do not revert to the previous practices and patterns. Concluding the DMAIC cycle, the control stage usually contains the following steps:

- *Implementing the measurement systems*, ensuring that appropriate measurement methods are in place and that the ways of working within the business have been amended.
- *Standardizing the solutions*, fixing the changes in the revised processes and standards.
- *Quantifying the improvement*, ascertaining that the project objectives have been achieved through measurement.
- *Closing the project*, bringing it to an end and looking for the next project.

RISK MANAGEMENT

As stated at the beginning of this chapter, quality problems pose a significant risk to all businesses. Quality risks have been characterized as operational risks, emphasizing that they arise within the business and can be addressed by managers. There are, of course, many other risks that managers must deal with. Such factors can relate to design decisions when product or services are created and to the behavior of the products and services as they go through their life cycle stages, as described in Chapter 6. Significant additional risks to businesses may arise from the macro-environment, as described in the introduction of the PESTEL method in Chapter 7.

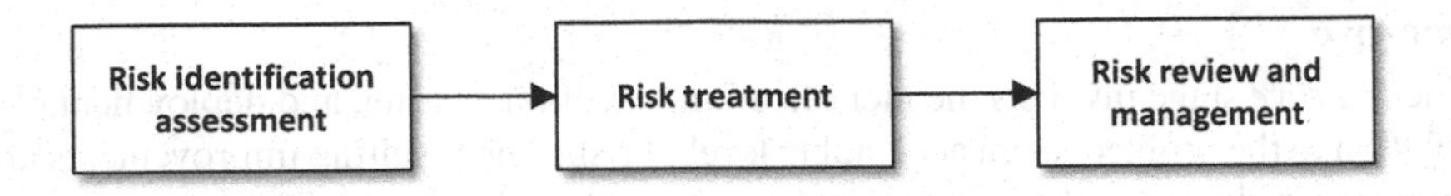

FIGURE 12.4 The structure of risk management

Addressing such risks is the task of risk management, as defined at the start of this chapter.

Generally, the professional risk management process is seen to consist of a sequence of distinct stages, with the initial elements aiming to identify and analyze the risk and the later elements focusing on addressing the risk in a sense that they are counteracted or accommodated for. Figure 12.4 summarizes the sequence of steps commonly associated with risk management. The remainder of this chapter will outline each stage.

Risk Identification and Assessment

Every kind of commercial activity, be it in the form of the day-to-day operations of a business or the completion of projects, is subject to risk. The starting point of the risk management process is to generate a wider awareness and visibility of specific and important risks among managers and other stakeholders. The identification of risks can be discussed in dedicated meetings, in which workers and other stakeholders can name specific risks. Such meetings are normally led by a facilitator acting as a referee.

The risk identification process usually aims to produce a document known as a *risk register*. The risk register names and briefly describes each identified relevant risk in a given context, allocates a category under which the risk falls, estimates the likelihood of the risk occurring, its expected impact should it occur, and its *severity*. The severity of a risk, also referred to as *risk level* or *risk rating*, is normally a combination of the risk likelihood and impact. The risk register also provides an indication of how a risk can be treated. Moreover, it names the person who is accountable for the risk and has the necessary authority to take action to address it, referred to as the *risk owner*.

Once the risk register has been created, it is updated on an ongoing basis to track the development of the identified risks over time. There are many different templates and formats for risk registers. Table 12.2 presents a simple example of a risk register for a fictitious project involving the construction of a powerboat. Risk registers often contain additional information such as the status of a risk and when it has last been updated or reviewed.

Both the likelihood and the impact of the risk normally enter risk registers in discrete categories that can be interpreted in a numerical way. In the example shown in Table 12.2, the "low" category is assigned a value of "1", the "medium" category is assigned a value of "2" and the "high" category is assigned a value of "3". This is done to calculate the severity of the risk as a numerical value by multiplication. It should be noted that judgment on the basis of subjective criteria flows into the determination of the likelihood and the impact of each risk.

TABLE 12.2
A Fictitious Example for a Risk Register

Risk name/ description	Risk category	Likelihood	Impact	Numerical expression of severity	Mitigating action	Owner
Engine has insufficient power	Technical risk	Low	High	3	Install a larger power supply	B. Gibbons
Hull design does not give required performance	Technical risk	Medium	High	6	Run a testing program for the proposed design	F. Beard
The target market for the powerboat is smaller than expected	Commercial risk	Low	Low	1	Conduct market research to ascertain potential market size	D. Hill
Molding chief technician retires early	Project risk	High	High	9	Organize a meeting with the head technician, offer a bonus for project participation	B. Ham

The next step in the risk assessment process is to impose an ordering on the identified risks to understand which risks managers should focus on. This process is referred to as *risk prioritization*. A common way to prioritize risks is by constructing a *risk map*, which is a matrix graphically representing the relative likelihood and impact of each risk. This provides intuitive insight into which risks are the most significant. It can also be used to group risks in terms of how they will be treated. Figure 12.5 presents a simple risk map with 3 × 3 cells. A common rule of thumb in risk prioritization is to first treat risks that have high impact and high likelihood, then treat risks that have high impact and low likelihood, and then treat risks that have low impact and high likelihood. All low-impact and low-likelihood risks should be treated after this.

A further way to prioritize risks identified in the risk register is through their expected cost impact, referred to in risk management as the *expected monetary value*. To specify such values, it is necessary to state a probability for an adverse outcome to occur and to quantify its cost impact. The expected monetary value is obtained by multiplying the cost impact with the probability. Table 12.3 lists the risks

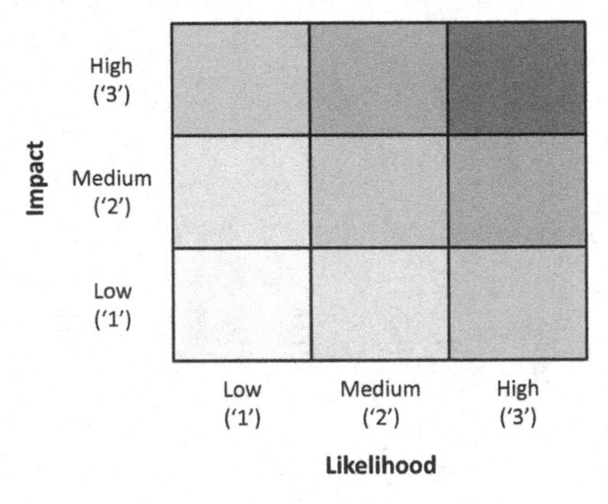

FIGURE 12.5 A simple risk map with 3 × 3 cells

TABLE 12.3
Calculation of Expected Monetary Impacts of Risks

Risk name/ description	Cost impact ($)	Probability (%)	Expected monetary value ($)
Molding chief technician retires early	50,000	50	25,000
Hull design does not give required performance	60,000	30	24,000
Engine has insufficient power	50,000	10	5,000
The target market for the powerboat is smaller than expected	2,000	10	200
Total technical contingency			$54,200

identified in the example with this additional information and ranks them according to expected monetary value. Additionally, the table sums the expected monetary impacts of all identified risks forming a value known as the *total technical contingency.* This value provides a simplistic yet useful indication to risk managers of how much money will likely be needed to address the impacts.

Information on expected monetary values and the total technical contingency can be used to perform a graphical method of analysis known as a Pareto analysis or a *Pareto chart.* To construct a Pareto chart, it is first necessary to express the expected monetary value of each risk as a percentage of the total technical contingency. This is done using the following equation:

$$\text{Risk share of total technical contingency} = \frac{\text{Expected monetary value}}{\text{Total technical contingency}} \times 100 \tag{12.8}$$

Once quantified this way, the risks are drawn as a column chart, with the risk with the largest share of the total technical contingency on the left, the second-largest share to the right, and so on. To further aid the visual impact of the chart, a line cumulating the risk shares of all risks is normally included in the chart. Figure 12.6 presents a Pareto chart for the simple powerboat construction example discussed in this section.

The presented risk prioritization methods include quantitative metrics of risk likelihood, probabilities, and impacts, both in terms of discrete categories ("low", "medium", and "high") and in the form of expected cost impacts. However, for most risks, the

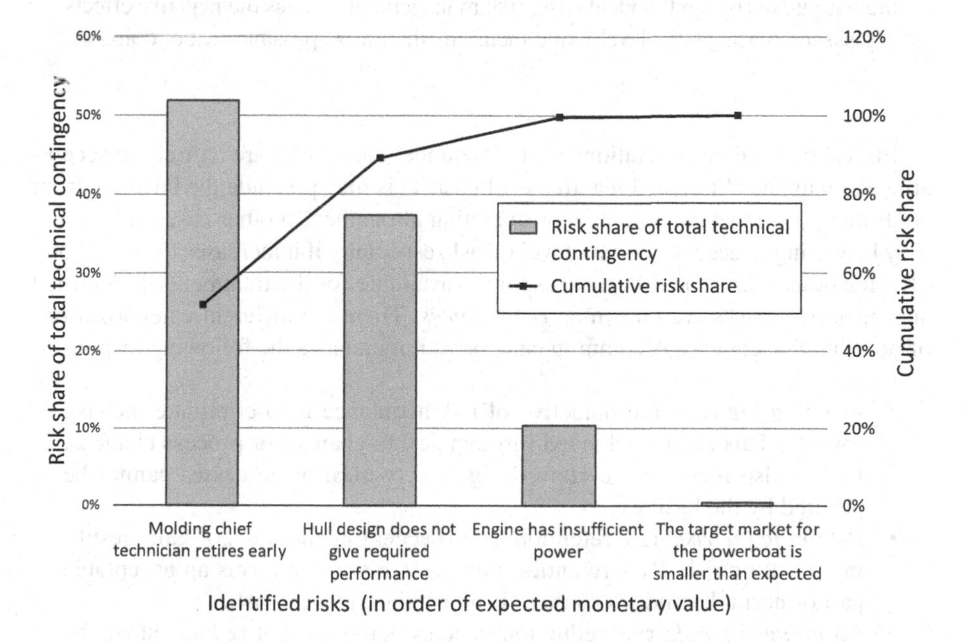

FIGURE 12.6 Pareto chart of risks in the example project

information required to determine such quantitative expressions through historical data or using statistical inference will be difficult or even impossible to obtain. This is the case especially for very unlikely events with catastrophic consequences, such as plane crashes or natural disasters. This means that informed but subjective guesses by experienced employees or stakeholders with specialist information will normally be needed.

Risk Treatment

Successful risk prioritization will determine an appropriate order in which the identified risks should be addressed. Due to limitations in resources, however, managers will not normally be able to treat all conceivable risks. This highlights the broader point that the treatment or elimination of all risks is not the objective of risk management. There will usually be a multitude of very low-likelihood risks that cannot realistically be treated, even if they have severe impacts.

CORE IDEA

Like most other areas of management, risk management is an activity in which benefits are weighed against costs. The concept of opportunity cost, defined in Chapter 3, forms the guiding principle for successful risk management. This insight has several consequences. Initially, opportunity cost should determine whether to allocate resources to risk management activities or to employ them elsewhere in the business. Moreover, as outlined in this section, it may be appropriate to choose alternative and less costly treatments other than eliminating or reducing a negative risk. In the ideal case, risk management reduces the negative effects of risks to an acceptable level while incurring the lowest possible cost of doing so.

Ethical or legal considerations will dictate that some risks are entirely unacceptable and must be eliminated regardless of cost. This may preclude the business from continuing with specific activities or operating altogether. In other cases, a business may be willing to accept a certain level of risk, especially if it increases the likelihood of some other gain. The different responses available for the treatment of identified and prioritized risks are known as *risk choices*. There are different categorizations of possible risk choices. A common categorization includes the following responses:

- *Avoiding the risk:* the objective of risk avoidance is to eliminate the risk entirely. This can be achieved through design changes or process changes. It may also mean that certain designs, activities, or processes cannot be pursued by the business.
- *Retaining the risk*: risk retention means accepting any loss, or gain, resulting from the risk. Risk retention implies that the risk forms an acceptable part of normal activities so there is no need to treat the risk.
- *Reducing the risk*: risk reduction denotes activities that reduce either the likelihood or the impact of a risk. Risk reduction is normally based on the

introduction of *risk controls*, which are changes to designs, activities, or processes in a business. To reduce financial risks, the business may decide to increase prices or accumulate cash reserves.

- *Transferring the risk*: risk transfer shifts the risk to another party, for example by taking out an insurance policy or by using an external contractor or supplier. Normally, risks cannot be fully transferred, so a residual level of risk is likely to remain. Analogous to the residual control rights discussed in Chapter 7, there will probably be aspects not covered by the risk transfer, no matter how comprehensive such an arrangement is. Often, risk transfer is limited to a degree of financial protection should an adverse event occur so there may still be negative non-financial impacts on the business.
- *Exploiting the risk*: building on the idea that not all risks are negative, there is often a possibility that by retaining or even increasing a certain risk a larger benefit can materialize to a business. Exploiting risks requires careful analysis of the possible impacts, and their costs, however.

Risk Review and Management

Due to the uncertainty involved in risk management, its outcomes will never be flawless. As the activities of the business proceed and information on risks and impacts accumulate over time, the way risks are assessed and treated by risk managers will change. For this reason, the risk register and risk choices should be periodically reviewed. This will allow managers to detect possible changes to the risk level faced by the business. If certain risk choices turn out to be inappropriate, they must be changed. The project monitoring technique presented in Chapter 16 is particularly helpful in identifying changes to some risks faced by the businesses as activities progress.

CONCLUSION

This chapter has introduced basic definitions and common methods from the fields of quality management and risk management and has shown how both areas are connected. The chapter has highlighted two core ideas from quality management that have been highly influential in most industries: the first is Taguchi's insight that the requirements of manufacturing systems are not the most important determinants of product quality. Instead, the objective of quality management should be the minimization of impacts on the product or service performance stemming from a wide variety of factors, some of which are entirely outside of the manufacturing setting. The second core idea is that quality management greatly benefits from the adoption of sampling techniques since they unlock the use of statistical methods even if the underlying distribution of the investigated parameter is unknown. In the presented summary of risk management, this chapter has highlighted the core idea that opportunity cost plays an important role in risk management and that, flowing from this, it may not be necessary (or even desirable) to attempt to eliminate every identified risk.

The areas of quality management and risk management are the subjects of intensive standardization. The ISO 9000 family of standards* is a comprehensive group of quality management standards that supports businesses and other organizations in meeting the needs of customers and other stakeholders related to the provision of products or services. Similarly, the ISO 31000 family of standards† provides a framework of principles and processes for the management of risk.

Several project management and operations management textbooks provide introductory chapters to quality management. The textbook *Operations Management* by John Naylor (2002) presents an overview of basic quality management techniques. The textbook *Creating Quality* by William Kolarik (1995) contains an in-depth treatment of quality management and statistical process control. The textbook *Process theory: The principles of operations management* by Matthias Holweg and colleagues (2018) provides a compact and accessible explanation of statistical process control and other relevant concepts. *Project Management* by Mike Field and Laurie Keller (1998) includes a brief introduction to risk management. For more information on the general attitude of managers towards risk, known as *risk appetite*, the textbook *The Psychology of Risk* by Glynis M. Breakwell (2014) is instructive. The economic disruption caused by the global coronavirus pandemic in the early 2020s has emphasized the importance of risk management and is certain to re-invigorate the field.

REVIEW QUESTIONS

1. Complete the below paragraph on quality management.
   (Question type: Fill in the blanks)
   "Failure to meet ________ by failing to deliver acceptable levels of quality is likely to have negative consequences for a business. In minor cases, such failure will result in ________. In more serious cases, ________.".

2. A manufacturing operation exhibits the following quality control chart:
   What does this chart suggest?

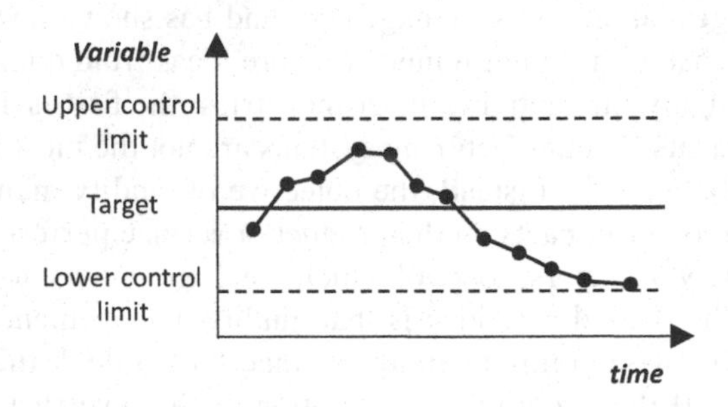

FIGURE 12.7 Control chart

* ISO (2015).
† ISO (2018).

(Question type: Multiple choice)

○ One-half of the items measured are acceptable
○ The plant is operating at one-sigma level
○ The manufacturing process cannot meet the required tolerance
○ There appears to be a trend in the data indicating a problem
○ The plant is operating at six-sigma level

3. Over-specifying dimensional tolerances will have which effect?

(Question type: Multiple choice)

○ Ensure the satisfactory operation of a part
○ Maximize productivity
○ Potentially lead to excessive cost increases
○ Make quality assurance easier to manage
○ Reduce short-term variability

4. Failure modes and effects analysis will uncover all possible ways in which products or services can fail if executed thoroughly.

(Question type: True/false)

Is this statement true or false?

○ True ○ False

5. Ignoring the deterioration of quality over time, a plant operating at Six Sigma level will produce on average:

(Question type: Multiple choice)

○ Three defective parts per thousand parts produced
○ Three defective parts per ten million parts produced
○ Three defective parts per ten thousand parts produced
○ Three defective parts per million parts produced
○ 0.002 defective parts per million parts produced
○ Three defective parts per hundred thousand parts produced

6. Order the provided steps of the main improvement cycle associated with Six Sigma.

(Question type: Ranking)

- Control
- Measure
- Analyze
- Define
- Improve

7. Which of the following are not included in a risk register?

(Question type: Multiple response)

☐ Status of risks
☐ Description of risks
☐ Likelihood and impacts of risks
☐ Ownership of risks

☐ List of attendance
☐ Frequencies and probabilities of risk
☐ Mitigations for risks

8. You are constructing a Pareto chart to analyze the risks in a project. You estimate the cost impact of a particular risk at $68,000. The probability of this impact to occur is estimated at 20%. The total technical contingency of the project is estimated at $54,400.
   (Question type: Calculation)
   Calculate the risk's share of the total technical contingency.

9. Risk is always bad and needs to be avoided.
   (Question type: True/false)
   Is this statement true or false?
   O True O False

10. Link the following risk choices to the situation in which they would be the most appropriate.
   (Question type: Matrix)

	Avoid	Retain	Exploit	Compensate	Transfer	Reduce
The risk is dangerous and must be eliminated	O	O	O	O	O	O
The risk is acceptable and part of normal business	O	O	O	O	O	O
It is possible to implement risk controls	O	O	O	O	O	O
The involvement of a third-party supplier allows this	O	O	O	O	O	O
There is an opportunity for greater return on investment	O	O	O	O	O	O
This is not a relevant risk choice	O	O	O	O	O	O

REFERENCES AND FURTHER READING

Boddy, D., 2017. *Management: An introduction.* 7th ed. New York: Pearson Education.

Breakwell, G.M., 2014. *The psychology of risk.* 2nd ed. Cambridge: Cambridge University Press.

Cook, S., 1997. *Practical benchmarking.* London: Kogan Page Publishers.

Field, M. and Keller, L., 1998. *Project management.* London: Thomson Learning.

Hauser, J.R. and Clausing, D., 1988. The house of quality. *Harvard Business Review,* May 1988.

Holweg, M., Davies, J., De Meyer, A., Lawson, B. and Schmenner, R.W., 2018. *Process theory: The principles of operations management.* Oxford: Oxford University Press.

ISO, 2015. *ISO 9000:2015 Quality management systems.* Geneva, Switzerland: International Organization for Standardization.

ISO, 2018. *ISO 31000:2018 Risk management.* Geneva, Switzerland: International Organization for Standardization.

Naylor, J., 2002. *Introduction to operations management.* London: Pearson Education.

Williams, R., Bertsch, B., Dale, B., Wiele, T.V.D., Iwaarden, J.V., Smith, M., and Visser, R., 2006. Quality and risk management: what are the key issues? *The TQM Magazine,* 18(1), pp.67–86.

Part III

Practical Management Techniques

13 Financial Planning and the Basics of Financial Accounting

OBJECTIVES AND LEARNING OUTCOMES

The objective of this chapter is to provide a first introduction to financial planning and financial accounting as important themes in management. The chapter begins with a summary of the structure of financial management and reflects how financial planning and financial accounting are situated in this field. This is followed by an introduction to basic financial planning techniques in the form of an extended example of a fictitious student-run business selling coffee at a university, the "Starbright Coffee Shop". The subsequent part of this chapter introduces the basics of financial accounting through a summary of the underlying financial accounting process, known as the accounting cycle, and presents simplified versions of three basic financial statements: the profit and loss statement, the balance sheet, and the cash flow statement. This chapter aims to achieve the following learning outcomes:

- an appreciation of the general field of financial management and its subfields, including the four major branches of accounting;
- an ability to define and use the concepts of assets, liabilities, depreciation, and amortization;
- an understanding of basic methods of financial planning and the ability to generate a simple profit and loss budget and a simple cash flow forecast;
- understanding the basic process of financial accounting in the form of the accounting cycle;
- an understanding of the importance of the accounting equation;
- knowledge of the three basic financial statements, which are the profit and loss statement, the balance sheet, and the cash flow statement; and
- understanding key terms, synonyms, and accepted acronyms.

THE ROLE OF FINANCIAL MANAGEMENT IN A BUSINESS

In any business, the activities concerned with the measurement and recording of flows of money in the forms of profit, expenses, cash, or credit are referred to collectively as *financial management*. The objective of financial management is to arrange financial resources such that a business can achieve its goals in the best possible way. This involves ensuring that the business has an adequate level of cash to carry out its

DOI: 10.1201/9781003222903-16

activities according to its plans and without disruption. It also helps the business to control its costs to promote competitiveness and profitability while maximizing the financial return to the owners and shareholders.

An important function within financial management is *accounting*. Accounting can be defined generally as the process of capturing information on the financial transactions of a business. This includes summarizing and analyzing financial information and reporting it to various stakeholders inside or outside of the business. Complicating matters somewhat, there are different branches of accounting within financial management. Figure 13.1 illustrates the relationships between four major branches, *managerial accounting*, *financial accosting*, *tax accounting*, and *cost accounting*. In the figure, solid arrows point toward areas generally considered subordinate to others.

Information flows from the business' records of transactions to the managerial accounting and financial accounting functions are shown as dashed arrows. The shaded boxes in Figure 13.1 highlight areas of financial management presented in this chapter and in Chapter 14 of this book. Figure 13.1 also expresses how the various parts of financial management are oriented in time, by showing which elements are backward-looking, forward-looking, or both.

The order in which this book treats the highlighted areas, financial planning, financial accounting, and financial analysis, is illustrated in Figure 13.2. This is based on the view of the business as a private venture with a life cycle, as discussed in Chapter 6. It is important to note that, in reality, these activities will be performed

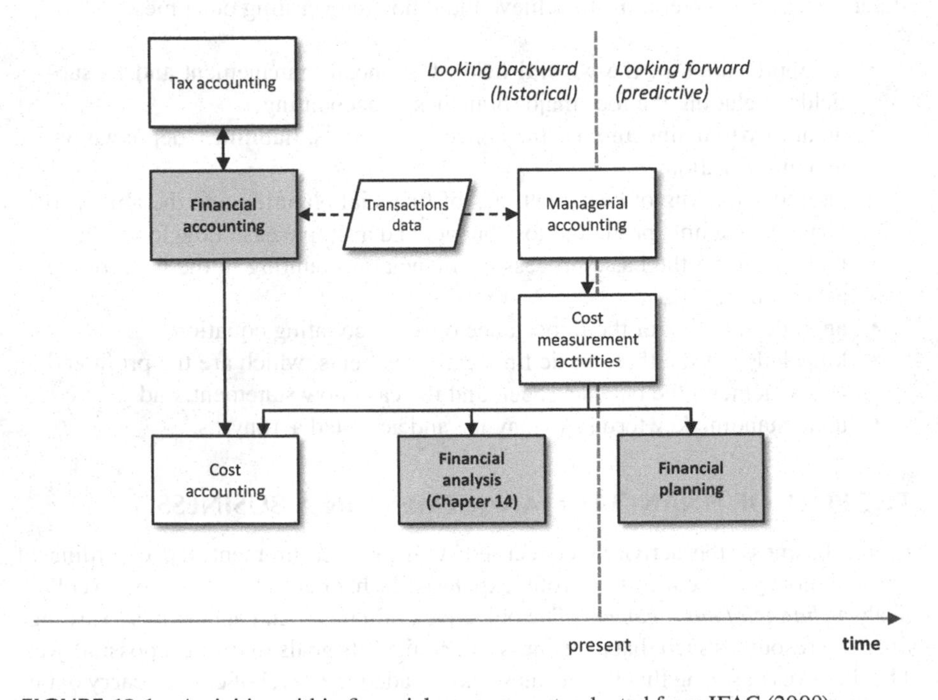

FIGURE 13.1 Activities within financial management, adapted from IFAC (2009)

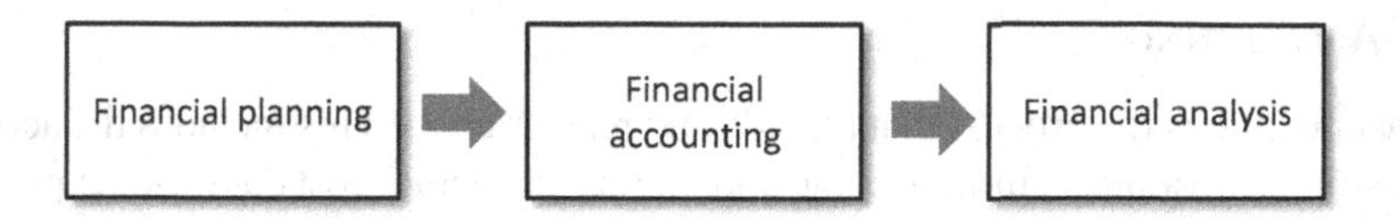

FIGURE 13.2 The sequence of financial management activities in a private venture

in parallel in most businesses, along with the other elements of financial management shown in Figure 13.1.

For completeness, and to avoid confusion, this chapter will open with brief disambiguation of the four major branches of accounting identified in Figure 13.1.

Managerial Accounting

To make sound decisions in line with the goals of a business, managers require detailed information on a broad range of financial aspects. It is the role of managerial accounting is to provide this information. Management accounting typically supports the allocation and control of budgets, the planning and control of cash flows, the measurement and analysis of various costs in the business, and the identification of optimal investment decisions, given the information available. An important additional role of management accounting is the financial planning of future business activities. Managerial accounting draws on historical transaction data recorded by the business and is both forward-looking, modeling and estimating possible future events, and backward-looking, analyzing past events.

Financial Accounting

Financial accounting is the process of collecting and recording information on past transactions resulting from the activities of the business over a defined period of time. This information is subjected to various forms of analysis and presented to managers and other stakeholders, including regulators and tax authorities, using a set of very specific reporting formats known as *financial statements*. The general objective of financial accounting is to provide detailed and accurate disclosure of the commercial performance and financial reserves in the business. Like managerial accounting, financial accounting draws on transaction data. Unlike management accounting, however, the information recorded and presented does not contain forecasts.

Tax Accounting

Tax accounting is the activity of accounting for taxation purposes. Tax accounting establishes how much tax must be paid by an individual or a business, known as the *tax burden*. Tax accounting must follow strict accounting codes and adhere to the laws and regulations determined by the relevant tax authorities. While some principles and practices in tax accounting differ from financial accounting, the information used in tax accounting is normally drawn from the financial accounting processes.

Cost Accounting

Cost accounting is the process that records and reports costs. This includes the activities of classifying, grouping, summarizing, and comparing actual cost data with expected or standard cost data. Cost accounting is subordinate to both management accounting and financial accounting. It provides information that informs managers on how to optimize business processes; it is, therefore, sometimes considered part of management accounting. However, cost accounting information is also used in financial accounting.

FINANCIAL PLANNING

The purpose of financial planning is to evaluate the future financial state of a business. This is done by using available information on financial variables to anticipate future income and expenses as the result of planned activities. To do this, financial plans specify how the income that will be received in the future will be allocated to costs incurred by the business, such as operating costs and raw material costs, as discussed in Chapter 2. Financial plans thus provide a projection of the cash flows and net profits expected to occur in a business, project, or other activity. For this reason, they form an important part of business plans, as introduced in Chapter 8. Another reason that managers regularly conduct financial planning is to understand the level of cash required by the business and whether the business needs to raise more funding to support its activities.

This section introduces the practical activity of constructing a financial plan. It does so by developing a simplified financial plan for a fictitious pop-up café owned and run by students at a university. The main outcomes of this activity are the *profit and loss budget* and the *cash flow forecast*. Both are aimed at projecting the future financial development of the proposed coffee business.

The Business Proposition

Imagine a straw poll indicating that most students do not like the coffee sold at a university-run coffee shop and think that it is too expensive. Sensing a commercial opportunity, a group of students plans to start their own business offering high-quality filter coffee at a reasonable price. While the objective is to generate some funds to reinvest, the students agree that the business need not make a high level of profit. In the following illustrative case study, the proposed (entirely fictitious!) business is called the "Starbright Coffee Shop".

The team involved in the business has conducted a series of surveys among students. On the basis of the information collected, the team has decided to construct a financial plan for the Starbright Coffee Shop making the following assumptions:

- coffee will be served in branded compostable paper cups at a price of $1.00;
- the business will offer a stamp-based loyalty program so that every tenth coffee will be provided free of charge;
- the business expects to serve a regular *customer base* of 60 students;
- each student will purchase, on average, two cups of coffee a day;
- the business will operate on weekdays for 30 weeks, distributed evenly throughout the year;

- the university will allow the business to operate on university premises and handle all processes such as permits and taxation under the condition that the students involved in the businesses are not paid a salary. The business is also required to contribute $50 per week to estates costs and purchasing services. Moreover, the business must pay for its own electricity costs, charged at a fixed rate of $15 per quarter; and
- the business will cover its own equipment costs and raw material expenses.

The first step in the financial planning process is to construct a *sales plan* and a *sales budget* for the business, covering the first year of operation. The sales plan is a forecast of the level of sales, measured in terms of the volume of product sold, that the business expects to achieve over a period of time, often measured in monthly or quarterly periods, usually denoted as *Q*1, *Q*2, and so on. The sales budget is an itemized list of the sales projected in the sales plan in the same time periods, usually aggregating sales according to product types or geographic markets. The sales budget also shows any sales discounts and allowances given to customers such as the planned loyalty program. On the basis of a sales rate of 120 cups per day and 150 days of operation per year, the sales plan covering Starbright Coffee Shop's first year of operations is shown in Table 13.1 and the sales budget covering the same period is shown in Table 13.2.

Projected Costs

Following the forecasting of the flows of money into the business, the next step in financial planning is to project flows of money out of the business arising in the form of costs.

TABLE 13.1
Sales Plan for the First Year of Operation for Starbright Coffee

	Annual sales in units	Q1 sales in units	Q2 sales in units	Q3 sales in units	Q4 sales in units
Forecast unit sales	18,000	4,500	4,500	4,500	4,500

TABLE 13.2
Sales Budget for the First Year of Operation for Starbright Coffee

	Annual sales	Q1 sales	Q2 sales	Q3 sales	Q4 sales
Forecast unit sales (units)	18,000	4,500	4,500	4,500	4,500
Price per unit ($)	1.00	1.00	1.00	1.00	1.00
Gross revenue ($)	18,000	4,500	4,500	4,500	4,500
Sales discounts and allowances ($)	1,800	450	450	450	450
Revenue ($)	16,200	4,050	4,050	4,050	4,050

TABLE 13.3
Projected Cost of Raw Materials and Consumables

Raw material/consumable	Purchase price ($)	Quantity in packaging units (units)	Materials/consumables cost per cup ($)
Coffee	6.00	50	0.12
Coffee filters	4.00	50	0.08
Milk	0.60	12	0.05
Branded paper cups and all other consumables	10.00	100	0.10
Projected total cost of materials and consumables per cup			0.35

For many businesses, a basic and important form of costs will be expenses incurred for the raw materials and consumables used in the operation. For Starbright Coffee Shop, it is assumed that coffee, coffee filters, milk, and paper cups will form the main raw material costs. Since materials are often purchased in fixed quantities or increments of multiple units, known as *packaging units*, material costs are often said to be "lumpy". Moreover, in many real businesses, the price paid for raw materials will decrease as the volume purchased increases. To avoid confusion originating from such aspects, material costs are usually broken down according to units of output, which is cups of coffee in the example. Table 13.3 summarizes the raw material purchase prices and packaging units for each raw material and breaks these values down to form costs per cup. The projected raw material costs per unit of output are assumed to be the only variable costs incurred by the business. As defined in Chapter 3, variable costs are costs that change in proportion to the total quantity of goods and services produced.

The next step in cost projection is to convert the identified costs into a projected quarterly cost rate, based on the sales budget. Such variable costs enter the financial plans as cost of goods sold (COGS), as defined in Chapter 2. Table 13.4 shows the variable cost budget for the Starbright Coffee Shop.

In contrast to variable costs, fixed costs are independent of sales quantity and remain constant whatever the sales, as introduced in Chapter 3. In the artificially simple example of the Starbright Coffee Shop, the two fixed costs incurred by the business are the fixed estates and purchasing charge and electricity costs. Constituting operating costs* both costs are charged at a different frequency, with estates and purchasing costs being charged weekly and the power costs being charged quarterly. This time span is referred to as the *billing period* or *billing cycle*. The estates and purchasing costs are only charged during 30 weeks of operation. Table 13.5 constructs the fixed costs budget for the Starbright Coffee Shop.

* Energy costs are fixed costs in this example and thus do not affect COGS. If energy costs arise as variable costs it would be appropriate to include them as process costs affecting COGS.

TABLE 13.4
Variable Costs Budget

Cost item	Cost per cup ($)	Annual sales in units	Annual variable costs ($)	Quarterly sales in units	Quarterly variable costs ($)
Coffee	0.12	18,000	2,160	4,500	540
Coffee filters	0.08	18,000	1,440	4,500	360
Milk	0.05	18,000	900	4,500	225
Branded paper cups and all other consumables	0.10	18,000	1,800	4,500	450
Projected total variable costs (equating to COGS in this example)	0.35	18,000	6,300	4,500	1,575

TABLE 13.5
Fixed Costs Budget

Fixed cost item	Cost ($)	Billing period	Annual fixed costs ($)	Quarterly fixed costs ($)
Estates and purchasing charge	50.00	Weekly (30 weeks only)	1,500	375
Power costs	15.00	Quarterly	60	15
Total fixed costs			$1,560	$390

Assets and Depreciation

One very important aspect that has been ignored in the financial planning process so far is the equipment needed by the business. After reviewing the performance of various available filter coffee machines, the team behind Starbright Coffee Shop has decided that four filter coffee machines of a particular type are needed. They have determined that a portable counter bar, signage, and a range of ancillary equipment are also required. Collectively, such objects form what is known in management and business as *assets*.

Assets can take many other forms beyond physical objects. Generally, an asset can be defined as "something valuable belonging to a person or organization that can be used for the payment of debts".* Unsurprisingly, the field of accountancy has a range of specialist definitions of assets, many of which are quite abstract. For example, an important standards-setting organization defines assets as "probable future

* https://dictionary.cambridge.org/dictionary/english/asset

economic benefits obtained or controlled by a particular entity as a result of past transactions or events".* For the purposes of this introductory textbook, it is possible to define assets more simply as follows:

An asset is any item of property, tangible or intangible, owned by a person or business that has value and can be sold or otherwise made available to meet commitments, debts, or obligations.

As contained in this definition, a *tangible asset* is an asset that is a physical object. Cash and cash equivalents, inventory, equipment, buildings, vehicles, and many forms of investments constitute tangible assets. Some readers may be surprised that financial assets such as shares and investments are classed as tangible assets. While there are some technical justifications for this, it is important to remember that many practices in accounting are shaped more or less arbitrarily by rules, standards, and laws.

Tangible assets are normally divided into the two main subclasses of *current assets* and *fixed assets*. Current assets are generally all objects of value that can be disposed of easily in return for cash; they include cash equivalents, raw materials, consumables, and stock†. Current assets are by convention expected to be used up or converted into cash within a year as a normal part of the operations of the business. In contrast, fixed assets, which are also known as *capital assets* or *property, plant, and equipment* (*PPE*) are tangible assets that are more difficult to convert into cash, such as equipment and buildings. Fixed assets normally include some financial assets such as long-term investments. Important techniques that can be used to evaluate financial assets are presented in Chapter 15.

Intangible assets are, as the name suggests, non-physical objects that can be used to meet the commitments, debts, or obligations of the business. Important forms of intangible assets are various forms of intellectual property introduced in Chapter 5 and valuable information.

An important, yet somewhat abstract, kind of intangible asset is *goodwill*. Goodwill is an accounting term for an asset that arises when an existing business, or a part of it, is sold to a new owner. More specifically, goodwill is the difference between the purchase price and the value of the assets owned by the acquired business minus other obligations and debts. It is abstract because it is so intertwined with the business that its value cannot be identified in isolation, unlike many other kinds of assets. Moreover, goodwill comes into existence solely as the result of the acquisition of the business, so it is not created by the business.

Examples of things that form goodwill include a business name, a validated business model, valuable customer relationships, and the culture within the business. The importance of goodwill becomes clear when considering, for example, the sale of ownership stakes in successful new businesses by venture capitalists, as discussed in Chapter 8. When successful startups are acquired by established businesses in this way, the buyers routinely pay many times the total value of the assets owned by the startup.

* FASB (1985). Statement of Financial Accounting Concepts No. 6

† In this usage, the word "stock" refers to inventory. This is not to be be confused with shares of ownership.

TABLE 13.6
List of Assets

Asset	Purchase price per unit	Quantity purchased	Total purchase cost
Coffee machine	$150	4	$600
Portable counter bar	$400	1	$400
Signage	$200	1	$200
Ancillary equipment (containers, small digital cash register, etc.)	$200	1	$200
Total assets			$1,400

To appropriately reflect assets in financial planning, it is necessary to first consider the costs of their purchase. The team at Starbright Coffee Shop has determined that it will be necessary to acquire the following assets before the business can begin operating: four filter coffee machines, a portable counter bar, signage, and a range of ancillary equipment, including various containers and a small cash register. Table 13.6 lists all assets required by the business.

Simply treating the purchase costs of the assets required by the business as expenses would not reflect the true profitability of the business since the assets themselves remain valuable and could be sold. However, it is essential to note that assets generally lose value over time. This process is known as *depreciation*. As an accounting term of great importance in all areas of management and commerce, depreciation refers to the reduction of the value of tangible assets over time due to wear and tear and other forms of obsolescence. It can be defined as follows:

Depreciation is the actual decrease of the value of tangible assets over the time they have been in use. This process is also referred to as *writing off* the asset. Intangible assets do not depreciate; instead, they are written off through *amortization*.

IMPORTANT DEFINITION

In practice, depreciation would not stop Starbright Coffee Shop from selling its assets. However, the money obtained from this sale would be smaller than the original purchase prices, reflecting the costs of depreciation incurred up to the time of sale. To reflect such *residual value*, the concept of net profit takes into account the costs of depreciation instead of the costs of purchasing the assets.

It is assumed that all assets owned by Starbright Coffee Shop will depreciate by 10% of their initial value each year, which is a reasonable level of depreciation associated with industrial equipment. Moreover, since a constant amount of value, measured in money terms at $140, is written off the total value of the assets each year, this form of depreciation is known as *straight line depreciation*. Figure 13.3 graphically expresses the depreciation of the assets listed in Table 13.6. As can be

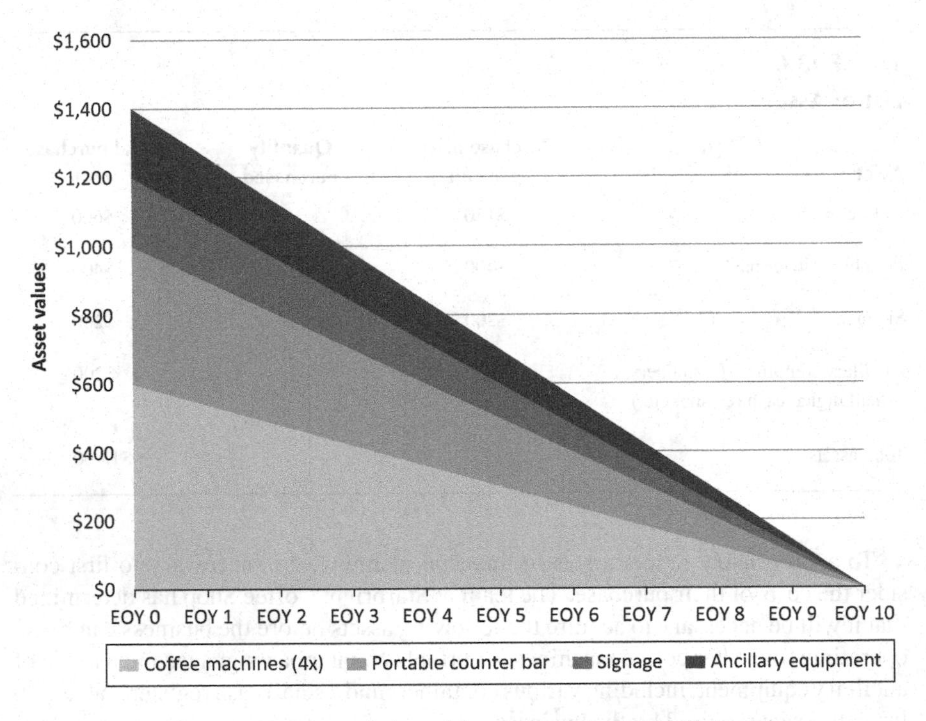

FIGURE 13.3 Straight line depreciation of the assets of Starbright Coffee Shop

seen, the assets will be written off fully after 10 years. Note that depreciation enters the financial plans in the *other costs* category, as defined in Chapter 2.

In some cases, it is not immediately clear if a cost incurred should enter the financial plans and accounts as an expense or as an asset that depreciates over time. Consider, for example, a piece of equipment that has a short useful life, such as tooling. In this case, managers may have the choice of recording this cost either as an expense when it occurs or as an asset depreciating over a period of time. Known as *expensing versus capitalizing*, this decision has an impact on the profit shown in financial plans and financial statements. It is important to note that there are rules and regulations about what can be expensed and what can be capitalized and that the decision will have tax implications.

The Profit and Loss Budget and the Cash Flow Forecast

The next stage in the financial planning process is the construction of the *profit and loss budget* and the *cash flow forecast*. The profit and loss budget is a financial plan setting out the amount of revenue the business is expecting to generate, the types and levels of cost that will be incurred, and how much profit or loss the business is planning to make. Also known as the *budgeted income statement*, it generally considers the previous financial performance of the business (if available) and projects the financial results of planned activities. The time horizon of a profit and loss budget usually ranges from one to five years into the future.

TABLE 13.7
Profit and Loss Budget for Starbright Coffee Shop, First Year of Operation

Starbright Coffee Shop		
Profit and loss budget for the first year of operation		
(All tangible assets are written off over 10 years)		
Sales revenue		**$16,200**
Cost of goods sold (COGS)		
Coffee	$2,160	
Coffee filters	$1,440	
Milk	$900	
Branded paper cups and other consumables	$1,800	
Total variable costs		**$6,300**
Gross profit		**$9,900**
Operating costs		
Estates and purchasing charge	$1,500	
Power costs	$60	
Total fixed costs		**$1,560**
Operating profit		**$8,340**
Other costs		
Depreciation	$140	
Total, other costs		**$140**
Net profit		**$8,200**

In existing businesses, the profit and loss budget is normally prepared annually – although the period can be shorter or longer depending on what the budget will be used for. It is normally presented in a vertical format with the expected income in the upper part and the expenses in the lower part, following the general logic and terminology presented in Chapter 2. Table 13.7 shows the profit and loss budget for the Starbright Coffee Shop. As can be seen from this projection covering the first year, the team at Starbright Coffee is planning to make a rather handsome net profit of $8,200 in their first year of operation.

Once the profit and loss budget has been constructed, the cash flow forecast covering the same period of time can be developed. The cash flow forecast, also known as *a cash flow projection*, is a plan that shows the amount of money a business expects to receive as revenue and to pay out as expenses over a given period of time. Like profit and loss budgets, cash flow forecasts are normally constructed with a time horizon of one to five years and normally forecast cash flows in monthly or quarterly time periods. If the business is experiencing a shortage of money, cash flow forecasts may be prepared for much shorter time periods, down to a single day. While there is no guarantee that cash flow forecasts will be accurate, they fulfill several important functions, including the following:

- they can be used to determine if the business holds enough cash. The importance of this point will be discussed in detail in Chapter 14;
- they allow the business to notice if its suppliers are paid on time and if its customers pay on time, which is not normally the case for all customers; and
- they are often required by investors in the business and other external stakeholders to see if the business is in a healthy financial position.

The cash flow forecast essentially breaks down the revenue and cost items listed in the profit and loss budget to shorter time periods. It drops the headers from the profit and budget and normally replaces them with the categories of *cash inflows* and *cash outflows*, or similar. Additionally, the cash flow forecast adds a further section at the bottom that summarizes the monthly cash inflows and outflows and presents a cumulative cash balance covering the entire forecasting period. It is important to note that only cash costs enter the cash flow forecasts. This means that depreciation is not shown – instead, the cost of the purchasing of the equipment is shown in a column reflecting cash flows prior to the business's start. This column also shows costs incurred for the purchasing of opening stock of coffee, coffee filters, milk, branded paper cups, and consumables. The team is planning to hold 10% of the projected annual requirement of these materials as opening stock.

The cash flow forecast for the first year of the Starbright Coffee Shop is shown in Table 13.8. Note that the cumulative net cash flow at the end of quarter 4 is $6,940, which equates to the net profit identified in the profit and loss budget plus depreciation and minus the purchasing cost for the assets.

While the example developed in this section is very simple and stylized, it conforms to the cash flow patterns associated with a successful project presented in Chapter 6, fully recovering the initial investment. The net quarterly cash flow indicates that it is projected that the Starbright Coffee Shop becomes cash flow positive in its first quarter of operations. The projected cumulative cash flow exceeds zero in the first quarter also, indicating that the business is expected to

TABLE 13.8
Cash flow Forecast for Starbright Coffee Shop, First Year of Operation

Starbright Coffee Shop

Cash flow forecast for the first year of operation, in quarterly time periods

	Start	Q1	Q2	Q3	Q4
Cash inflows					
Sales revenue	$0	$4,050	$4,050	$4,050	$4,050
Total cash inflow	**$0**	**$4,050**	**$4,050**	**$4,050**	**$4,050**
Cash outflows					
Coffee	$216	$486	$486	$486	$486
Coffee filters	$144	$324	$324	$324	$324
Milk	$90	$202.50	$202.50	$202.50	$202.50
Branded paper cups and other consumables	$180	$405	$405	$405	$405
Estates and purchasing charge	$0	$375	$375	$375	$375
Power costs	$0	$15	$15	$15	$15
Coffee machines (4x)	$600	$0.00	$0.00	$0.00	$0.00
Portable counter bar	$400	$0.00	$0.00	$0.00	$0.00
Signage	$200	$0.00	$0.00	$0.00	$0.00
Ancillary equipment (containers, small digital cash register, etc.)	$200	$0.00	$0.00	$0.00	$0.00
Total cash outflow	**$2,030**	**$1,807.50**	**$1,807.50**	**$1,807.50**	**$1,807.50**
Net cash flow	**–$2,030**	**$2,242.50**	**$2,242.50**	**$2,242.50**	**$2,242.50**
Cumulative net cash	**–$2,030**	**$212.50**	**$2,455**	**$4,697.50**	**$6,940**

break even very quickly. This is likely due to the low operating expenses incurred by the company as a consequence of the workers not receiving a salary. The addition of staff costs would have pushed back the breakeven for the company considerably. Nevertheless, the financial planning for Starbright Coffee Shop suggests a lucrative business opportunity.

While specialized software tools are available for the construction of profit and loss budgets and cash flow forecasts, financial plans are mostly constructed using standard spreadsheet tools, as done in the illustrative case study presented in this chapter.

THE BASICS OF FINANCIAL ACCOUNTING

The financial plans presented in the previous section of this chapter contain prospective financial statements. In the introduction to the various types of accounting presented at the beginning of this chapter, it was stated that financial accounting and cost accounting are not forward-looking but deal exclusively with past, historical transactions. This means that financial accounting is fundamentally different from financial planning. Financial accounting can be characterized as the process of collecting, recording, summarizing past transactions, and generating formal reports of these.

In financial accounting, the transactions undertaken by a business (or another entity) are reported in a set of highly specific documents known as financial statements. Across the globe, there are many different accounting standards that define how such statements should be presented and what information they must contain.* It is important to note that compliance with these standards is normally required by law.

In many cases, the financial statements prepared by financial accountants must be made available to the certain stakeholders and the wider public. To ensure that these statements are useful they must satisfy two main criteria, which are:

- *relevance*, referring to the criterion that the financial statements must be able to influence and inform the decisions of stakeholders. Though not containing any forward-looking information, they must have value for the prediction of future events, and they must be useful in confirming past events; and
- *faithful representation*, referring to the requirement that financial statements must describe the recorded transactions in a truthful way. More specifically, the records must be complete, neutral, and free from errors.

* The main set of accounting principles accepted in most jurisdictions around the world is the International Financial Reporting Standards (IFRS) issued by the IFRS Foundation, which is a non-profit organization committed to the standardization of accounting processes. In the United States, the main set of accepted accounting principles is defined in the Generally Accepted Accounting Principles, issued by the Financial Accounting Standards Board (FASB), which is also a non-profit organization.

Further criteria for the acceptability of financial statements are that their contents should be verifiable by outside parties, such as auditors, they should have a format that allows comparisons to the financial statements produced by other businesses, they should be comprehensible to those for whom the information is relevant, and they should be made available in a timely fashion.

The Accounting Cycle

It is difficult to reach a practical understanding of financial accounting without basic knowledge of how the process of accounting works. The most important model used to introduce the process is the so-called *accounting cycle*. There are various formats for the accounting cycle; the most common format is an eight-stage cycle presented as a circle, shown in Figure 13.4.

This section briefly outlines each stage of the accounting cycle and introduces important concepts and definitions. While accounting is very much a practical activity requiring hands-on training, this overview is nevertheless useful because it introduces the basic knowledge needed to grasp the meaning and structure of the financial statements summarized at the end of this chapter.

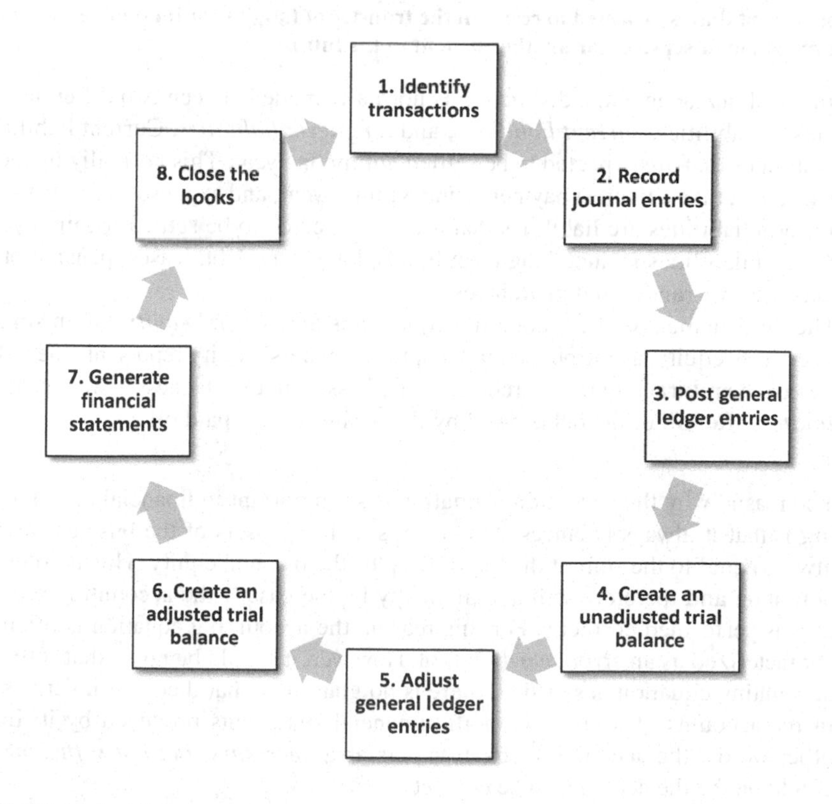

FIGURE 13.4 The accounting cycle

1. Identify Transactions

The first stage in the accounting cycle is the identification of events that may have an effect on the finances of the business. Financial accounting employs a very specific criterion for such events. Only events that have an effect on the *accounting equation* are of interest to financial accounting. In this sense, the accounting equation forms the conceptual core of financial accounting. In its basic form, this equation states that the assets in the business equate to the liabilities plus the equity the owners hold:

$$\text{Assets} = \text{Liabilities} + \text{Owners}' \text{ equity} \tag{13.1}$$

The first element is the assets of various types owned by the business. As described in this chapter there are different kinds of assets that fall into this category.

The second element is the *liabilities* of the business. In accounting terminology, the word "liability" is used in a very specific sense. It can be characterized as an obligation (i.e. a commitment to take some action) connected to past events that are expected to lead to some form of payment in the future. Liabilities can thus be defined as follows:

In accounting, a liability is an obligation of a business or person arising from a past event that is expected to result in the transfer of tangible or intangible assets, a provision of services, or another benefit in the future.

In financial management, a distinction is normally made between two different categories of liabilities, *current liabilities*, and *long-term liabilities.* Current liabilities are liabilities that are expected to be settled within one year. This normally includes salaries, fees, taxes, interest payments due within a year, and expenses for purchases. Long-term liabilities are liabilities that are not expected to be settled within a year. Long-term liabilities include long-term bonds, long-term debt, leases, pension obligations, and warranties and guarantees.

The third element in the accounting equation is the *owners' equity.* Often simply referred to as equity, as introduced in Chapter 2, owners' equity represents the value that would be returned to the shareholders if all assets in the businesses were sold, or *liquidated*, after all of the debts owed by the business were paid off.

The reason why the accounting equation is so important in financial accounting is that it always balances. This means that the assets of the business are always equal to the sum of the liabilities plus the owners' equity. This is so by definition and therefore will automatically be the case if the accounting process is performed correctly. For this reason, the accounting equation is often characterized as an error detection tool. However, it should be noted that if the accounting equation is satisfied, there is no guarantee that there are no errors in the accounts of a business or the financial statements produced by it. In other words, the accounting equation acts as a *necessary, but not sufficient,* condition for the accounts to be correct.

2. Record Journal Entries

The second stage in the accounting cycle following the identification of relevant transactions is the capturing of such transactions in an accounting document* referred to as the *general journal*. The general journal records every eligible business transaction, for example, relating to sales, receipts of money, changes in inventory, and purchases. It records the transactions in chronological order. The general journal usually has a format containing columns for reference numbers or serial numbers, descriptions, dates, and indicates the accounting records into which the entries are transferred.

3. Post General Ledger Entries

The third stage of the accounting cycle is the posting of the journal entries into a further book of accounts referred to as the *general ledger*. The general ledger, also known as the *nominal ledger*, is the main and most important information repository in any accounting system. It forms the collection of all accounts maintained by the accounting system of the business. In businesses running an enterprise resource planning (ERP) system, as described in Chapter 9, the general ledger is likely to form a central element of such a system. The general ledger is not organized chronologically but on the basis of *accounts* that are affected by transactions. In this context, accounts are defined as follows:

> **In accounting and bookkeeping, an account is a record that tracks the financial activities and transactions involving a specific asset, liability, equity, revenue, or expense.**

The general ledger follows a T-shaped format in which *debits* are recorded on the left-hand side and *credits* are recorded on the right-hand side. Whether an entry constitutes a debit or a credit depends on the nature of the account affected. Moreover, every transaction is entered into two accounts. This process is fundamental to modern accounting practice and is known as *double-entry bookkeeping*. The process of posting journal entries to the general ledger and correctly identifying in which two accounts a transaction enters as a debit or a credit is considered a specialist task in basic accounting that exceeds the scope of this book. Recommendations for financial accounting textbooks are provided in the conclusion.

4. Create an Unadjusted Trial Balance

Following the posting of entries to the general ledger, the next step is to construct an *unadjusted trial balance*. A trial balance is an accounting report that summarizes the account balances from the general ledger in two columns, debits and credits, and shows if the debits and credits are equal to one another.† If the two totals do not equate, it is clear that an accounting error has occurred and that some adjustments are

---

* In financial management, accounting documents are referred to as *books of accounts*. Such documents may exist in paper or digital forms.

† The trial balance should not be confused with the balance sheet. While both concepts are driven by the accounting equation, the former is a report that is part of the accounting cycle and the latter is a financial statement as the outcome of the accoutring cycle.

required to balance the accounts. The differences may be due to incorrectly posted ledger entries, the omission of accounts, simple miscalculations, or other errors.

Trial balances are normally constructed toward the end of an accounting period. It is common that the debits do not equate to the credits in the first trial balance constructed. The purpose of the trial balance is to detect such errors and to resolve them so the accounting books and financial statements can be made accurate.

5. Adjust General Ledger Entries

At the end of the accounting period, some expenses may have been incurred without payments having been settled yet. Likewise, some invoices issued to customers may still be unpaid. Such occurrences require an adjustment to general ledger entries to capture in the accounts that some transactions have occurred but have not yet been recorded. A further example of such a misalignment could be the interest earned on a bank account balance. While the interest payment might not have been recorded in the business accounts, it may already appear on the bank statements. In such cases, an adjusted entry will be used to identify the interest payment correctly in the accounts. Again, adjusting general ledger entries is considered a non-trivial activity done by professional accountants.

6. Create an Adjusted Trial Balance

The final step before the financial statements are created is the construction of an *adjusted trial balance*. The adjusted trial balance verifies that the debits and credits match after making the adjustments and serves as an indication that the accounting records are accurately prepared and up to date.

7. Generate Financial Statements

The previous stages in the accounting cycle lead up to the point at which the financial statements are prepared. As introduced in this chapter, there are three main financial statements for businesses: the *profit and loss statement*, the *balance sheet*, and the *cash flow statement*. Collectively, these statements should provide a detailed and accurate representation of the financial situation of the business. They can be used to determine how the business can be improved, to set appropriate goals, and to analyze performance, as discussed in Chapter 14. Financial statements are also used to provide insight into the current situation of the business to shareholders and prospective investors. The final part of this chapter will provide a brief characterization of the main financial statements.

8. Closing the Books

The last stage in the accounting cycle is to finalize the accounting process in a step referred to as *closing the books*. This involves tasks such as updating and reconciling accounts, reviewing small, discretionary cash resources, referred to as *petty cash*, and measuring stock levels. An important additional function of closing the books is to prepare the accounting system for the next accounting period by filing financial documents and discarding documents that are no longer needed.

Three Major Financial Statements

The three major financial statements produced in the accounting cycle are the profit and loss statement, the balance sheet, and the cash flow statement. This section will briefly introduce these financial statements and present simplified examples for each. It is important to note that all financial statements are subject to stringent legal requirements, both in terms of their content and in terms of appearance. Often a fourth financial statement is included in such summaries, which is the *statement of changes in equity*. This additional financial statement presents changes in the share capital, accumulated reserves, and retained earnings in the business. It is omitted from this chapter for brevity.

The Profit and Loss Statement

The purpose of the profit and loss statement is to show the financial results of the business over a period of time, normally reflecting an accounting period of one year. The profit and loss statement is known under several other names, including *income statement*, *profit and loss account*, *statement of profit or loss*, *revenue statement*, and *earnings statement*. It reports the total sales revenues achieved by the business and shows how this financial inflow is transformed into net profit. Table 13.9 provides a simplified profit and loss statement for a fictitious business ("Example Company A Inc."). It should be noted that the entries found in a profit and loss statement depend on the nature of the business and the accounting regulations it must meet.

The headers used in the example profit and loss statement are in line with the financial terminology introduced in Chapter 2, with the exception of the terms "operating expenses" and "other expenses" instead of "operating costs" and "other costs". It should be noted how depreciation enters the profit and loss account as an operating expense. As can be seen, the business reports a net profit of $4,850 over the accounting period, resulting from a total reported revenue of $135,100.

The Balance Sheet

The *balance sheet* shows the assets and liabilities of a business and from which sources finance has been raised. It is also known as the *statement of financial position* or the *statement of financial condition*. Unlike the other financial statements, the balance sheet describes the finances of a business at a specific point in time, which is normally the end of the accounting period reflected in a profit and loss statement. For this reason, the balance sheet is often described as a *snapshot of the financial condition of a business*. Because a balance sheet is extracted from the general ledger, its structure is determined by the accounting equation such that the total assets of the business must equate to the sum of the total liabilities plus the owner's equity.

There are several accepted formats for balance sheets. An intuitively accessible format is the so-called *horizontal* balance sheet format as shown in the simplified example presented in Table 13.10 for another fictitious business ("Example Company B Inc."). The appearance and entries in balance sheets vary depending on the accounting regulations and the nature of the business. Moreover, as is the convention

TABLE 13.9
Simplified Example of a Profit and Loss Statement

Example Company A Inc.

Profit and loss statement for the year ended December 31, 2020

Sales Revenues		
Cash sales	$84,300	
Credit sales	$50,800	
Total sales revenue		**$135,100**
Cost of goods sold	$86,300	
Gross profit		**$48,800**
Operating expenses		
Salaries	$25,600	
Advertising	$2,700	
Office Rent	$2,200	
Utilities	$1,500	
Office supplies and consumables	$450	
Depreciation	$1,300	
Other expenses	$3,400	
Total operating expenses		**$37,150**
Operating profit		**$11,650**
Other expenses		
Interest expenses	$2,300	
Income tax expenses	$4,500	
Net profit		**$4,850**

TABLE 13.10

Simplified Example of a Balance Sheet in the Horizontal Format

Example Company B Inc.

Balance sheet, December 31, 2020

Assets			*Liabilities and owners' equity*		
Current assets			*Current liabilities*		
Cash and cash equivalents	$100,000		Accounts payable	$45,000	
Accounts receivable	$30,000		Notes payable	$12,000	
Inventory	$16,000		Accrued expenses	$6,000	
Prepaid expenses	$50,000		Deferred revenue	$1,500	
Investments	$8,000		**Total current liabilities**		**$64,500**
Total current assets		**$204,000**			
			Long-term liabilities		
Fixed assets			Long-term debt	$252,800	
Land	$25,000				
Buildings and improvements	$310,000		**Total liabilities**		**$317,300**
Equipment	$45,000				
Less accumulated depreciation	($4,500)		*Owners' equity*		
			Common shares	$20,000	
Other assets			Additional paid-in capital	$14,000	
Intangible assets	$3,500		Retained earnings	$234,000	
Less accumulated amortization	($200)		Treasury shares	($2,500)	
Total assets		**$582,800**	**Total liabilities and owner's equity**		**$582,800**

in accounting practice, items that are subtracted from a balance are shown in parentheses and not as negative values.*

As can be seen in Table 13.10, the sum of the assets in the business shown on the left-hand side equates to the sum of all liabilities plus all forms of owners' equity shown on the right-hand side. This example contains a few terms that have not been introduced elsewhere in this book. *Accounts receivable* refers to outstanding payments from customers for goods or services that have already been delivered. *Accounts payable* and *notes payable* are outgoing payments that were promised but have not yet been made by the business. Similarly, *accrued expenses* are costs that have been recognized in the accounting system but have not yet been paid. *Deferred revenue* refers to prepayments from customers that have not yet received their goods or services. *Retained earnings* are the profits made in previous accounting periods that have been kept in the business and not been paid out to the owners in the form of a dividend. Finally, *treasury shares* are shares in the business that the business has not yet sold to investors. Because they have not yet been sold, they enter the owners' equity section as a "negative" and are shown in parentheses.

The Cash Flow Statement

The cash flow statement, also known as the *statement of cash flows*, reports how changes in the balance sheet entries and financial flows in and out of the business over a period of time change the cash and cash equivalents present in the business. In the most common format, these flows of money are broken down into *operating activities*, *investing activities*, and *financing activities*. For managers and other stakeholders, the cash flow statement, therefore, indicates whether or not the business can operate in the short run and will be able to pay its bills. A simplified example of a cash flow statement is provided in Table 13.11 for a further fictitious business ("Example Company C Inc."). Again, values that are subtracted are shown in parentheses. As can be seen from the example, the business has started the accounting period with a cash balance of $115,000 and ended the accounting period with a cash balance of $527,000, largely due to financial inflows received from issuing shares.

CONCLUSION

This chapter has provided an overview of the activities of financial planning and financial accounting, as part of the overarching field of financial management. The chapter introduced financial planning as a practical activity, developing a simple financial plan consisting of a profit and loss budget and a cash flow forecast for the proposed Starbright Coffee Shop. It is hoped that this brief introduction will be useful for the development of financial plans for businesses and projects. The remainder of the chapter introduced the field of financial accounting. As discussed at the beginning of the chapter, it is important to first grasp that there are various distinct disciplines in accounting. The chapter outlined four such disciplines, each with its own goals and conventions and subject to accounting practices that are difficult to

* In fact, due to the structure of double-entry bookkeeping, negative values are *never* entered into accounting systems.

TABLE 13.11
Simplified Example of a Cash Flow Statement

Example Company C Inc.

Cash flow statement for the year ended December 31, 2020

Cash and cash equivalents, beginning of the period		**$115,000**
Operating activities		
Cash receipt from		
Customers	$900,000	
Other operations	$30,000	
Cash paid for		
Inventory purchases	($450,000)	
General operating expenses	($112,000)	
Wage expenses	($220,000)	
Interest	($34,000)	
Income taxes	($20,000)	
Net cash provided by operations		**$94,000**
Investing activities		
Sale of property and equipment	$38,000	
Purchase of investments	($45,000)	
Net cash flow provided by investing activities		**($7,000)**
Financing activities		
Issuance of shares	$350,000	
Issuance of long-term liabilities	$20,000	
Payment of dividends	($45,000)	
Net cash provided by financing activities		**$325,000**
Net increase (decrease) in cash and cash equivalents		**$412,000**
Cash and cash equivalents, end of period		**$527,000**

navigate by non-accountants. Nevertheless, an understanding of the underlying principles, especially of the accounting cycle and the accounting equation, which was presented as a core idea in this chapter, is important for all managers. Moreover, as will be seen in Chapter 14, financial statements form an excellent, and often very accessible, source of reliable information for further analysis.

There are many useful textbooks dedicated to managerial and financial accounting. A relevant textbook is *Principles of Managerial Finance* by Chad Zutter and Scott Smart (2019). The textbook *Financial Accounting* by Robert Libby and colleagues (2020) provides an accessible introduction to the accounting cycle. Other introductory management textbooks provide a more general description of financial statements and other aspects of managerial accounting and financial accounting, such as *Management – An Introduction* by David Boddy (2017) and *Management for Engineers, Scientists and Technologists* by John Chelsom and colleagues (2005).

REVIEW QUESTIONS

1. Complete the below paragraph introducing financial management.
   (Question type: Fill in the blanks)
   "The objective of financial management is to arrange ________ such that the business can ________. Financial management involves activities such as ensuring that the business has an adequate level of ________.

2. A piece of construction machinery is assumed to depreciate by 15% in its first year of operation. A straight line model of depreciation is applied.
   (Question type: Calculation)
   Calculate the integer number of years after which the asset will be fully written off.

3. Which effect does depreciation have?
   (Question type: Multiple choice)
   - O After the asset has depreciated, it is useless
   - O After it has depreciated fully it must be sold
   - O Depreciation creates a cost for the company that diminishes net profit
   - O Depreciation decreases the value of intangible assets
   - O Depreciating assets are undesirable

4. Match the following asset categories to specific assets.
   (Question type: Matrix)

	Fixed assets	Current assets	Long-term investments	This is not an asset
An office building	O	O	O	O
Land owned for investment purposes	O	O	O	O

Deferred revenue	○	○	○	○
Inventory	○	○	○	○

5. You are planning the finances of a new business manufacturing a chocolate product. It is planned that the business will achieve a sales revenue of $45,000 in its first year of operation. You are assuming a quarterly COGS of $5,000 and an annual operating cost of $7,500. You expect the equipment owned by the business to depreciate in value by $1,300.
   (Question type: Calculation)
   Calculate the operating profit in the first year.

6. Arrange the provided stages in the accounting cycle in the correct order.
   (Question type: Ranking)
   - Create an unadjusted trial balance
   - Close the books
   - Post general ledger entries
   - Adjust ledger entries
   - Create an adjusted trial balance
   - Generate financial statements
   - Identify transactions
   - Record journal entries

7. What is the significance of the accounting equation?
   (Question type: Multiple choice)
   ○ It guarantees that there are no errors in the accounting system
   ○ It forms the basis for double-entry bookkeeping and the general ledger
   ○ It was widely used before digital accounting systems emerged
   ○ It can sometimes be incorrect, yet provides orientation
   ○ It helps in making entries into the journal

8. Which of the following are not found in a profit and loss statement?
   (Question type: Multiple response)
   ☐ Current assets
   ☐ Sales revenue
   ☐ Operating expenses
   ☐ Common shares
   ☐ Tax expenses
   ☐ Added value

9. Which of the following statements about balance sheets are true and which are false?

   (Question type: Dichotomous)

   True False

   O O It is incorrect if the sum of assets and liabilities equates to owners' equity
   O O Only incorporated legal forms produce balance sheets
   O O It provides a snapshot of the financial situation in a business
   O O Assets are always on the left side and liabilities are on the right side
   O O Intangible assets are impossible to value and are not shown
   O O Inventory is shown as an asset

10. Complete the below paragraph on cash flow statements.

    (Question type: Fill in the blanks)

    "In the most common formats of cash flow statements, the ________ are broken down into ________, ________, and ________ activities. The cash flow statement can thus be summarized as a report of the flow of ________ into and out of the business. The cash flow statement indicates whether or not the business can operate in the ________ and will be able to ________".

REFERENCES AND FURTHER READING

Boddy, D., 2017. *Management: An introduction*. 7th ed. New York: Pearson Education.

Chelsom, J.V., Payne, A.C., and Reavill, L.R.P., 2005. *Management for engineers, scientists and technologists*. 2nd ed. Chichester: Wiley.

IFAC, 2009. *International good practice guidance, evaluating and improving costing in organizations*. New York: International Federation of Accountants, July, 2009, IFAC.

Libby, R., Libby, P.A., and Hodge, F., 2020. *Financial accounting*. 10th ed. New York: McGraw-Hill Education.

Zutter, C.J. and Smart, S.B., 2019. *Principles of managerial finance*. 15th ed. London: Pearson Higher Education.

14 Financial Analysis

OBJECTIVES AND LEARNING OUTCOMES

This chapter introduces basic methods of financial analysis which are used by managers to draw insight from financial data about a business and its activities. Two major families of methods of financial analysis are introduced in this chapter: the first is the application of financial ratios to data typically contained in the financial statements introduced in Chapter 13. The second is cost-volume-profit analysis, also known as breakeven analysis, which forms a versatile methodology to model the relationship between the scale or level of activity and financial parameters such as revenue, cost, and profit. This chapter aims to achieve the following specific learning outcomes:

- an appreciation of the general activity of financial analysis and the major aspects it seeks to identify in the situation of a business;
- knowledge of basic financial ratios, their categories, and the ability to apply them to available financial data;
- understanding the method of cost-volume-profit analysis and an ability to construct simple linear models of this kind;
- an ability to calculate and interpret the contribution, the contribution margin ratio, and the margin of safety; and
- understanding key terms, synonyms, and accepted acronyms.

INTRODUCTION TO FINANCIAL ANALYSIS

The topic of *financial analysis* concerns the assessment of the commercial viability, stability, and profitability of a business, a part of a business, or any other project of a commercial nature. Where financial analysis draws information from the financial statements introduced in Chapter 13, it is also referred to as *financial statement analysis* or *analysis of finance*. Typical questions addressed by financial analysis include whether to continue or discontinue an operation, whether to invest to expand the scale of an activity, which new technologies to adopt, whether to engage in a research and development project, or to understand if additional funding for the business in the form of loans or outside investment should be sought.

Since a business is likely to have competitors, financial analysis is frequently used to compare different businesses against each other. Moreover, the results of financial analysis are likely to play an important role in encouraging external investors to invest in a business. Given that many large and powerful

DOI: 10.1201/9781003222903-17

businesses are obliged by law to disclose their financial statements,* for example, because they are listed on the stock exchange, a surprising amount of information and insight can be extracted from such statements by outside parties using financial analysis techniques.

The following will briefly introduce these aspects. When applied to the business as a whole, financial analysis is often directed at identifying four aspects: *profitability*, *solvency*, *liquidity*, and *stability*.

Profitability and Solvency

The profitability of a business constitutes its ability to generate income and grow in the short term and long term. Various measures of profit have been introduced in Chapter 2. Chapter 13 has discussed where profit-related information can be found in the financial statements issued by a business.

A precondition of profitability is solvency. In financial management, solvency is the degree to which the current assets of a business exceed the current liabilities of that business. Solvency thus reflects the ability to pay bills, meet commitments, and pursue activities directed at growth. Conversely, *insolvency* is a state of financial distress in which a business (or a person!) is not able to pay bills or meet other financial obligations. In this situation, a business may be forced to wind up, involving a process of voluntary or compulsory liquidation in which the assets the company owns are sold to pay for the business's debt, as outlined in Chapter 8. This usually leads to an acrimonious legal process in which multiple stakeholders lose part or all of their investment.

Liquidity

Liquidity is a property that affects all types of assets, as introduced in Chapter 13. It describes the readiness with which assets can be converted into cash. Holding *liquid assets* is important for businesses (and individuals) because they are the source of the cash flows required to meet any current and future commitments. If a business does not have sufficient access to liquid assets, including cash, it cannot buy raw materials, rent its premises and pay its workforce, despite possibly owning valuable illiquid assets. Liquidity can be defined as follows:

IMPORTANT DEFINITION

Liquidity is the ease with which any asset can be converted into cash without leading to changes in its value. Cash is by definition the most liquid asset.

Fixed assets are generally considered relatively *illiquid*. Such assets may be difficult to sell rapidly due to a low volume of trading activity in the markets for such assets or simply because the number of potential buyers is small. This does not imply that

* Many businesses make financial statements or excerpts of their financial statement freely available on their corporate websites.

illiquid assets are less valuable than liquid assets. Typical examples for illiquid assets include real estate, vehicles, antiques, works of art, collectibles, and shareholdings in businesses that are not traded on a stock exchange.

While this may not be immediately obvious to a non-businessperson, maintaining an appropriate level of liquidity is of the utmost importance for any business. If a business does not hold sufficient liquid assets and it finds itself in a situation in which more money is flowing out than is flowing in, a *cash flow shortage* is inevitable. In consequence, the business may not have enough money to cover payroll and other essential operating expenses. This will cripple its ability to further generate revenues. If the cash flow shortage is not dealt with swiftly and decisively by managers, such as through the prompt liquidation of some assets or by taking out a loan, a *cash flow crisis* will occur. This may lead to the demise of an otherwise viable business. Many healthy businesses have failed due to cash flow crises.

Stability

The stability of a business is its ability to continue operating in the long term. Stability refers to the ability to survive temporary problems such as periods of low sales, lower than ideal levels of funding, the unavailability of important equipment, or the departure of important employees. Where such problems result in financial losses in the long term, the business is considered *unstable*.

In financial terms, the characteristics of stable businesses are that they do not maintain excessive levels of debt, use assets efficiently in their operations, and will return acceptable levels of profit. To be able to respond to emergencies or unforeseen events, such as the global coronavirus pandemic of the early 2020s, stable businesses will usually hold liquid assets in reserve. Having a large number of long-term customers is considered an important contributing factor to the stability of businesses.

FINANCIAL RATIOS

Financial ratios form a common method of analyzing the financial information presented in the financial statements issued by a business. By doing so, it is possible to understand the strengths and weaknesses of the business and how these change over time. This section presents a selection of widely used financial ratios grouped in three major categories: *profitability ratios*, *activity ratios*, and *liquidity ratios*. Since this section will build on the terminology of financial flows introduced earlier in this book, it may be a good idea to briefly revisit the relevant sections in Chapter 2.

As the name suggests, financial ratios are a set of ratios expressing relationships between basic financial parameters that are usually available in the financial statements produced by businesses, as presented in Chapter 13. Financial ratios thus give information on the relative magnitude of two or more financial values. There are many different financial ratios and it should be noted that some are known under alternative names. There is also considerable variation in the way some financial ratios are specified; this chapter aims to present the most common versions.

Profitability Ratios

As the name suggests, profitability ratios measure the use of assets and the control of expenses to generate an acceptable level of profit. Profitability ratios give insight into the question of whether a business produces an attractive financial return from the sales it generates.

A basic ratio evaluating gross profit is the *gross profit to sales ratio*, also known as the *gross margin*. Customarily shown in percentage terms, like many other financial ratios, the gross profit to sales ratio expresses the gross profit as a percentage of sales revenue. If the ratio is positive, the activity of the business is least potentially profitable if other costs can be controlled. A negative ratio indicates that the business will always make a loss unless something can be done about the cost of goods sold (COGS), which is the cost incurred for inputs and direct process costs. A negative gross profit to sales ratio is considered highly problematic for any established business. The gross profit to sales ratio is calculated as follows:

$$\text{Gross profit to sales ratio} = \frac{\text{Gross profit}}{\text{Sales revenue}} \times 100 \qquad (14.1)$$

The *net profit to sales ratio*, also known as the *net margin*, is a measure of the net profit relative to the sales revenue achieved. Since it is based on net profit, it takes operating costs and other costs into account. If this ratio is positive, then the business is profitable. If not, the business is making a loss. Such a financial situation may be recoverable if the costs of operating the company can be controlled, for example, by downscaling operations or cutting research and development expenditure. A change in the net profit to sales ratio over time indicates either a change in the gross profit to sales ratio or a change in the expenses involved in running the company. Usually expressed as a percentage, it is calculated as follows:

$$\text{Net profit to sales ratio} = \frac{\text{Net profit}}{\text{Sales revenue}} \times 100 \qquad (14.2)$$

The *expenses to sales ratio* represents, as the name suggests, the cost of running the company relative to the sales revenue achieved. High values are not desirable in this ratio. Typically, the operating expenses of a business are used to calculate this ratio. An increase in this ratio over time indicates either an increase in the costs of running the business or a deterioration in sales. Both aspects are alarming for managers and suggest that action is urgently required. Again, usually expressed as a percentage, it is calculated as follows:

$$\text{Expenses to sales ratio} = \frac{\text{Expenses}}{\text{Sales revenue}} \times 100 \qquad (14.3)$$

TABLE 14.1
Example Calculation of ROCE

Item from financial statements	Company A	Company B
EBIT ($)	3,837	13,955
Total assets ($)	11,123	115,406
Current liabilities ($)	3,200	29,210
Capital employed (*Total assets – Current liabilities*) ($)	7,923	86,196
Return on capital employed (%)	48.43	16.19

Return on capital employed (*ROCE*) is a group of financial ratios comparing the profit earned by a business against the investment employed to earn this profit. A common version of ROCE reflects the level of capital employed at a particular point of time. The capital employed can be established by subtracting the current liabilities from the total assets, as identified in the balance sheet. Using the term EBIT (see Chapter 2), ROCE is frequently expressed as a percentage and calculated as follows:

$$\text{Return on capital employed} = \frac{\text{EBIT}}{\text{Total assets} - \text{Current liabilities}} \times 100 \quad (14.4)$$

Table 14.1 shows information extracted from the financial statements of two fictitious businesses and illustrates the usefulness of ROCE in judging the performance of a business. As shown, company B shows a far higher profit than company A. However, due to the much smaller level of capital employed, company A exhibits a far higher return on capital employed, indicating a substantial operational advantage.

Activity Ratios

Also known as *efficiency ratios*, activity ratios measure the effectiveness with which a business uses the resources available to it. Only one activity ratio will be considered in this chapter, the *stock turnover ratio*, also known as the *stock turn*. This ratio is a reflection of the efficiency of an operation. More specifically, it shows how many times the inventory held ("stock") is being sold and re-purchased over a given time period. It is calculated by dividing COGS by the average value of the inventory held. Note that this ratio is customarily expressed as a simple number, not as a percentage.

$$\text{Stock turnover ratio} = \frac{\text{Cost of goods sold (COGS)}}{\text{Average inventory}} \quad (14.5)$$

A low stock turnover ratio may indicate *overstocking*, *obsolescence*, or problems with a product line or marketing activities. An excessively high turnover may indicate inadequate inventory levels, which may lead to a loss in business. Generally, a high stock turnover ratio is seen as favorable. This reflects a desire to avoid holding unnecessary inventory and to sell goods as soon as possible.

The appropriate level of stock turn depends heavily on the industry and the nature of the business model adopted. This means that comparisons should only be made against businesses in the same industry and those pursuing a similar approach. For example, a supermarket predominantly selling fast-moving consumer goods (FMCGs) is likely to exhibit a high stock turnover ratio, possibly around 50. Compared to this, a *white goods* retailer, selling durable household goods such as washing machines, would expect a lower stock turnover ratio of around 5. For manufacturing businesses in general, a reasonable stock turn would normally be in the region of 2–8. Nevertheless, many managers believe that stock turn should be increased as far as safely possible.

WORKED EXAMPLE

For further illustration, consider the following numerical example: a small business sells a particular product line, incurring a *COGS* of $50,000 over an accounting period. Its accounts show an opening inventory of $4,000 and a closing inventory of $6,000 in the accounting period. The average inventory level can be obtained by forming the average of the opening and closing inventory ($4,000 + $6,000)/2 = $5,000. By using equation 14.5, the resulting stock turn is calculated at 10, which would be reasonable for a manufacturing business.

Liquidity Ratios

Liquidity ratios shed light on whether a company has sufficient liquid assets to meet its short-term liabilities. A company holding enough liquid assets, generally referred to as *working capital*, is said to have satisfactory liquidity. The *current ratio* is the most commonly used liquidity ratio; it indicates whether the business has a sufficiently large cushion of cash or cash equivalent assets to pay its bills on time. The current ratio can be calculated by simply forming the ratio of current assets over current liabilities:

$$\text{Current Ratio} = \frac{\text{Current assets}}{\text{Current liabilities}} \tag{14.6}$$

For example, if current assets in a business are $150,000 and current liabilities are $50,000, then the current ratio is 3. The current ratio may fluctuate significantly over time as a business pays large bills, for example, to settle debt or to fund new product development. Businesses generally try to have a current ratio of between 1.2 and 2. A current ratio lower than 1 is a warning sign that the business does not have the necessary liquid assets to meet its short-term liabilities. This situation is referred to as *negative working capital* or a *working capital deficit*. Resulting in a cash flow shortage, this requires urgent action to avoid a cash flow crisis. However, businesses will not want to hold too much cash and cash equivalents either since investing this money elsewhere may yield a higher return. To reduce the level of cash held by a business, it can also be paid out to the owners or shareholders as a dividend, further increasing the attractiveness of the business to investors.

A special liquidity ratio is the *acid test*, also known as the *quick ratio*. This ratio takes into account only current assets that are cash or quickly convertible to cash, excluding the inventory held by the business. Thus, the acid test shows if the business can meet its short-term obligations if it cannot sell its inventory. It is calculated as follows:

$$\text{Acid Test} = \frac{\text{Current assets} - \text{Inventory}}{\text{Current liabilities}} \tag{14.7}$$

Quickly convertible assets include marketable securities and accounts receivable. The acid test thus shows whether there are enough liquid assets to be able to pay current liabilities rapidly. As with the current ratio, if its value decreases below 1, the business may find it difficult to operate normally. This would obviously be a clear warning sign to managers, investors, and the creditors of the business.

A Worked Example of Financial Analysis Using Financial Ratios

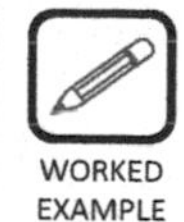

This section provides a worked example of the application of the presented financial ratios to the financial data of two fictitious businesses, Crazy Things Inc. and Spaced-Out Inc. The nature of these businesses is not important but it is assumed that they operate in the same industry. Table 14.2 compares excerpts from the profit and loss statements of both businesses and Table 14.3 compares excerpts from the balance sheets. Note that unlike the balance sheet example presented in Chapter 13, the balance sheets are presented in a vertical format, showing assets above the liabilities and owners' equity.

Further to the information provided in the profit and loss statements and balance sheets, it is known that Crazy Things Inc. had an opening inventory of $2,000 and a closing inventory of $4,000 in the relevant accounting period. Over the same time period, Spaced-Out Inc. had an opening inventory of $7,000 and a closing inventory of $6,000. Table 14.4 presents the calculation for all financial ratios introduced in this chapter. After calculating the financial ratios for both businesses, the next step toward an inter-business financial analysis is to interpret the results.

TABLE 14.2
Profit and Loss Statement Excerpts (Worked Example)

Profit and loss statement entries for the year ended December 31, 2020		
	Crazy Things Inc. ($)	**Spaced-Out Inc. ($)**
Sales Revenues		
Cash sales	25,000	22,500
Credit sales	50,000	91,000
Total sales revenue	75,000	113,500
Cost of goods sold	15,000	17,500
Gross profit	60,000	96,000
Operating expenses		
Salaries	9,000	6,000
Advertising	4,000	5,000
Office Rent	4,000	6,000
Utilities	300	600
Office supplies and consumables	400	200
Depreciation	100	700
Other expenses	300	200
Total operating expenses	18,100	18,700
Operating profit	41,900	77,300
Other expenses		
Interest expenses	3,400	2,000
Income tax expenses	4,500	7,000
Net profit	34,000	68,300

TABLE 14.3
Balance Sheet Excerpts (Worked Example)

	Balance sheet entries, December 31, 2020	
	Crazy Things Inc. ($)	**Spaced-Out Inc. ($)**
Assets		
Current assets		
Cash and cash equivalents	3,000	7,000
Accounts receivable	200	500
Inventory	4,000	6,000
Prepaid expenses	1,500	500
Investment	4,000	2,000
Total current assets	**12,700**	**16,000**
Fixed assets		
Land	10,000	1,200
Buildings and improvements	10,500	1,000
Equipment	3,300	4,000
Less accumulated depreciation	(1,000)	(500)
Other assets		
Intangible assets	3,500	4,000
Less accumulated amortization	(200)	(300)
Total assets	**38,800**	**25,400**
Liabilities and owners' equity		
Current liabilities		
Accounts payable	500	1,000
Notes payable	100	100
Accrued expenses	2,800	500
Deferred revenue	1,200	200

(*Continued*)

TABLE 14.3 (CONTINUED)
Balance Sheet Excerpts (Worked Example)

Balance sheet entries, December 31, 2020	Crazy Things Inc. ($)	Spaced-Out Inc. ($)
Total current liabilities	**4,600**	**1,800**
Long-term liabilities		
Long-term debt	15,000	1,200
Total liabilities	**19,600**	**3,000**
Owners' equity		
Common shares	17,400	15,000
Additional paid-in capital	2,000	3,000
Retained earnings	2,300	4,900
Treasury shares	(2,500)	(500)
Total liabilities and owner's equity	**38,800**	**25,400**

TABLE 14.4
Calculation of Financial Ratios (Worked Example)

Financial ratio	Crazy Things Inc.	Spaced-Out Inc.
Gross profit to sales ratio	$\frac{\$60,000}{\$75,000} = 80\%$	$\frac{\$96,000}{\$113,500} = 85\%$
Net profit to sales ratio	$\frac{\$34,000}{\$75,000} = 45\%$	$\frac{\$68,300}{\$113,500} = 60\%$
Expense to sales ratio	$\frac{\$18,100}{\$75,000} = 24\%$	$\frac{\$18,700}{\$113,500} = 16\%$
ROCE	$\frac{\$34,000 + \$3,400 + \$4,500}{\$38,800 - \$4,600} = 123\%$	$\frac{\$68,300 + \$2,000 + \$7,000}{\$25,400 - \$1,800} = 328\%$
Stock turnover ratio	$\frac{\$15,000}{(\$2,000 + \$4,000)/2} = 5.0$	$\frac{\$17,500}{(\$7,000 + \$6,000)/2} = 2.7$
Current ratio	$\frac{\$12,700}{\$4,600} = 2.8$	$\frac{\$16,000}{\$1,800} = 8.9$
Acid test	$\frac{\$12,700 - \$4,000}{\$4,600} = 1.9$	$\frac{\$16,000 - \$6,000}{\$1,800} = 5.6$

The financial metrics calculated for the worked example presented in Table 14.4 suggest the following conclusions:

- Spaced-Out Inc. is more profitable than Crazy Things Inc., both in terms of net profit and in terms of ROCE. The very high values of ROCE for both businesses are typical of small companies requiring only very limited capital.
- Crazy Things Inc. exhibits a higher stock turnover ratio than Spaced-Out Inc., coupled with a lower level of sales. On this basis, it could be speculated that Crazy Things Inc. is pricing its products too low. The stock turnover ratio for both businesses is not high.
- Crazy Things Inc. has far greater fixed assets than Spaced-Out Inc. and exhibits a higher expense to sales ratio. This suggests that Crazy Things Inc. should consider re-structuring it operations, perhaps by reducing asset ownership.
- Spaced-Out Inc. exhibits a very high current ratio and acid test. This suggests that too much cash is held in the business. This money could be employed more productively elsewhere. It could be reinvested within the business to expand the operations or be paid out as a dividend to the owners.

COST-VOLUME-PROFIT ANALYSIS

A recurring theme in financial analysis is how the various costs to a business respond to the volume of its activity. Answering this question is important in order to develop an understanding of how the profits generated by the business respond to changes in scale. Such aspects are of interest when a business is compared against its direct competitors or benchmarked against businesses in other industries.

The fundamental relationships between revenue, cost, and profit were introduced in Chapter 2. For the analysis of factors relating to profit, revenue, quantity, and cost as a part of financial management, this chapter will introduce a framework known in finance and management as cost-volume-profit analysis or breakeven analysis. This method can be characterized generally as comparing relevant financial relationships that share the same variables, i.e., as mathematical functions. In practice, the objective of such investigations is to identify points at which these functions intersect, i.e. *breakeven*, and to see how rapidly different characteristics, such as costs and revenues, converge or diverge as the volume of an activity changes. As seen in Chapter 5, this form of investigation has significant applications beyond financial analysis, such as in the comparison of alternative technologies. The following sections of this chapter describe how such methods are used in financial analysis. It is also shown how special financial ratios can be applied in this context.

THE COST SIDE

Recall from the section on cost theory presented in Chapter 3 that in the production of a good or service it is possible to distinguish between fixed costs FC that are independent of quantity q and variable costs VC that are dependent on quantity and can be expressed as a function of quantity $VC(q)$. Also, recall that the total cost function $TC(q)$ denotes is the sum of these costs such that:

$$TC(q) = FC + VC(q) \tag{14.8}$$

Labor costs often enter such analyses as a fixed cost since it is assumed that employees are not normally paid according to the level of output they generate. However, if labor costs are treated as process costs, they should enter as variable costs.

In the following, it will be assumed for simplicity that the variable cost function $VC(q)$ is linear, which is likely to be a justifiable assumption if the scope of the investigation is limited. More advanced models can employ quadratic or cubic cost functions. A linear variable cost function $VC(q)$ running through the origin is one in which variable cost is the product of some cost increment b and quantity q:

$$VC(q) = b \times q \tag{14.9}$$

Combined with fixed cost FC, which enters the simple model simply as a constant a, the linear total cost function $TC(q)$ can be stated as:

$$TC(q) = FC + VC(q) = a + b \times q \tag{14.10}$$

Two further cost functions are of interest, average cost $AC(q)$, also known as unit cost, and marginal cost $MC(q)$, which reflects the cost of the last unit of output, as defined in Chapter 3. AC can be obtained by dividing total cost over q:

$$AC(q) = \frac{TC(q)}{q} = \frac{a + b \times q}{q} = a \times q^{-1} + b \tag{14.11}$$

The marginal cost function $MC(q)$ reflects the rate with which the total cost function $TC(q)$ changes as the quantity changes. Thus, the cost contribution of the qth unit of output, where $q \in \mathbb{N}$, can be obtained by forming the derivative of the total cost function, $TC'(q)$, assuming that $TC(q)$ is differentiable:

$$MC(q) = TC'(q) = \frac{d}{dq}\left[TC(q)\right] \tag{14.12}$$

Using the linear total cost model, marginal cost is, therefore, simply a constant reflecting the variable cost element b for each additional unit:

$$MC(q) = \frac{d}{dq}\left[a + b \times q\right] = b \tag{14.13}$$

The Revenue Side

To construct a simple revenue model, it is assumed that total revenue TR is, like total cost, a linear function of quantity q. Linearity in the total revenue function $TR(q)$ implies that the price does not change as quantity chances. This can only be the case if the business under investigation is a *price taker*, i.e. that it faces a horizontal demand curve and has no way of affecting the market price p it faces. Recall from Chapter 3 that the average revenue function AR equates to the demand curve in the market in this situation. If average revenue is constant, it also equates to the marginal revenue MR obtained from the sale of the last unit of output. In the simple model presented in this chapter, $TR(q)$ does not have a constant term since zero units transacted will not yield any revenue for the business. Under the stated conditions, $TR(q)$ is simply a function of q and p:

$$TR(q) = p \times q \tag{14.14}$$

Analogous to the cost side, it is possible to specify two further revenue functions of interest, average revenue $AR(q)$ and marginal revenue $MR(q)$, which reflects the revenue obtained of the last unit of output, as defined in Chapter 3. $AR(q)$ can be obtained by dividing $TR(q)$ over q:

$$AR(q) = \frac{TR(q)}{q} = \frac{p \times q}{q} = p \tag{14.15}$$

The marginal revenue function $MR(q)$, which is the revenue contribution of the qth unit of output, where $q \in \mathbb{N}$, is obtained by forming the derivative of the total revenue function, $TR'(q)$, assuming it is differentiable:

$$MR(q) = TR'(q) = \frac{d}{dq}\left[TR(q)\right] \tag{14.16}$$

In the simple model with linear revenue, $MR(q)$ is, therefore:

$$MR(q) = TR'(q) = \frac{d}{dq}\left[p \times q\right] = p = AR(q) \tag{14.17}$$

Bringing the Cost Side and the Revenue Side Together

Once the total cost function and the total revenue function are understood, they can be brought together in a model. Showing $TC(q)$ and $TR(q)$ in the same chart with q on the horizontal axis and a money term reflecting cost and revenue as the vertical axis, results in the *breakeven chart.* Figure 14.1 (a) shows the simple linear model developed in the previous sections graphically. Figure 14.1 (b) shows MC, MR, and AR, which are all constant in this simple model and are therefore shown as horizontal lines. Note that $MR = AR$, which also equates to the demand curve as the business is a price taker. This implies that if the business sets prices lower or equal to p it will sell as much output as it can generate. If it prices higher it will sell nothing. Likewise, the marginal cost equates to the slope of the total cost function $TC(q)$.

The point at which the $TC(q)$ and $TR(q)$ intersect yields the point of breakeven q^*. Below this level of activity, the business is making a loss, shown as the shaded area in Figure 14.1 (a). Above this level of activity, the business is making a profit, shown

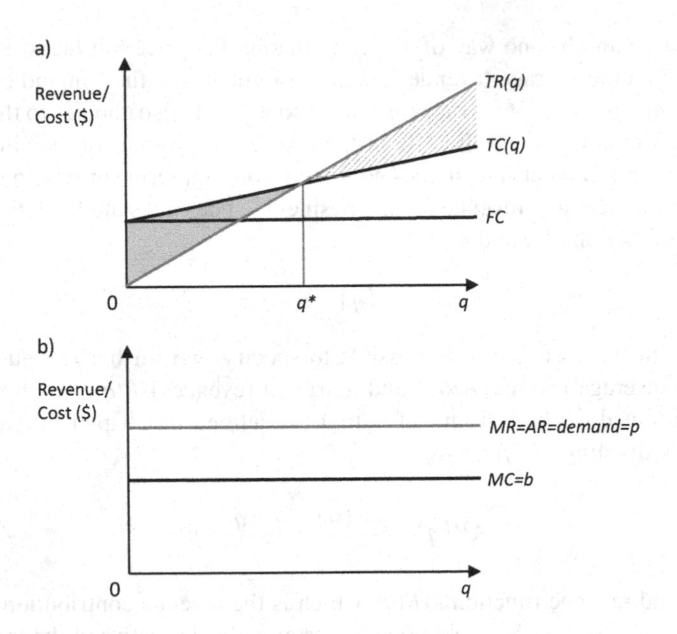

FIGURE 14.1 Breakeven chart (a) and constant model elements (b)

as the hatched area. At quantity q^*, the following condition is satisfied, which can be used to identify q^* mathematically:

$$TR\left(q^*\right) = TC\left(q^*\right) \tag{14.18}$$

In the simple example with linear costs and revenue, solving for q^* yields:

$$q^* = \frac{a}{p-b} \tag{14.19}$$

When comparing the constructed model to the economic theory of the firm introduced in Chapter 3, it becomes clear that this case is rather special. This is because MC and MR are both constants and $MC < MR$ if $b < p$. This means that the functions will never intersect at any quantity and that the first-order condition of profit maximization discussed in Chapter 3, $MR = MC$, will never be satisfied, no matter how large the quantity produced becomes. This situation is apparent in Figure 14.1 (b). Coupled with the assumption of the business being a price taker, which is implicit in the linearity of the total revenue function, this suggests that this model is limited to companies that are relatively small compared to an overall market. It also suggests that rational businesses should be expanding the quantity as far as possible and that products should be priced at market price p.

FINANCIAL METRICS AND RATIOS USED IN COST-VOLUME-PROFIT ANALYSIS

Two financial ratios are applied routinely to breakeven analyses of the sort presented in this chapter: *the contribution margin ratio* and the *margin of safety*. This section will discuss each in turn.

Often managers will want to know what *contribution* the sale of the marginal unit has made to overall profit. This metric, also known as the *contribution to margin*, can be obtained through the *marginal profit function*. First, recall from Chapter 3 that profit π of the firm results from the difference between total revenue and total cost. Therefore:

$$\pi(q) = TR(q) - TC(q) \tag{14.20}$$

The next step is to form the derivative of the profit function, $\pi'(q)$, which yields the contribution of the qth unit of output to profit, assuming $q \subset \mathbb{N}$ and that $\pi(q)$ is differentiable:

$$\text{Contribution} = \pi'(q) = \frac{d}{dq}\left[TR(q) - TC(q)\right] \tag{14.21}$$

Proceeding with the simple linear cost-volume-profit model introduced in the previous section, the profit function can be specified as:

$$\pi(q) = p \times q - (a + b \times q) = q(p - b) - a \tag{14.22}$$

The marginal profit function, and with it the contribution of the qth unit, can thus be expressed as follows, which is constant and simply the difference between the price p and the variable cost term b:

$$\pi'(q) = \frac{d}{dq}\left[q(p - b) - a\right] = p - b \tag{14.23}$$

The contribution can also be expressed as the *contribution margin ratio.* In this case, the contribution is divided by the price, thereby expressing the extent to which revenue is eaten away by the variable cost. Usually expressed as a percentage, the contribution margin ratio is specified as follows:

$$\text{Contribution margin ratio} = \frac{\text{Contribution}}{\text{Price}} \times 100 \tag{14.24}$$

For the linear cost-volume-profit model, the contribution margin ratio is thus:

$$\text{Contribution margin ratio} = \frac{p - b}{p} \times 100 \tag{14.25}$$

A further financial ratio of interest is the *margin of safety*. This metric is indicative of the overall strength of a current or planned position of the business and allows managers to identify what amount is or can be gained over or below the point of breakeven q^*. It is calculated by first forming the difference of total revenue arising from an actual or planned quantity q_{margin} and the total revenue arising at q^*. This expression is divided over the total revenue at q_{margin}. It is normally expressed in percentage terms as follows:

$$\text{Margin of Safety} = \frac{TR(q_{margin}) - TR(q^*)}{TR(q_{margin})} \times 100 \tag{14.26}$$

A Worked Example of Cost-Volume-Profit Analysis

WORKED EXAMPLE

A new and growing business in the medical sector, Meditech Inc., is being analyzed financially. The data shown in Table 14.5 have been identified. Additionally, it has been reported that the total revenue and total cost functions are linear. You are asked to establish the breakeven quantity, breakeven revenue, contribution, and margin of safety.

The first step is to identify the variable cost VC per unit. It is known that price p = \$10 and that total revenue TR = \$500,000. Therefore, the quantity $q_{reported}$ is given by \$500,000/\$10 = 50,000. Hence, $VC = TVC/q_{reported}$ = \$350,000/50,000 = \$7. This information allows the definition of the total cost function $TC(q)$ and the total revenue function $TR(q)$:

$$TC(q) = a + b \times q = \$100{,}000 + \$7 \times q \tag{14.27}$$

$$TR(q) = p \times q = \$10 \times q \tag{14.28}$$

At breakeven, where $TR(q^*) = TC(q^*)$, the breakeven quantity q^* is obtained by solving for q^*:

TABLE 14.5
Meditech Inc. Key Financial Data

Sales revenue, $TR(q_{reported})$	\$500,000 per year
Total variable costs, TVC	\$350,000 per year
Fixed costs, FC	\$100,000 per year
Price, p	\$10

$$\$10 \times q^* = \$100{,}000 + \$7 \times q^* \quad (14.29)$$

$$q^* = 33{,}333 \quad (14.30)$$

Breakeven revenue is simply $TR\left(q^*\right)$, which is:

$$p \times q^* = \$10 \times 33{,}333 = \$333{,}330 \quad (14.31)$$

As defined, for linear cost-volume-profit models, the contribution is obtained by subtracting the unit variable cost VC from the price p:

$$p - VC = \$10 - \$7 = \$3 \quad (14.32)$$

As the final step, the margin of safety is obtained using the reported total revenue $TR(q_{reported})$ and $TR(q^*)$:

$$\frac{\$500{,}000 - \$333{,}330}{\$500{,}000} \times 100 = 33.33\% \quad (14.33)$$

Following this, you are asked to evaluate, the effect on both the breakeven quantity and the margin of safety of a technical upgrade to the manufacturing line. The new machinery will increase annual fixed costs by \$100,000 while reducing unit variable costs to \$2.

To develop an answer, the first step is to note that the $TR(q)$ remains unchanged. $TC(q)$, however, changes to $TC_{\text{new}}\left(q\right) = \$200{,}000 + \$2 \times q$. This is solved for the new breakeven quantity q^*_{new}, which also allows the calculation of *margin of safety*$_{\text{new}}$:

$$q^*_{\text{new}} = 25{,}000 \quad (14.34)$$

$$TR\left(q^*_{\text{new}}\right) = p \times q^*_{\text{new}} = \$10 \times 25{,}000 = \$250{,}000 \quad (14.35)$$

$$\text{Margin of safety}_{\text{new}} = \frac{\$500{,}000 - \$250{,}000}{\$500{,}000} \times 100 = 50\% \quad (14.36)$$

As can be seen, the technological improvement would lead to an improvement in the margin of safety from 33% to 50%, which indicates that the stability of the business could be increased through the upgrade.

Limitations in the Presented Methods of Cost-Volume-Profit Analysis

This chapter has treated the cost functions and revenue functions as linear. In reality, there are forces in business and economics that cause nonlinearities in such functions. A particularly powerful force shaping cost functions are economies of scale (as outlined in Chapter 3) and economies of scope (outlined in Chapter 4). Both are likely to reduce the slope of the total cost function as output expands at low levels. At higher levels of output, beyond some point, the opposite will be the case if there are diseconomies of scale. In this situation, the slope of the total cost function will increase as output expands beyond a certain level. Similarly, as discussed in this chapter, the conditions that lead to a linear total revenue function are quite specific. For example, if the business has any kind of market power, as discussed in Chapter 3, and thus faces a downward-sloping demand curve, the slope of the total revenue function will decrease as total quantity increases.

A further limitation of breakeven analyses is their reliance on a clear distinction between fixed costs and variable costs. In reality, many supposedly fixed costs increase as certain capacity or output thresholds are crossed. For example, if labor enters as a fixed cost and an increase in output requires the business to hire an additional employee then the fixed cost faced by the business increases. This leads to so-called *stepped fixed costs*.

Finally, cost-volume-profit analyses are difficult to carry out in operations that generate multiple products or services alongside each other. In this case, the sale of one product might affect the sale of another if there are relationships of complementarity or substitutability, as discussed in Chapter 4. Moreover, it may be difficult to apportion fixed costs incurred by the whole operation to individual products. In combination, these limitations can make the development of accurate cost-volume-profit analyses challenging.

CONCLUSION

After introducing the objectives of financial analysis and highlighting both the free availability of high-quality data in financial statements and the concept of liquidity as core ideas, this chapter has provided an overview of important methods used in financial analysis and has also frequently shown how they relate to the other topics covered in this book. Two broad methodologies were presented, the application of financial ratios to financial data and the construction cost-volume-profit models. Financial ratios are a widely used form of financial analysis that can be applied to the published financial statements, which are available for many businesses. A surprising amount of information can be extracted from financial statements in this way. Cost-volume-profit models, on the other hand, are extraordinarily versatile and can be used in many ways to understand the financial implications of decisions in business.

Most textbooks on managerial accounting cover the methods presented in this chapter. An extensive and useful treatment of financial ratios is provided in *Principles of Managerial Finance* by Chad Zutter and Scott Smart (2019). Most management and finance textbooks introduce cost-volume-profit analysis. An accessible yet discerning introduction is presented in *Management Accounting for Decision Makers* by Pater Atrill and Eddie McLaney (2018).

REVIEW QUESTIONS

1. Who can use financial analysis techniques to assess the financial situation of a business?

   (Question type: Multiple choice)

   ○ The managers of the business
   ○ Potential investors, but only after signing a non-disclosure agreement
   ○ Potential investors, without signing a non-disclosure agreement
   ○ Managers and auditors
   ○ Anyone with access to the financial statements issued by the business

2. Which of the following aspects is financial analysis normally used to investigate?

   (Question type: Multiple response)

   ☐ Financial budgets
   ☐ Competitive performance
   ☐ Stability
   ☐ Strategy
   ☐ Solvency
   ☐ Accountability
   ☐ Performance

3. Which of the following best characterizes the property of liquidity?

   (Question type: Multiple choice)

   ○ The value of an asset
   ○ The importance of an asset to meet financial obligations
   ○ The ease with which an asset can be converted into cash
   ○ The extent of financial commitments of a business
   ○ The financial reserves of an organization

4. Match the below financial ratios to the appropriate descriptions.

   (Question type: Matrix)

	Gross profit to sales ratio	Current Ratio	Return on capital employed	Expenses to sales ratio
Expresses how well resources are utilized	○	○	○	○
Measures how well the costs are controlled	○	○	○	○
Expresses gross profit as a percentage of revenue	○	○	○	○
Measures the level of liquid assets in the business	○	○	○	○

5. You are analyzing a business that produces car seats. You are concerned that the design of the product sold by the company is outdated. You have been provided the following information about the latest accounting period:
   - The business has sold car seats worth $750,000
   - It has incurred input costs of $50,000
   - It has three warehouses with a total area of 500,000 square feet
   - It has made a gross profit of $290,000
   - The accounts show an opening stock of $14,000
   - The accounts show a closing stock of $22,000

   (Question type: Calculation)
   Calculate the financial metric that shows how frequently inventory is being sold and purchased over a given time period (ignoring irrelevant data).

6. You are investigating a company that shows the following balance sheet excerpt:

Balance sheet entry	Value
Fixed assets	
Equipment at cost	$10,000
Less depreciation to date	$8,000
Current assets	
Inventory	$64,000
Debtors	$60,000
Cash in bank	$7,000
Current liabilities	
Creditors	$10,000

   (Question type: Calculation)
   Calculate the acid test.

7. Complete the following figure showing a break even chart by inserting the correct labels.
   (Question type: Labeling)

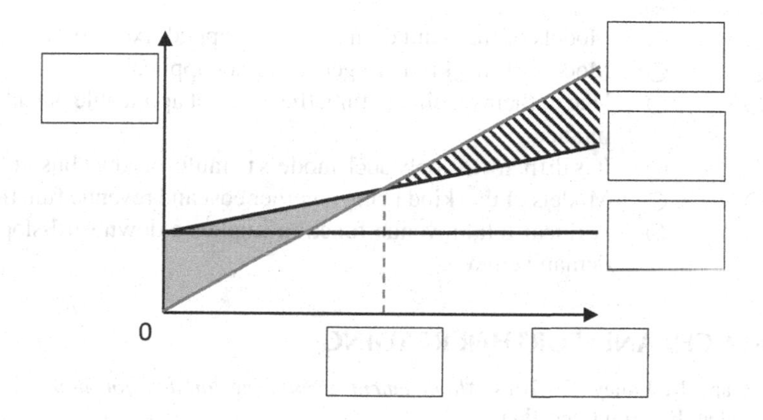

FIGURE 14.2 Identify the appropriate labels

8. A financial analysis is performed on a business. So far, the analysis has shown that the total revenue and total cost functions are linear. The information is available for the most recent accounting period:
   - Sales revenue of $2,500,000
   - Total variable costs of $750,000 and fixed costs of $125,000
   - The product has been sold at a price of $500

(Question type: Calculation)
Calculate the contribution of each unit sold.

9. You have been hired as an engineering consultant by a printing business that manufactures note pads. The business provided the following financial information for the most recent accounting period:
   - Sales of $300,000
   - Total variable costs of $40,000 and fixed costs of $125,000
   - The product has been sold at an average price of $0.60
   - The business reports that the total revenue and total cost functions are linear

You are proposing a technical improvement that will allow the business to halve its total variable costs.
(Question type: Calculation)
Compute the margin of safety after the improvement.

10. Which of the following statements about linear cost-volume-profit models as presented in this chapter are true or false?
    (Question type: Dichotomous)

True	False	
O	O	Models of this kind do not reflect stepped fixed costs
O	O	Models of this kind are generally not applicable
O	O	Due to their specific nature, they are not applicable outside of finance
O	O	It is difficult to apply such models to multi-product businesses
O	O	Models of this kind bring together cost and revenue functions
O	O	A linear total revenue function implies a downward-sloping demand curve

REFERENCES AND FURTHER READING

Atrill, P. and McLaney, E., 2018. *Management accounting for decision makers.* 9th ed. London: Pearson Education.

Zutter, C.J. and Smart, S.B., 2019. *Principles of managerial finance.* 15th ed. London: Pearson Higher Education.

15 Evaluating Flows of Money

OBJECTIVES AND LEARNING OUTCOMES

This chapter introduces methods that can be used to evaluate cash flows arising over time from business activity, investments, or projects. After presenting a range of simplistic methods, the chapter introduces a group of extremely widely used techniques known as compounding methods. This is accompanied by an introduction to a range of relevant concepts, including time preference, interest rates, compound interest, discount rates, net present value, and various rates of return. This chapter aims to achieve the following learning outcomes:

- an appreciation of how cash flows can be expressed as a series of flows of money occurring over time;
- the ability to apply simplistic, shape-based techniques to assess cash flows;
- understanding basic financial concepts, such as time preference, interest rates, compound interest, discount rates, and present value;
- the ability to discount sequences of simple cash flows to net present value and to determine attractiveness based on the minimum acceptable rate of return;
- the ability to construct the internal rate of return from simple sequences of cash flows and to be able to interpret its meaning;
- the ability to model the effect of constant rates of inflation on compounding methods; and
- understanding key terms, synonyms, and accepted acronyms.

MONEY AND TIME

As explored at the start of this book, most managers and shareholders would agree that the main objective of a business is to be as profitable as possible. At the same time, managers have only limited resources at their disposal. This leads to the problem that in a business – and in fact in most other situations in life – rational decision-making is governed by opportunity costs. Recall from Chapter 3 that opportunity costs represent the loss of forgoing the benefit attached to the next best alternative choice or course of action.

To successfully deal with such decisions, the aspect of time often plays a decisive role. This can be illustrated by asking a simple question: should one prefer to receive $1,000 now or $1,000 in five years' time? By the standards of rationality used in business, every sensible manager would want to be given this sum immediately. After all, who can guarantee that this promise will be honored five years into the

DOI: 10.1201/9781003222903-18

future? As economist John Maynard Keynes famously wrote, "in the long run we are all dead".*

Nowhere is this idea more important than in the world of finance and in related disciplines such as insurance, financial economics, and actuarial science, which is the discipline that uses mathematics and statistics to assess risk. Questions of this kind are also prominent in all areas of business. For example, managers frequently face the problem of allocating scarce financial resources to specific activities or projects in order to ensure the survival of the business or to pay an attractive dividend to shareholders. In this, it is not only important how large expected benefits are, but crucially also when they will materialize. This is captured by the concept of *time preference*, which can be defined as follows:

IMPORTANT DEFINITION

Time preference is the current relative value assigned by a person to receiving a good, service, or object at an earlier time compared with receiving it at a later time.

Time preference must thus be taken into account in the evaluation of activities, transactions, or projects if they promise to generate cash flows in the future, especially if these benefits arise in the distant future. Known as the *time value of money*, changes to the valuation of money based on time preference are of particular importance for life cycle-based views of various processes and activities, as presented in Chapter 6.

SIMPLE METHODS OF EVALUATING CASH FLOWS OVER TIME

In many cases, it will be necessary for managers to compare options that will result in different flows of money over time. To illustrate this challenge, Table 15.1

TABLE 15.1
Three Alternative Projects with Identical Cumulative Cash Flows after Year 5

	Project A		Project B		Project C	
EOY	Cash flow ($)	Cumulative cash flow ($)	Cash flow ($)	Cumulative cash flow ($)	Cash flow ($)	Cumulative cash flow ($)
1	–20,000	–20,000	–32,500	–32,500	–15,000	–15,000
2	6,250	–13,750	7,500	–25,000	–15,000	–30,000
3	8,750	–5,000	12,500	–12,500	20,000	–10,000
4	10,000	5,000	12,000	–500	12,500	2,500
5	8,750	13,750	10,500	10,000	10,000	12,500
6	6,250	20,000	10,000	20,000	7,500	20,000

* Keynes (1923, p. 80).

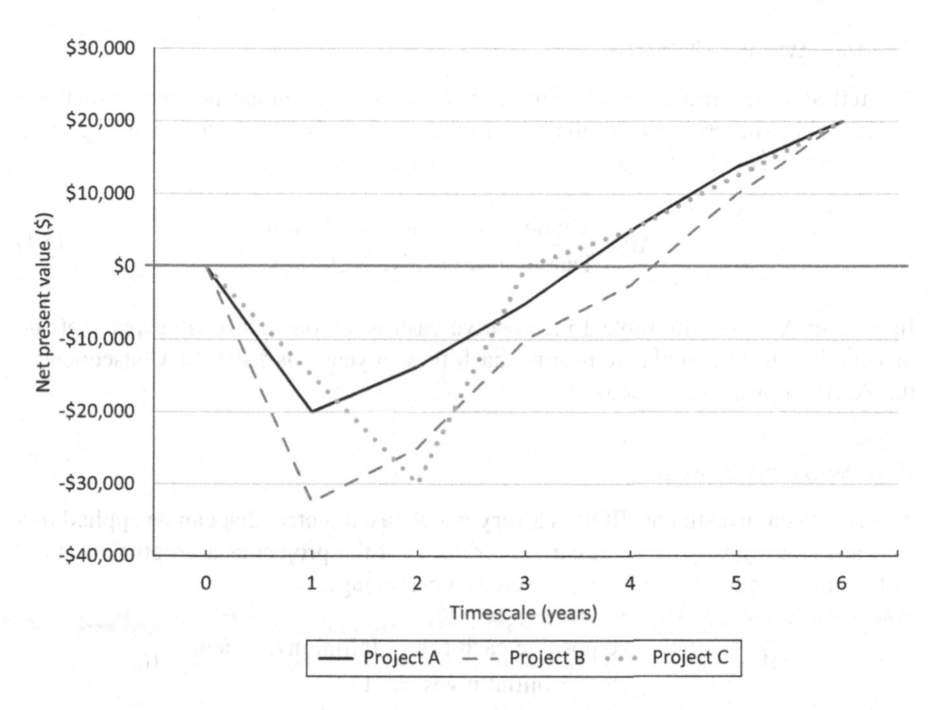

FIGURE 15.1 Cumulative cash flow profiles of the three projects presented in Table 5.1

summarizes the cash flows exhibited by three alternative projects, each with a duration of six years. The cash flows are assumed to occur at the end of each year and start with an investment at the end of year 1 (EOY 1). This example is contrived so that all projects show a total cumulative cash flow of $20,000 at the end of year 6. Figure 15.1 shows the cumulative cash flows arising from the three project options in graphical form.

Chapter 6 has presented a group of shape techniques that can be used to interpret such cash flow data, using metrics such as I_{MAX}, R_{MAX}, the end ratio, the average rate of investment, the average rate of recovery, and the average rate of return. Going beyond the metrics presented in Chapter 6, additional simplistic methods that can be employed to analyze the presented projects are the *payback period*, *annual average proceeds* (*AAP*), and the *return on investment* (*ROI*).

Payback Period

The payback period assumes that there is an initial investment, i.e. a negative cash flow, which will be recouped from the subsequent revenues that are positive cash flows. In project A shown in Table 15.1, the initial investment is $20,000. The initial outlay is recovered at EOY 4, when the cumulative cash flow exceeds zero. Note that in project C, the initial investment takes place in the first two years, so the start of the payback period is delayed by one year. Conceptually, the payback period is similar to *time to breakeven* t_{BE} presented in Chapter 6, with the difference that the payback period is normally expressed in years.

Annual Average Proceeds

Annual average proceeds (AAP) measures the average annual positive cash flows generated by the project after the initial investment. It is similar to the average rate of return presented in Chapter 6 and is calculated as follows:

$$\text{AAP} = \frac{\text{Cumulative positive cash flow}}{\text{Duration of positive cash flows}} \tag{15.1}$$

In project A shown in Table 15.1, positive cash flows occur over five years of the project. The total cumulative positive cash flow in year 6 is $40,000. Consequently, the AAP for project A is $8,000.

Return on Investment

The return on investment (ROI) is a very widely used metric that can be applied in a forward-looking way by expressing the returns of the project as a proportion of the initial investment*, normally expressed as a percentage:

$$\text{ROI} = \frac{\text{Cumulative positive cash flow} - \text{Initial investment}}{\text{Initial investment}} \times 100 \tag{15.2}$$

In project A in the example, the cumulative positive cash flows amount to $40,000. The initial investment is $20,000, so the ROI amounts to 100%.

Making a Decision Based on Different Metrics

A manager faced with the task of evaluating different cash flow profiles will first select an appropriate range of metrics, such as those defined. A simple way to identify the most attractive option is to assign a numerical score based on rank for each metric, for example, by assigning a value of "3" to the best result, "2" to the next lower result, etc. Assuming that all metrics carry equal weight, an overall score as a measure of attractiveness can be calculated by summing up these scores. This is done in Table 15.2, identifying project C as the most attractive option, due to the shortest payback period and the highest AAP.

The decision to pick project C in this example has been affected by the two-year initial investment period, resulting in a shortened payback period and a shorter duration of positive cash flows, in turn resulting in higher AAP. While methods of this kind are easy to apply and may be effective as tools for managerial decision-making

* ROI is also frequently applied in a backward-looking way to evaluate historical cash flows. In this case, the following, more generally termed, specification is used:

$$\text{ROI} = \frac{\text{Current value of an investment} - \text{Cost of the investment}}{\text{Cost of the investment}} \times 100$$

TABLE 15.2
Three Alternative Projects with Identical Cumulative Cash Flows after Year 5

	Project A		Project B		Project C	
Metric	**Value**	**Score**	**Value**	**Score**	**Value**	**Score**
Payback period	3 years	2	4 years	1	2 years	3
AAP	$8,000	1	$10,500	2	$12,500	3
ROI	1.00	3	0.62	1	0.67	2
Total score	–	6	–	4	–	8

in some situations, their weakness is that they do not address the issue of time preference. In fact, in the example provided, the time preference exhibited by the decision-maker is implicit in the weighting attached to the payback period when scoring the available options.

EVALUATING CASH FLOWS USING COMPOUNDING METHODS

More advanced analyses can be constructed by acknowledging that negative cash flows result in opportunity costs, since the business will have to use cash that could be invested in some other way, thereby forgoing another return. Moreover, there may be financial costs due to borrowing that must be factored in. Two concepts are indispensable when reasoning about such costs, the *interest rate*, and the *discount rate*.

THE INTEREST RATE

Stated generally, an interest rate is an amount of money – the *interest* – due per period of time, as a proportion of an amount loaned, borrowed, or deposited. The original amount of money loaned, borrowed, or deposited is known as the *principal sum* or simply the *principal*. Interest rates can thus be defined as follows:

An interest rate is the proportion of a principal sum that is charged as interest to the recipient of the principal sum, typically expressed as an annual percentage.

IMPORTANT DEFINITION

Following from this, the total interest payable depends on the principal sum, the interest rate, the frequency with which interest is allocated, and the length of time over which the principal sum is loaned, borrowed, or deposited. Most frequently, interest rates are expressed as annual interest rates stating the interest accruing over one year.

It is crucial to note that in many fields such as banking, finance, and economics, the received interest is normally assumed to be reinvested, rather than paid out. This implies any interest received is earned on the principal sum and all previously accumulated interest. This practice is known as *compound interest* and it is an essential practice in the world of commerce. Calculating the future value of current sums in this way is also known as *compounding forward.*

The most important technique for the calculation of compound interest is referred to as *periodic compounding.** Here, the total accumulated value of the investment V is formed by the principal sum P, the annual interest rate r, and the duration of the investment in years t:

$$V = P(1+r)^t \tag{15.3}$$

Correspondingly, the total interest generated I is the final value V minus the initial principal sum, so that:

$$I = P(1+r)^t - P \tag{15.4}$$

A numerical example for the calculation of V is straightforward. If \$100 is invested for five years at an annual interest rate of 8%, the investment will return a final value of $V = \$100 \times (1+0.08)^5 = \$146.93$. The total interest amounts to $I = \$100 \times (1+0.08)^5 - \$100 = \$46.93$. This example shows that compound interest can lead to rapid, nonlinear, gains in the value of investments.

If interest is charged on a loan that is repaid to the lender in regular payments over time, for example monthly, the loan is referred to as *installment debt.* The process of repaying debt through principal and interest payments over time, lowering the balance of the debt, is known as generally known as *amortization.* Note that in financial accounting, *amortization* has a different meaning, referring to the loss of value of intangible assets over time, as presented in Chapter 13.

The Discount Rate

In finance, there is a conceptual "opposite" to interest rates. Instead of looking at investing a principal P and eventually obtaining a total accumulated value V, it is often essential to be able to calculate how much money needs to be invested now in order to receive a certain total accumulated amount at a specified time in the future. For example, it might be necessary to know how much money needs to be invested now to obtain \$100 in five years' time.

* The compounding methods presented in this chapter can also be constructed in their non-periodic, continuous forms. These methods are mathematically more complex, exceeding the scope of this book.

In finance, such questions are treated using the technique of *discounting to present value* (*PV*). The analog to the interest rate used in this practice is known as the discount rate. Unlike compound interest, discounting to PV *compounds backward*, such that the discount rate reflects the return that could be obtained per unit of time, normally per year, on an investment with equal risk. It can be defined as follows:

The discount rate is the interest rate used to determine the present value of future cash flows.

Because it forms a variation of the concept of the interest rate, the symbol used for the discount rate is also r. Thus, the present value of a sum S, denoted by S_{PV}, can be calculated in a periodic technique using the discount rate r and the time t at which S is paid out:

$$S_{PV} = S\frac{1}{(1+r)^t} = S(1+r)^{-t} \quad (15.5)$$

It is helpful to note that the term $(1 + r)^{-t}$ is known as the *discount factor*. Specifying the the discount factor as a separate term avoids confusion when calculating the present value of cash flows.

Again providing a simple numerical example, the present value of $100 receivable in three years' time subject to a discount rate of 6% amounts to $S_{PV} = \$100 \times (1+0.06)^{-3} = \$83.96$. The interpretation of this result is that a rational individual applying a discount rate of 6% would be indifferent between receiving $100 in three years' time or $83.96 now.

DISCOUNTING TO NET PRESENT VALUE AND THE MINIMUM ACCEPTABLE RATE OF RETURN

A family of techniques used by managers and financial analysts for the evaluation of flows of money is based on the concept of the rate of return. The rate of return is an expression of the benefits received from activities or investments, normally arising annually. The rate of return can be used to compound forward historical cash flows to establish their present value. By compounding forward, it is, for example, possible to calculate the present value of a sum obtained in the past or to measure the opportunity cost of not having invested a specific sum of money at a specific point. Conversely, the rate of return can also be used to compound backward cash flows occurring in the future to establish their value in the present.

It is important to note that the idea of discounting to PV applies to both inflows and outflows of cash. For example, if an investor knows that they will be invoiced for $100 in 5 years' time then they could meet that requirement now by investing $62.09 at 10%

interest. Vice versa, if they were to be paid $100 in 5 years, that would be equivalent to being paid $62 now and investing this amount at 10% for the same period.

WORKED EXAMPLE

To illustrate the process of discounting future cash flows to present value, Table 15.3 presents the calculation of the discounted annual cash flows of a modified version of project A discussed in Table 15.1. In this example, 18% is used as a discounting rate. For simplicity, it is also assumed that the investment takes place immediately (EOY 0). In this case, the initial investment, as a cash flow that takes place immediately, is not discounted because the discount factor is 1 (since $(1 + r)^{-0} = 1$). As can be seen from Table 15.3, the project is assumed to generate positive cash flows in every year after the initial investment in Year 0.

TABLE 15.3
Discounting of Cash Flows in a Modified Version of the Project A, Based on a Discount Rate of 18%

EOY	Cash flow ($)	Discount factor	Cash flow at PV ($)	Cumulative PV ($)
0	–20,000	1.00	–20,000	–20,000
1	6,250	0.85	5,297	–14,703
2	8,750	0.72	6,284	–8,419
3	10,000	0.61	6,086	–2,333
4	8,750	0.52	4,513	2,180
5	6,250	0.44	2,732	4,912

Subject to the given discount rate (18%) and the cash flow profile shown in Table 15.3, the project cash flows, including the initial investment, can be added up to form a cumulative present value of $4,912. This value is referred to as the *net present value* (*NPV*). It is worth highlighting that the amount of information that has flowed into this value is significant. It contains information on the magnitude of specific cash flows, their direction, and their timing. Armed with this technique, managers can proceed with the estimation of the net present value of the cash flows associated with similar projects in order to select the most attractive option. However, the wide applicability of discounting to NPV means that it can also be used to compare cash flows resulting from very different investments with dissimilar durations.

A frequent criticism of the method of discounting to NPV presented in this section is that it does not yield an absolute indication of the attractiveness of an investment but offers a comparative measure against an arbitrary interest rate *r*.

This thought forms the basis for a project selection technique in which *r* is specified as the *minimum attractive rate of return* (*MARR*), also known as the *minimum acceptable rate of return*, the *required rate of return*, the *hurdle rate*, or the *cut-off rate*. In this technique, it is assumed that managers are able to identify a discount rate at which all conditions required for the project to be worthwhile are satisfied. This includes the repayment of the money invested, covering the costs of the investment, compensating for uncertainty, and the generation of a sufficient additional profit.

Returning to the example shown in Table 15.3, it is assumed that for the project to be deemed attractive, the MARR is set at 20%. Table 15.4 presents the data used to calculate the NPV of the specified cash flows evaluated at the MARR. Since the discounted value of the cash flows is positive, and quite significantly so at $3,803, managers would take this as an indication that the project is promising. If it is the only option considered, managers would likely go ahead with the project on this basis.

THE INTERNAL RATE OF RETURN

A method that does away with the need to determine a rate of return outside of such models is the *internal rate of return* (*IRR*). While the method may initially appear complex and is difficult to solve mathematically, its logic is closely related to that of NPV. Alongside discounting to NPV, the calculation of IRR is a frequently used tool in the evaluation of cash flows arising from projects or investments. The most intuitive way to explain the IRR is by providing a numerical example.

TABLE 15.4
NPV of the Cash Flows of Project A Evaluated at a MARR of 20% (Worked Example)

EOY	Cash flow ($)	Discount factor	Cash flow at PV ($)	Cumulative PV ($)
0	–20,000	1.00	–20,000	–20,000
1	6,250	0.83	5,208	–14,792
2	8,750	0.69	6,076	–8,715
3	10,000	0.58	5,787	–2,928
4	8,750	0.48	4,220	1,291
5	6,250	0.40	2,512	3,803

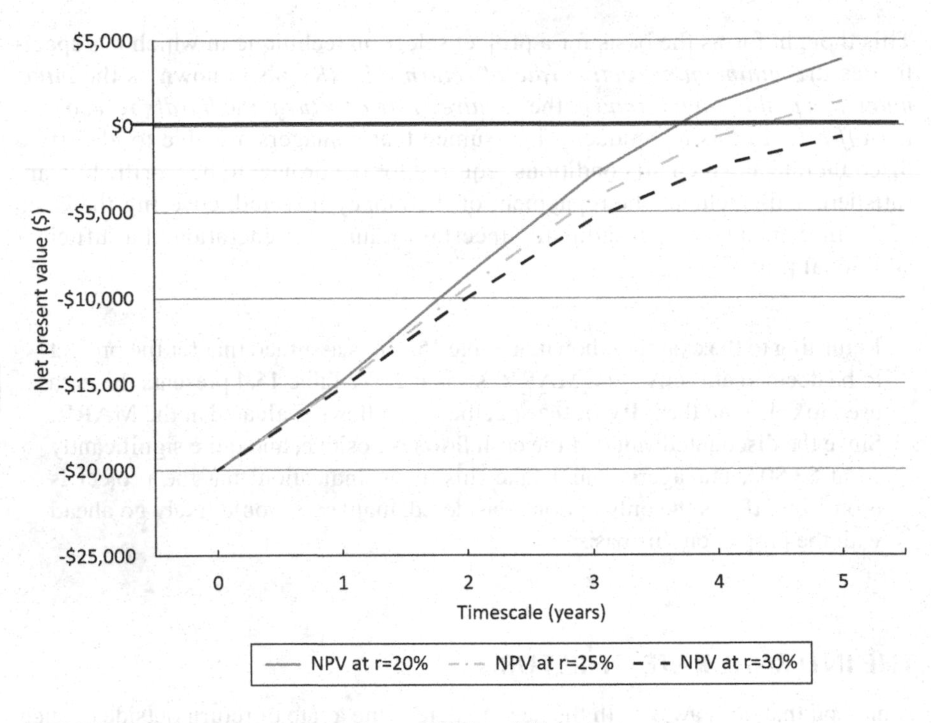

FIGURE 15.2 Graphical comparison of NPV at different rates of return

As discussed, the net present value at the end of a project or investment can be positive, zero, or negative. To illustrate this point, Figure 15.2 shows how changing the discount rate r can impact the NPV for the cash flows presented in Table 15.4.

IRR is defined as the rate of return at which the NPV of the cash flows is zero. As illustrated in Figure 15.2, setting $r = 30\%$ generates a negative NPV of –$716. Setting $r = 25\%$ results in a positive NPV of $1,352. In the absence of unusual cash flow sequences,* this indicates that the IRR is between the values of 25% and 30%. Setting $r = 20\%$, as in the example of MARR, is also shown in Figure 15.2 for illustration.

As illustrated by this example, the IRR is the rate of return at which the sum of any discounted negative cash flows is equal to the sum of any discounted positive cash flows. In other words, at this rate r, the outgoing and incoming discounted cash flows of the investment or project are in balance. If different investments or projects are being compared using IRR and it is the sole criterion used, the investment or project with the highest possible IRR should be chosen.

* This is subject to some mathematical conditions that exceed the scope of this book. Lanigan (1992), provides a brief and useful explanation.

In the example provided, the IRR is approximately 28.17%. Many spreadsheet applications have easily usable functions to determine the IRR.* Alternatively, if the cash flows are evaluated by constructing a table such as Table 15.4, it is usually possible to hone in on the IRR by iteratively guessing values for r until the NPV of the project is close to zero.

As explained, with periodic compounding across T time periods, IRR is the discounting rate r at which the sum of cash flows S_t at time t discounted to net present value equals zero. Finding a mathematical solution to IRR is difficult since the equation that must be solved is a polynomial of degree n in the variable $(1 + r)$, which has multiple roots:

$$S_0 + S_1(1+r)^{-1} + S_2(1+r)^{-2} + \ldots + S_T(1+r)^{-T} = \sum_{t=0}^{T} S_t(1+r)^{-t} = 0 \quad (15.6)$$

Without spreadsheet software, it is, therefore, advisable to use a graphical solution method to identify IRR. The first step in the graphical method is to draw a graph of NPV values at three different levels of r. Ideally, these will produce both positive and negative values of NPV that are close to zero. Figure 15.3 uses r values of 20%, 25%, and 30%. IRR is then identified by the intersection of the graph and the x-axis. This method can be approximated mathematically by interpolation or extrapolation

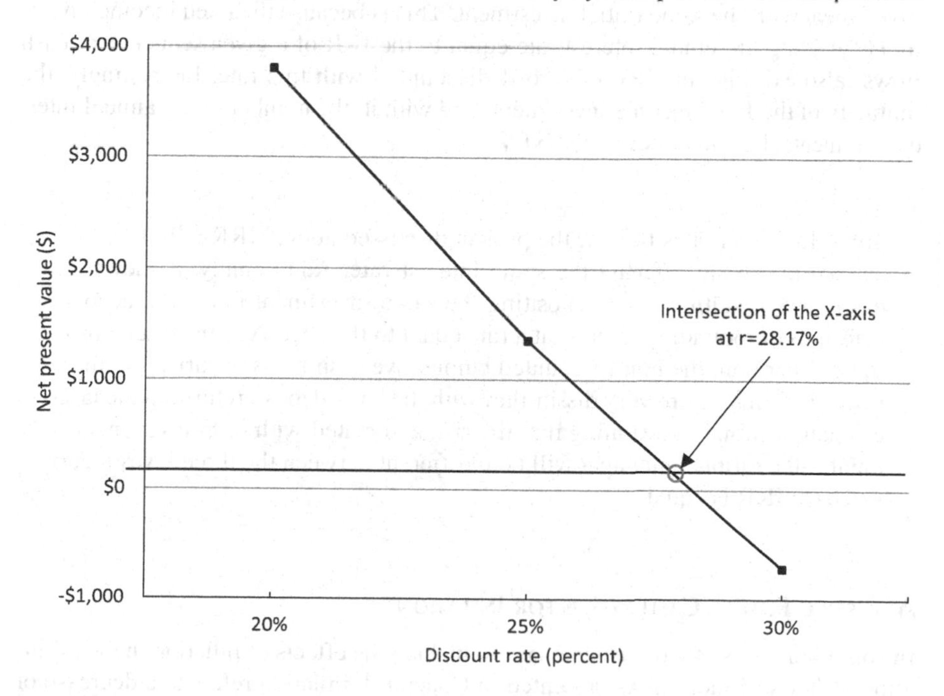

FIGURE 15.3 Graphical determination of IRR (r = 20%, r = 25%, and r = 30%)

* For example, the syntax for the IRR function in Microsoft Excel is "=IRR(range of values, optional starting value)".

using two interest rates and corresponding levels of NPV to identify at which r the condition $NPV = 0$ is satisfied.

COMPARING INVESTMENTS BASED ON THE INTERNAL RATE OF RETURN

The cash flows discussed in this chapter represent project-type investments in which costs are incurred more or less immediately and are repaid through revenues generated by the project in the future. Such initiatives were characterized in Chapter 6 as private ventures. An important alternative class of investments in which interest is non-compounding is *fixed income investments*, the most important type of which are *fixed rate bonds*.

When a bond is issued, a sum is loaned by the lender, or *bondholder*, to the issuer of the bond. The issuer is obliged to pay a fixed level of interest to the bondholder in agreed time intervals, normally quarterly, semi-annually, or annually. The duration of these payments, which is called *maturity*, is normally greater than one year. Upon maturity, the sum, also known as the face value, is returned. In the world of bonds, fixed interest payments are referred to as *coupon* payments.

Once the IRR of a given sequence of cash flows starting with an initial investment has been calculated, it is possible to use it to construct an equivalent fixed income investment with the same initial investment. This is because the fixed income investment, yielding an annual interest rate equal to the IRR of a given sequence of cash flows, also exhibits an NPV of zero if discounted with this rate. Interestingly, the maturity of the fixed income investment, and with it, the number of the annual interest payments, has no effect on the NPV.

WORKED EXAMPLE

Table 15.5 illustrates this for the project discussed above (IRR = 28.17%) and a fixed rate bond offering the same interest rate. Additionally, it shows the cash flows resulting from depositing the same investment in a bank account offering compounding interest at a rate equal to the IRR. As can be seen from the comparison, the non-discounted cumulative cash flows occurring in these three investments are very dissimilar, with the bank deposit returning the largest sum of money. Assuming that the risk associated with each investment is identical, a rational manager will be indifferent between the three investments since the IRR is equal.

ADJUSTING FUTURE CASH FLOWS FOR INFLATION

In some situations, it will be necessary to include the effects of inflation in the evaluation of flows of money. As presented in Chapter 3, inflation refers to a decrease of the value of money over time. In this section, the inflation rate is captured by the annual inflation rate f. There are two common ways in which cash flow models can

TABLE 15.5
Three Alternative Projects with Identical Cumulative Cash Flows at PV after Year 5 (Worked Example)

	Project A		Fixed income investment ($r = IRR$)		Bank deposit ($r = IRR$)	
EOY	Cash flow (non-discounted) ($)	Cumulative PV ($)	Cash flow (non-discounted) ($)	Cumulative PV ($)	Cash flow (non-discounted) ($)	Cumulative PV ($)
0	–20,000	–20,000	–20,000	–20,000	–20,000	–20,000
1	6,250	–15,124	5,635	–15,604	–	–
2	8,750	–9,798	5,635	–12,174	–	–
3	10,000	–5,049	5,635	–9,498	–	–
4	8,750	–1,807	5,635	–7,410	–	–
5	6,250	0	25,635	0	69,188	0
Cumulative non-discounted cash flow	20,000	–	28,174	–	49,187	–

incorporate inflation, the *constant-value method*, and the *composite-rate method.* This section will briefly introduce both. For simplicity, it is assumed in the following that the inflation rate remains constant over time.

The *constant-value method* of handling inflation is based on the separation of the effect of inflation from time preference. This is done by first deflating all then-current cash flows $S_{then\text{-}current}$ to constant-value cash flows $S_{constant\text{-}value}$ using the inflation rate f and the time t at which the cash flows occur:

$$S_{constant\text{-}value} = S_{then\text{-}current}\frac{1}{(1+f)^t} = S_{then\text{-}current}(1+f)^{-t} \quad (15.7)$$

Note that this equation is analogous to the periodic discounting equation (15.5), with the term $(1+f)^{-t}$ referred to as the *inflation factor.* Once future cash flows have been adjusted for inflation in this way, they can be discounted to their present values. This means that the future cash flows are modified twice, once to account for inflation and once to discount to PV.

The *composite-rate method* combines the steps performed in the constant-value method into a single step by forming the *composite discount factor* by multiplying the *discount factor* with *the inflation factor*:

$$(1+f)^{-t}(1+r)^{-t} = (1+r+f+f\times r)^{-t} \quad (15.8)$$

Following this, the constant-values cash flow discounted to present value S_{CVPV} can be calculated by applying the composite discount factor to all then-current cash flows $S_{then\text{-}current}$ as follows:

$$S_{CVPV} = S_{then\text{-}current}(1+r+f+f\times r)^{-t} \quad (15.8)$$

WORKED EXAMPLE

Table 15.6 summarizes the calculation of the constant-values NPV in the example project using both methods, subject to an inflation rate of $f = 3\%$ and a discount rate of $r = 18\%$. As can be seen from this example, both methods produce the same results. Despite inflation, the project still exhibits a positive NPV and is therefore an attractive option.

CONCLUSION

This chapter has presented an overview of methods that can be used to evaluate future flows of money or, indeed, of any measurable quantity that is subject to time preference. After introducing some simplistic methods building on the shape techniques presented in Chapter 6, the chapter has presented a group of methods based on the concept of the interest rate and its conceptual counterpart,

TABLE 15.6
Discounting of Then-Current Cash Flows in Project A, Based on an Inflation Rate of 3% and a Discount Rate of 18%, Using Both Methods (Worked Example)

EOY	Then-current cash flow ($)	Inflation factor	Constant-values cash flow ($)	Discount factor	Constant-values cumulative PV (calculated using the constant-value method) ($)	Composite factor	Constant-values cumulative PV (calculated using the composite-rate method) ($)
0	–20,000	1.00	–20,000	1.00	–20,000	1.00	–20,000
1	6,250	0.97	6,068	0.85	–14,858	0.82	–14,858
2	8,750	0.94	8,248	0.72	–8,934	0.68	–8,934
3	10,000	0.92	9,151	0.61	–3,364	0.56	–3,364
4	8,750	0.89	7,774	0.52	645	0.46	645
5	6,250	0.86	5,391	0.44	3,002	0.38	3,002

the discount rate, both of which are presented as core ideas. Time preference can be modeled head-on through a discount factor that can be used to adjust the value of future flows of money to establish their present value. Due to their fundamental relevance to virtually all aspects of finance, knowledge of these concepts is essential to professional managers. The calculation of IRR is particularly interesting since it does not require the manager to estimate a discount rate or to state a MARR, either of which is a complex task. Once the manager has evaluated cash flows using the provided methods, some form of ranking technique is normally applied to identify the most promising investment or project. In real-world settings, the benefit of performing more quantitative analysis on proposed investments projects will be exhausted at some point. This means that the manager's subjective judgment will ultimately be called upon to make a decision.

Discounting techniques are required for many approaches based on life cycle thinking, such as life cycle costing and cost-benefit analyses. However, the reader should note that the logic of discounting to net present value is not limited to money. Focusing on the carbon impact of a process, for example, it is likely that the reduction of carbon emissions in the near future is more valuable than reductions of carbon emissions in the more distant future, due to uncertainty. This example shows that, to reach appropriate conclusions about the sustainability of processes creating environmental impacts over time, the use of discounting techniques is essential. This assumes, of course, that an appropriate discount rate is known.

Most managerial finance and managerial accounting textbooks introduce the presented methods, including *Principles of Managerial Finance* by Chad Zutter and Scott Smart (2019) and *Management Accounting for Decision Makers* by Peter Atrill and Eddie McLaney (2018). A number of project management textbooks provide practical introductions to cash flow evaluation techniques, each highlighting different aspects. The authors find the textbook *Engineers in Business* by Mike Lanigan (1992) particularly illuminating. This book discusses the construction of ranking techniques that can be used to pick the most attractive option in detail. An engineering management textbook introducing these methods is *Management for Engineers, Scientists and Technologists* by John Chelsom, Andrew Payne, and Lawrence Revill (2005).

REVIEW QUESTIONS

1. Complete the below sentence on the concept of a time preference.

   (Question type: Fill in the blanks)

   "Time preference is the ________ assigned by a person to receiving a good or an object at ________ rather than receiving it ________."

2. A proposed project exhibits the cash flow profile shown below, with an investment required immediately:

End of year	Cash flow
0	–$38,500
1	$7,500
2	$12,500
3	$10,500
4	$6,000
5	$10,000

(Question type: Calculation)
Calculate the annual average proceeds of the project.

3. An interest rate is the proportion of a principal sum that is charged over a period of time as interest to the recipient of the principal sum, typically expressed as an annual percentage.
   (Question type: True/false)
   Is this statement true or false?
   O True O False

4. Which of the following characterizes the "principal sum" in a financial investment?
   (Question type: Multiple response)
   ☐ It is a sum that can be borrowed
   ☐ It is a sum that can be lent
   ☐ It is the result of interest payments
   ☐ It is a sum that can be deposited
   ☐ It will always be returned to the investor
   ☐ It is determined by time preference
   ☐ It is a liability to the business

5. Given a particular interest rate, which of the following does the total interest received through an investment not depend on?
   (Question type: Multiple choice)
   O The principal sum
   O The frequency of interest payments
   O The duration of the investment
   O The discount factor

6. A saver has a bank account with a fixed annual interest rate of 5%. The saver deposits $2,000 into the account.
   (Question type: Calculation)
   How much money will be in the account after 3 years?

7. Cash flows occurring in the future are discounted because of which reason?
   (Question type: Multiple choice)
   - O Discounts encourage high-quality contractors
   - O An organization can obtain government discounts on its costs
   - O It minimizes the interest paid at start-up
   - O It allows for the variation in the value of money in different time periods
   - O The discount rate encourages long-term investment

8. An investor is offered a cash payment in the future. The offered amount is $9,000, it will be paid in 12 years, and it is discounted annually at 4%.
   (Question type: Calculation)
   What is the present value of the payment?

9. You are investigating the internal rate of return of a project. In your analysis, you have identified the net present value of the cash flows in the project at three different discount rates. Your estimations are as follows:

Discount rate	Net present value
22%	–$600
15%	–$160
10%	$1,100

(Question type: Calculation)
Use a graphical or numerical method to estimate the internal rate of return.

10. A potential buyer of a sportscar is interested in the future costs of a vehicle per mile traveled at net present value. The buyer is planning to sell the car after three years at the residual value. The car costs $32,000 to purchase, due immediately. The buyer expects to incur costs for insurance, fuel, and maintenance of $1,600 per year. Moreover, the car is expected to depreciate in value by $3,000 per year and the buyer is planning to travel 15,500 miles per year.
    (Question type: Calculation)
    What is the whole life cost per mile traveled at net present value, assuming that all payments arise annually, are subject to an annual discount rate of 8%, and that the miles traveled are not subject to discounting?

REFERENCES AND FURTHER READING

Atrill, P. and McLaney, E., 2018. *Management accounting for decision makers.* 9th ed. London: Pearson Education.

Chelsom, J.V., Payne, A.C., and Reavill, L.R.P., 2005. *Management for engineers, scientists and technologists.* 2nd ed. Chichester: Wiley.

Keynes, J.M., 1923. *A tract on monetary reform.* London: Macmillan.

Lanigan, M., 1992. *Engineers in business: The principles of management and product design.* Boston: Addison-Wesley Publishing Company.

Zutter, C.J. and Smart, S.B., 2019. *Principles of managerial finance.* 15th ed. London: Pearson Higher Education.

16 Planning and Monitoring Projects

OBJECTIVE AND LEARNING OUTCOMES

The final chapter in this book provides an overview of important and practically applicable techniques used in project management. After defining what a project is and outlining the area of project management, the chapter introduces two important topics within project management, the planning of projects and the monitoring of projects. Using simple example projects, important practical techniques are presented for both topics, focusing on network analysis and earned value management. This chapter aims to achieve the following learning outcomes:

- an appreciation of the general field of project management;
- an ability to define what a project is and how project management is different from day-to-day management activities;
- an ability to use basic techniques in project planning, including task lists and logic networks;
- understanding the difference between activity-on-node networks and activity-on-arrow networks;
- the ability to perform simple network analysis on a small project, identifying float (or slack) for individual activities and the critical path;
- understanding the basic principles of project monitoring, including the use of project milestones;
- an understanding of the fundamentals of earned value management, including an appreciation of the significance of budgeted cost of work performed and other relevant metrics;
- the ability to perform a simple earned value analysis of a small project, including the interpretation of results and an appreciation of its limitations; and
- understanding key terms, synonyms, and accepted acronyms.

THE CONTEXT OF PROJECT MANAGEMENT

Some activities undertaken in a business can and should be classified as *projects*. For example, the research and development of a new product, as introduced in Chapter 5, is often characterized as a project. Generally, for an activity to take the status of a project, a number of criteria must be satisfied. These include the following:

- a project must have a clear purpose and well-defined goals;
- it must be novel and unique in a sense that it has not been carried out before;

DOI: 10.1201/9781003222903-19

- it is a temporary activity that ends after it has achieved its goals or failed;
- it pulls together people and resources from different departments, functional units or organizations; and
- it will always involve some degree of uncertainty and risk.

The process of devising and executing projects is the remit of *project management*. The objective of project management is to reach all project goals within the specified time and resource constraints. The characteristics of projects set them apart from ongoing operational activities, as introduced in Chapter 9, also referred to as *business-as-usual*.

The requirements of project management have led to a set of specific methods that are applied wherever projects are undertaken. This chapter introduces two specific activities within project management, the planning of projects and the monitoring of projects. Figure 16.1 shows how these two activities, shown as shaded boxes, relate to other major phases in project management, including project initiation, work execution, and closing the project.

PROJECT PLANNING

The first step in a project is to arrange the elements of the project into a well-configured sequence. This process is referred to as *project planning* or *project scheduling*. Given that some projects can be very large, incorporating many hundreds of elements, or highly complex, a range of established methods has emerged to configure

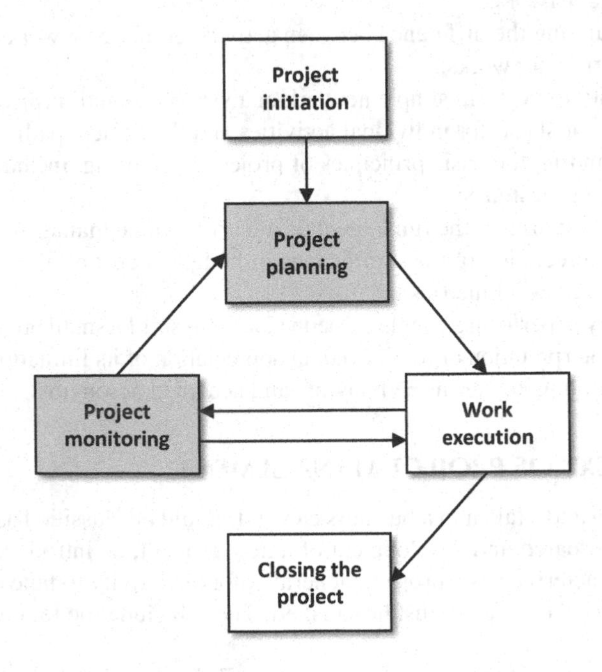

FIGURE 16.1 Project planning and project monitoring (shaded boxes) in the context of project management

projects appropriately. The activity of determining an appropriate structure of project activities is also known as *network scheduling*.

Henry Gantt (1861–1919)

FAMOUS THINKER

One of the most important early contributors to project planning techniques was the mechanical engineer and management consultant Henry Gantt in the late years of the 19th century and the early 20th century. His almost universally known contribution is the *Gantt chart*, which is a graphical tool expressing the schedule of work that allows the recording of progress against this schedule.

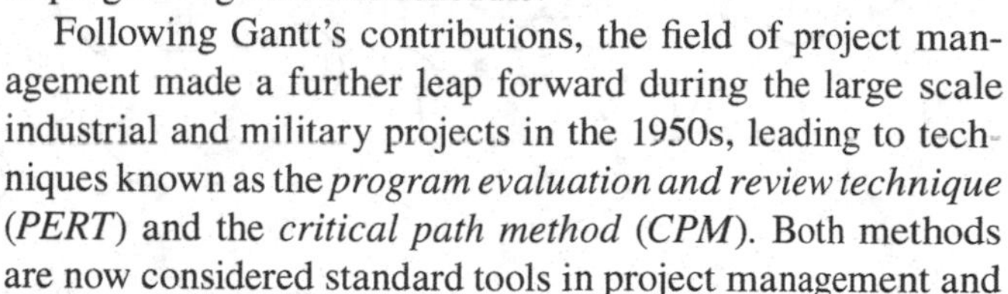

Following Gantt's contributions, the field of project management made a further leap forward during the large scale industrial and military projects in the 1950s, leading to techniques known as the *program evaluation and review technique* (*PERT*) and the *critical path method* (*CPM*). Both methods are now considered standard tools in project management and form the basis for numerous project management approaches and software packages.

An Introductory Example

Almost any project, from the simplest to the most complex, will require an element of project management. Large projects, such as the construction of ships, office buildings, and the development of new products will all have substantial and professional project management teams behind the scenes.

In project management, the term *deliverable* is often used to refer to the quantifiable products or services generated during or on completion of a project. Deliverables can be tangible or intangible objects the project may or may not be contractually obliged to produce. Often, deliverables are associated with the completion of portions or work. Deliverables can be defined as follows:

In project management, a deliverable is any measurable product, service, information, or result that must be completed to finish a project or a part of a project.

To introduce basic aspects of project planning and scheduling, this chapter will return once more to preparing hot beverages. It will discuss a very simple project, the making of tea following the English method, as illustrated in Figure 16.2. The main project deliverable is said cup of tea, correctly brewed.*

Any description of a project should start with the description of activities arranged in a logical sequence, known as a *task list*. Ideally, this should include information on the duration of each activity. For the tea-making project, the task list is as follows:

- fill kettle, which takes 0.5 minutes;
- boil kettle, which takes 10 minutes;

* In fact, the order of inserting milk and tea is a hotly contested topic in the United Kingdom. The standard ISO 3103:2019 / BSI 6008:1980 defines that the milk should be poured into the cup before the tea.

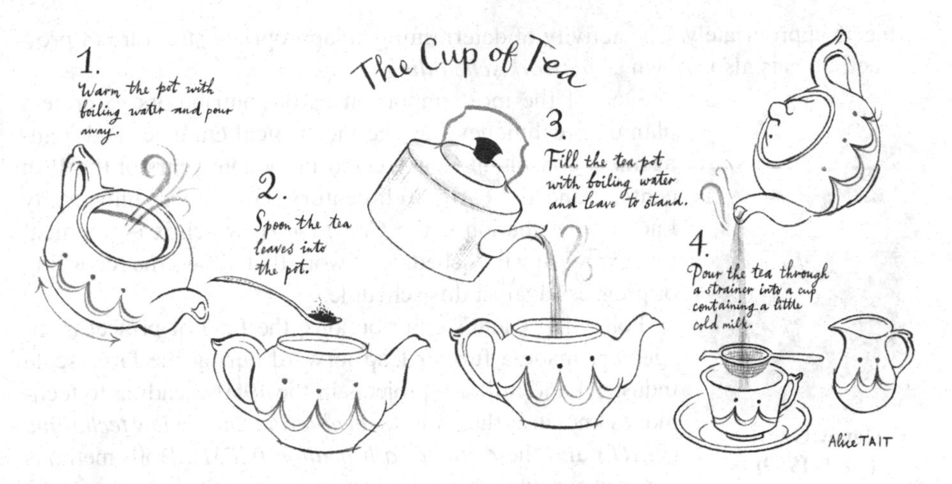

FIGURE 16.2 Making tea the English way (illustration by Alice Tait)

- warm pot, which takes 2 minutes;
- spoon tea into pot, which takes 0.5 minutes;
- pour water into pot, which takes 1 minute;
- pour cold milk into cup, which takes 0.5 minutes;
- pour tea into cup, which takes 0.5 minutes; and
- serve to customer, which takes 1 minute.

In most cases, simply executing projects as one-by-one sequences of activities is not acceptable, however. In the example of tea-making, a more efficient way of completing the work would be to carry out activities in parallel by splitting activities up into two groups, as done in Table 16.1. By parallelizing work in this way, the pot can be warmed, tea can be put into the pot and the milk can be poured while the kettle is brought to a boil.

TABLE 16.1
Making Tea in Parallel Groups of Activities

First group of activities	Task 1: Fill kettle
	Task 2: Boil kettle
	Task 3: Pour water into pot
	Task 4: Pour tea into cup
	Task 5: Serve to customer
Second group of activities	Task 1: Warm pot
	Task 2: Spoon tea in pot
	Task 3: Pour cold milk in cup

Work Breakdown Structure and Work Packages

A preliminary step in the planning of any project is to create a *work breakdown structure (WBS)*. Work breakdown structures are fundamental to project management because they provide the underlying framework for the organization and control of the work and specify the logical hierarchy of the work to be performed in order to reach the project objectives. Work breakdown structures can be defined as follows:

A work breakdown structure is a hierarchical decomposition of the work to be performed in the project to reach project objectives and produce the required deliverables.

A common way to show a work breakdown structure is in the form of a hierarchical tree of activities, with high-level activities shown at the top and specific activities and tasks shown at the bottom. Figure 16.3 shows a three-level work breakdown structure for the tea-making process described in this chapter.

One important function of work breakdown structures is that they can be used to define groups of activities of a more manageable size and complexity. Such groups of activities are referred to by project managers as *work packages*. They can be defined as follows:

A work package is a unit of related activities, tasks, products, or functions in a project. Work packages are defined in a work breakdown structure to break projects up into smaller sections of work.

Work packages are normally seen to be the smallest units in a project according to which resources can be formally allocated and which can incur costs. Normally, it is possible to break down the work contained in work packages into sub-work packages, individual activities, or tasks, as shown in Figure 16.3.

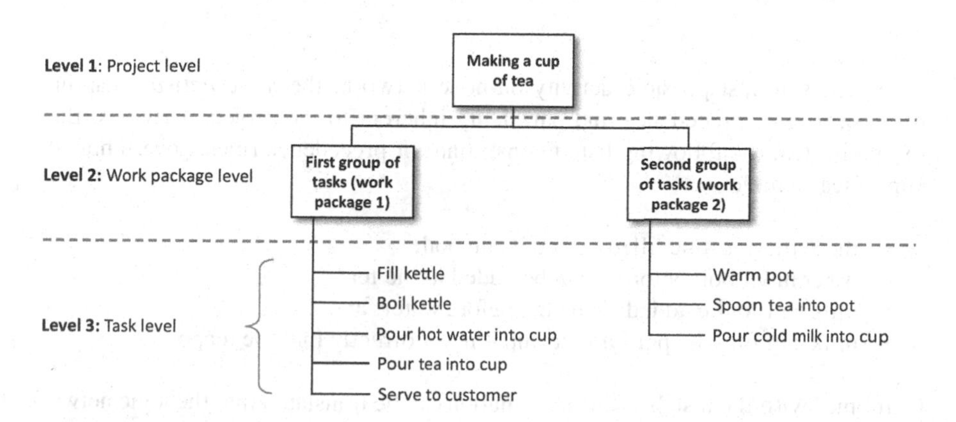

FIGURE 16.3 Work breakdown structure of making tea

The Logic Network and the Critical Path

Once the project scope and the relevant activities have been defined, the emphasis of project management shifts to systematically capturing the relationships between activities, tasks, and work packages in terms of their *parallelity* and *precedence*. Parallelity refers to the simultaneous way in which work on some activities can be progressed. Precedence refers to constraints dictating which activity must be completed before another activity can begin. In project management, these aspects are analyzed using a group of methods known as *network techniques*.

There are several variations of network techniques such as PERT and CPM, as outlined in this chapter. The fundamental reason for representing a project as a network of activities is to represent the parallelity and precedence relationships between the various project activities. An initial and important distinction is between network techniques that generate *activity-on-node networks* and those that generate *activity-on-arrow networks:*

- Activity-on-node networks are networks that show activities as nodes, connected by arrows representing the links between activities. The activities carry durations and the links represent relationships of precedence between activities.
- Activity-on-arrow networks are networks representing activities as arrows, each starting and finishing at a node. In this type of network, nodes are characterized as *events* with zero time duration, occurring only if all necessary activities leading to that node are complete.

Important insight can be gained about projects if they are analyzed in this way. In particular, it is possible to identify which activities are of special importance in a project and where there might be spare time or capacity that could be used otherwise by project managers.

As the simplest possible activity-on-node network, the *logic network* captures the sequence of precedence and parallelity inherent to the project activities. Back to making tea, the following list of constraints, or precedence rules, govern how the cup of tea is made:

- the kettle must be filled before it can boil;
- water must boil before it can be added to the tea;
- milk cannot be added to the tea before water; and
- milk can only be put into the cup – never directly into the teapot.

Combined with the task list, this information can be translated into the logic network shown in Figure 16.4, capturing activity precedence and the required sequence.

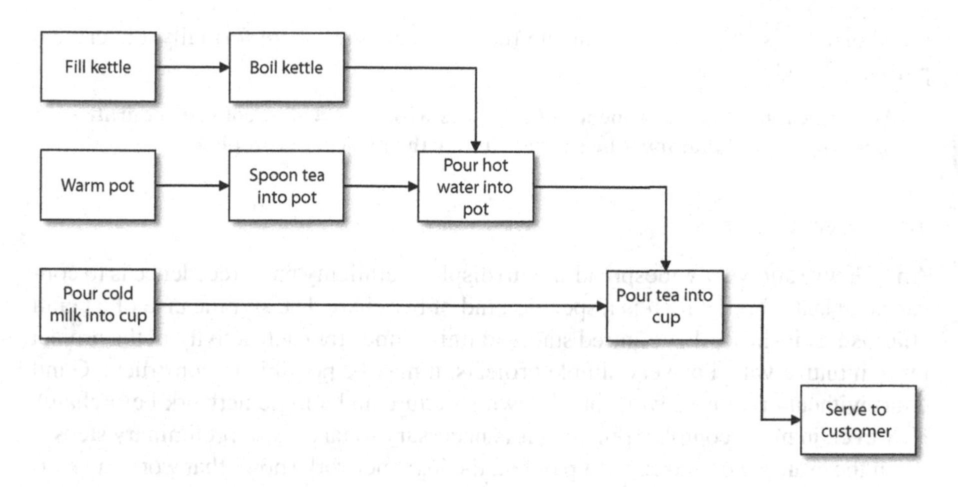

FIGURE 16.4 Logic network for the example project

On the basis of the logic network, it is possible to assess the time necessary to complete the full project. The duration of each step in the process is indicated on the task list. Including the durations in the logic network allows the identification of the minimum possible duration of the project from start to finish. The activities involved in this chain are shown as shaded boxes in Figure 16.5. As can be seen, the activities on this path define the minimum duration required to complete the entire project. In the example of making a cup of tea, the project has a rather short duration of 13 minutes. If any activity shown as a shaded box is held up, the overall completion of the project will be delayed. Because project managers will normally want to avoid any kind of delay, or *slippage*, the activities on this path carry special significance. Collectively, they form the *critical path* of the project, which represents

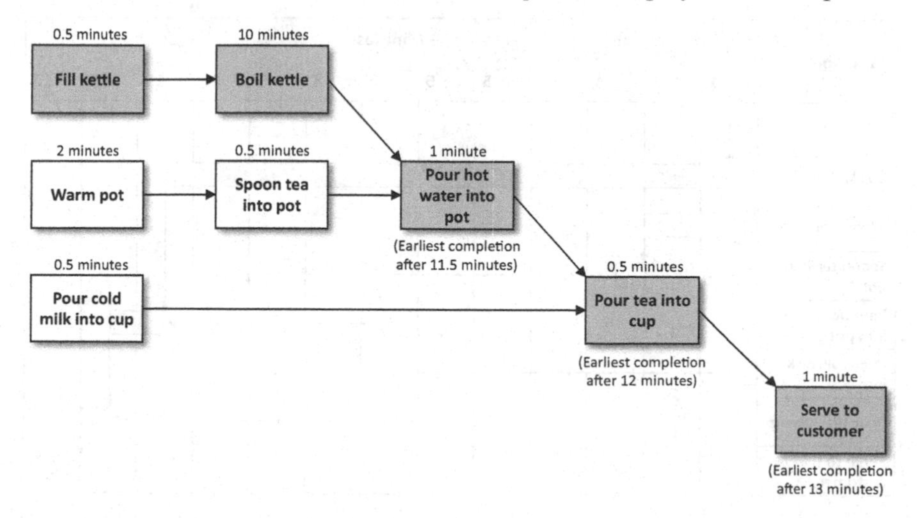

FIGURE 16.5 Modified logic network for the example project

the shortest possible way to complete the logic network. More formally, the critical path is defined as follows:

The critical path is the sequence of activities with the greatest collective duration in a project plan that must be carried out for the project to complete.

The Gantt Chart

An effective and very widespread way to display parallelity and precedence is to construct a Gantt chart, which is a specific kind of bar chart. The significance of a Gantt chart is that it shows the required start and finish times for each activity in the project in an intuitive way. For very simple projects, it may be possible to construct a Gantt chart without creating a work breakdown structure and a logic network beforehand. However, in more complex projects, it is necessary to take these preliminary steps.

In the example of making a cup of tea, the logic network shows that work can start with three activities in parallel: fill kettle, warm pot, and pour milk in cup. Using these three activities as starting points, the logic network, and the timings shown, it is possible to construct the Gantt chart presented in Figure 16.6. The activities on the critical path are shown as shaded bars.

Starting a Simple Network Analysis

To analyze the structure of a project in more detail, *network analysis* is required. As introduced, the logic network is based on activities, such as "boil kettle" or "warm pot". The first step in network analysis is to switch from an activity-on-node view of the project to an activity-on-arrow view. Recall that an event has zero time duration and occurs only if all necessary activities leading to it are completed. In this sense, an event signals that an activity is complete. It is separated from a previous event

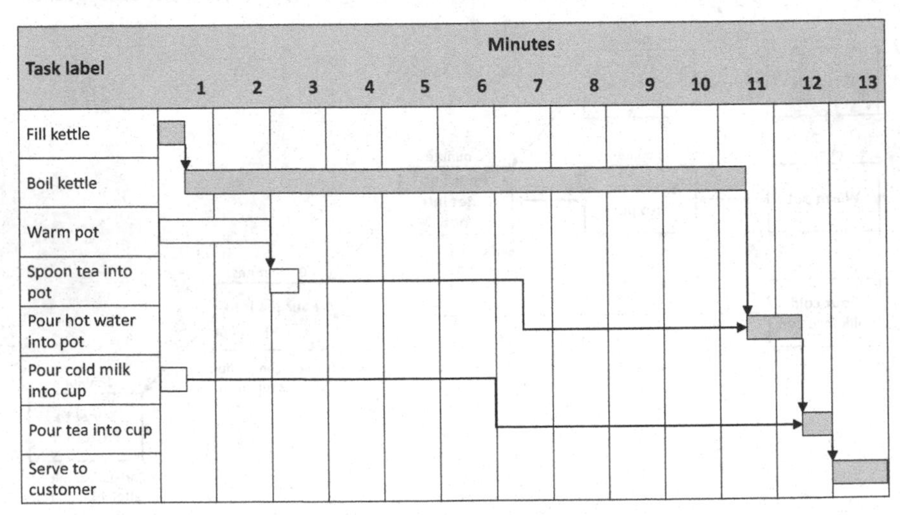

FIGURE 16.6 The Gantt chart for the example project

through the duration of the preceding activity. Each event in the project is assigned a unique number for identification, such as *Event 1* or *E*1.

In the tea making project, event *E*1 can be defined as occurring if and when the kettle has been filled. Since the duration of filling the kettle is 0.5 minutes, *E*1 occurs 0.5 minutes after the preceding event *E*0, which forms the starting event in the project. An event can require the completion of more than one activity to occur. For example, another event *E*2 is defined as occurring if and when the kettle has been boiled and the tea is in the pot.

Capturing the events occurring in a project plan can be achieved by drawing an additional form of project network, the *events network*. In its basic form, the events network is determined by the information contained in the logic network. Figure 16.7 shows the events network of the tea-making project, identifying the duration of each activity bookended by events. Note that event numbers are arbitrarily assigned and that *E*0 represents the start of the project and *E*5 represents the end of the project.

To more clearly identify each activity relative to events, they can be labeled using the notation *E*X,Y, where X is the number of the preceding event and Y is the number of the following event. Using this notation, "boil kettle" is identified by the notation *E*1,2. To avoid confusion, it is often a good idea to construct a table identifying each activity in this way, as shown in Table 16.2.

Earliest Activity Start and Latest Activity Finish

The next step in network analysis is to evaluate the starting and finishing times for the various activities in the project that can occur without holding up project completion. This requires the introduction of two additional variables for each activity: *earliest activity start* (*EAS*) and the latest *activity finish* (*LAF*).

EAS is the earliest time at which an activity can start, i.e., the earliest time at which the preceding event can occur. If there are multiple paths to an event, as is the case with *E*2 and *E*3 in the example project, EAS is the latest of the possible times. The EAS for each activity can be obtained through a technique called a *forward pass*. The forward pass is done by assessing each possible pathway through the project in

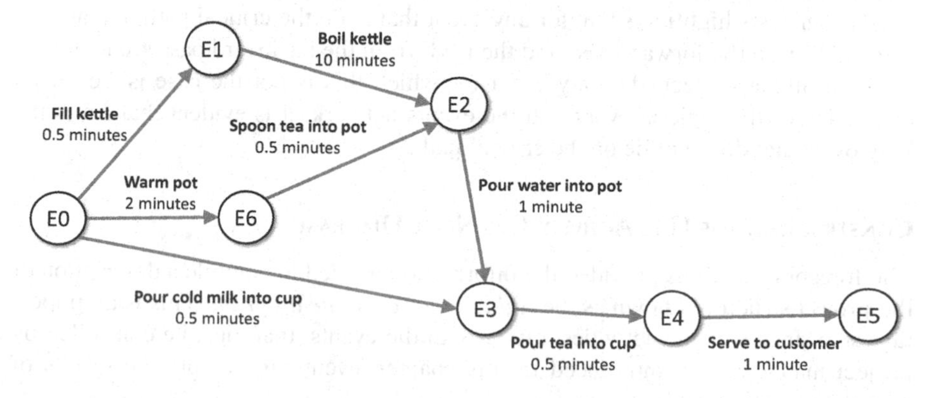

FIGURE 16.7 Events network with additional descriptions for the example project

TABLE 16.2
Definition of Labels for Activities for the Example Project

Activity	Activity duration	Preceding event	Following event	Activity label
Fill kettle	0.5	*E*0	*E*1	*E*0,1
Boil kettle	10	*E*1	*E*2	*E*1,2
Warm pot	2	*E*0	*E*6	*E*0,6
Spoon tea into pot	0.5	*E*6	*E*2	*E*6,2
Pour water into pot	1	*E*2	*E*3	*E*2,3
Pour cold milk into cup	0.5	*E*0	*E*3	*E*0,3
Pour tea into cup	0.5	*E*3	*E*4	*E*3,4
Serve tea to customer	1	*E*4	*E*5	*E*4,5

the events network, assuming no waiting or delay. For each path, beginning with the initial event and ending with the finishing event, the times as which the events occur are recorded, as shown in Table 16.3 for the three possible pathways of the example project, denoted as path A, path B, and path C. Note that the path with the longest duration is the critical path and defines the minimum duration of the project.

Analogous to this, LAF is the latest time at which an activity can finish (i.e., when the following event can occur) without causing the whole project to slip. This is obtained through a *backward pass* in which times of occurrence of events are identified by subtracting activity durations. This process begins at the end of the project, and is done for all alternative pathways through the project, as shown in Table 16.4. The LAF for each activity is obtained by recording the earliest time of occurrence for each event following the respective activity.

This analysis highlights that for any event that is on the critical path, the associated EAS from the forward pass and the LAF from the backward pass are identical. In the example project, the only event for which this is not the case is *E*6. From inspection of the logic network and the events network, it is evident that *E*6 is the only event that does not lie on the critical path.

Constructing the Full Activity-on-Node Diagram

The foregoing sections provided the information needed to compile a description of the project structure known as the *full activity-on-node diagram.* This is an important step since it is the activities, rather than the events, that must be controlled by project managers. As introduced in this chapter, events are simply the results of completing activities.

TABLE 16.3
Forward Pass, with Time in Minutes, and EAS Shown in Bold for the Example Project

Path A (critical path)		Path B		Path C	
Event	Time of occurrence	Event	Time of occurrence	Event	Time of occurrence
E0	**0**	*E0*	**0**	*E0*	**0**
E1	**0.5**	*E6*	**2**	*E3*	0.5
E2	**10.5**	*E2*	2.5	*E4*	1
E3	**11.5**	*E3*	3.5	*E5*	**2**
E4	**12**	*E4*	4	–	–
E5	**13**	*E5*	5	–	–

TABLE 16.4
Backward Pass with Time in Minutes, and LAF Shown in Bold for the Example Project

Path A (critical path)		Path B		Path C	
Event	Time of occurrence	Event	Time of occurrence	Event	Time of occurrence
E5	13	*E5*	13	*E5*	13
E4	12	*E4*	**12**	*E4*	**12**
E3	11.5	*E3*	**11.5**	*E3*	**11.5**
E2	10.5	*E2*	**10.5**	*E0*	11
E1	**0.5**	*E6*	**10**	–	–
E0	**0**	*E0*	8	–	–

Activity-on-node diagrams are based on the formal representation of activities as *modules*, with each module forming a node in the diagram. It should be noted that different types of modules exist and these are subject to various standards. A common activity representation is shown in Figure 16.8.

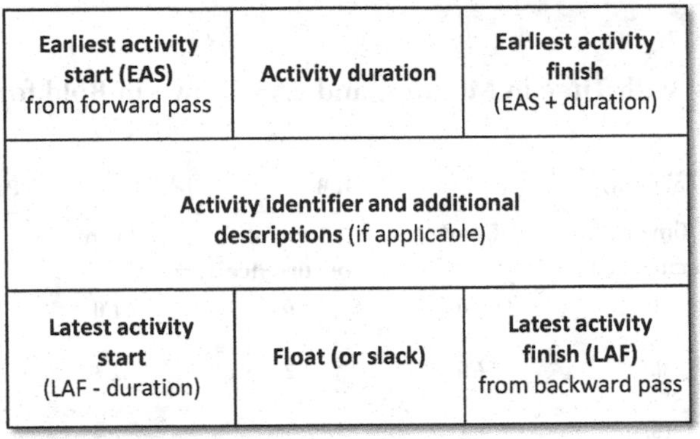

Earliest activity start (EAS) from forward pass	**Activity duration**	**Earliest activity finish** (EAS + duration)
	Activity identifier and additional descriptions (if applicable)	
Latest activity start (LAF - duration)	**Float (or slack)**	**Latest activity finish (LAF)** from backward pass

FIGURE 16.8 Activity representation in activity-on-node diagrams, adapted from BS 6046, BSI (1992)

This activity representation summarizes the information gained from the task list, the events network, and the forward and backward passes. The middle section identifies the activity by label or name and provides other relevant information if necessary. The other elements report the EAS, the duration of the activity, and the earliest activity finish, which is the sum of the EAS and the duration. The module also reports LAF and the latest activity start, which is obtained by subtracting the duration from LAF.

The final, and arguably most interesting, element contained in the module is activity *float*, also known as *slack*. Float represents the spare time around an activity in addition to its estimated duration. Float is calculated as follows:

$$\text{Float} = \text{LAF} - \text{EAS} - \text{duration} \tag{16.1}$$

An activity that has float can be lengthened, for example by reducing its personnel resource, or delayed without causing the project to slip. This gives the project manager flexibility to manage the flow of the project should unforeseen events occur or if the project hits difficulties. It should be noted that only activities with zero float lie on the critical path of the project.

The final step to be carried out in the simple network analysis is to replace all nodes in the logic network with the filled-in activity representations, as done in Figure 16.9. The completed activity-on-node diagram gives the project manager information on:

- the earliest and latest time each activity can start;
- the earliest and latest time each activity can finish;
- the float (or slack) of each activity; and
- whether an activity lies on the critical path (activities with zero float).

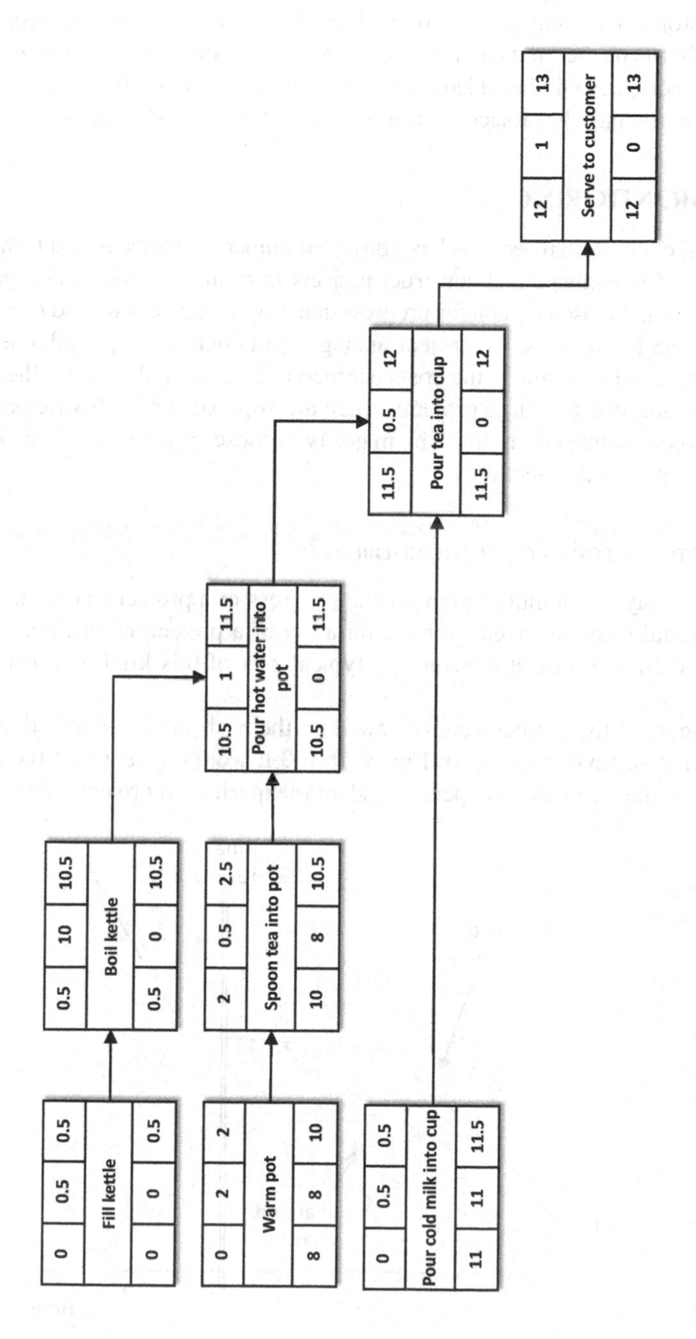

FIGURE 16.9 The completed activity-on-node diagram for the example project

By showing this information, the full activity-on-node diagram provides an extremely useful representation of the project structure to managers. It should also be stressed that this form of project planning is not a "do and forget" process. As a project progresses, plans are likely to change. In the real world, some activities will slip or sometimes finish early. New activities will be added, and some will be removed. The project manager will continually monitor the project and modify the activity-on-node diagram as needed.

PROJECT MONITORING

The preceding chapters and sections have provided numerous methods and techniques that can be used to evaluate and construct projects in business. This section presents techniques that can be used to monitor progress in a project once it is up and running.

In any project in business, the project manager will continuously need to monitor the progress that is being made, the costs accrued over time, and whether the activities meet their targets. For large projects, there are sophisticated software packages to support project managers in this. The majority of these systems are aligned to the methods presented in this section.

Using Milestones for Project Monitoring

A very simple way to attempt to monitor the progress of a project against targets is to plot the actual costs incurred over the duration of a project against the planned costs expected by the project manager. A typical plot of this kind is illustrated in Figure 16.10.

What happens if the project were reviewed on the basis of the information available at the time of review shown in Figure 16.10? It would appear that the project has not cost as much money as expected and an inexperienced project manager may

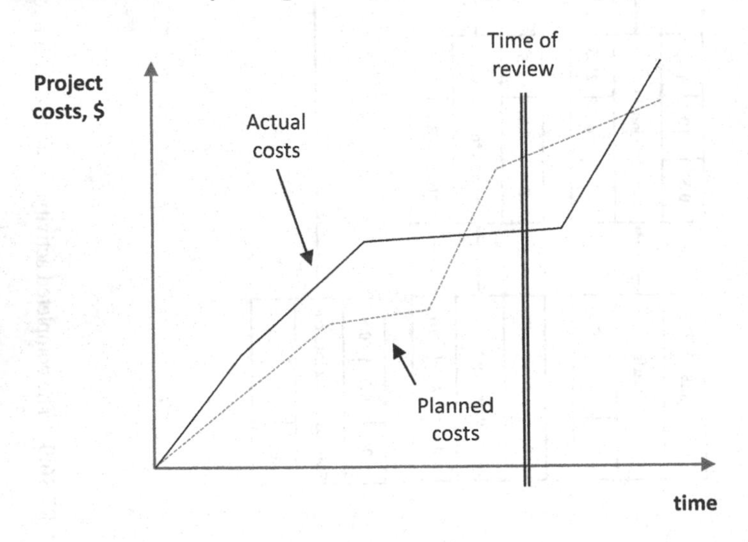

FIGURE 16.10 An example of actual spend plotted against planned spend over time

see this as an encouraging sign that the project is progressing well. However, there may be several underlying reasons for the under-spend at the time of review. Possible reasons include:

- the project is in fact running on time and is costing less than expected;
- the project is costing less than expected but this is because less work has been done than was planned and the project is running late; and
- the project is severely delayed *and* the work performed so far is costing more than expected.

A simple method to overcome this problem is to introduce *milestones* into the project at the project planning stage. Milestones represent some tasks or events that represent significant achievements of progress. Milestones can be defined as follows:

A milestone is a specific point in time during the course of a project that can be used to measure the progress of the project toward its ultimate goal.

Milestones must be connected to identifiable events, such as the completion of a preliminary design, the submission of a final report, or the end of the review phase. As such, milestones can form part of deliverables. Milestones must not be fuzzy or indefinite points such as scheduling a meeting or beginning work on a report.

Revisiting the example project, a series of milestones has been added in Figure 16.11, adopting the notation *M*1, *M*2, and *M*3. The milestones are marked on the lines showing planned costs and actual costs. By adding milestones, significantly more insight on the status of the project can be obtained:

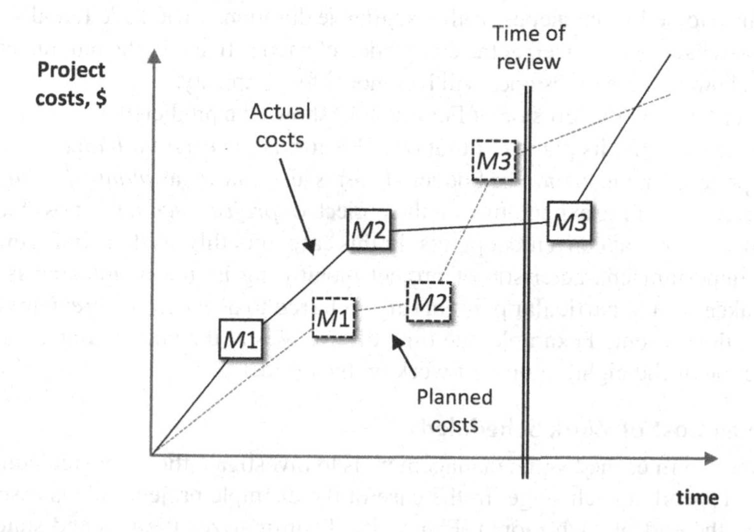

FIGURE 16.11 An example of actual spend plotted against planned spend over time, including milestones

- *M*1 is achieved ahead of schedule and below predicted spend. At this stage, the project is doing very well;
- *M*2 is again achieved ahead of schedule but this time above cost. At this stage, the project manager would consider a number of options, such as reducing personnel resources since the project is running faster than expected; and
- *M*3 has not been completed at the time of review and is overdue. Since the actual project spend before the review is below the planned spending rate, the project manager should consider using more resources, perhaps by subcontracting work out.

Earned Value Management

A more advanced and very well-known technique used by managers to monitor project progress is *earned value management (EVM)*, also known as *earned value analysis (EVA)*. It allows the monitoring of the progress of work and project spend against original plans by using a set of specific project monitoring metrics. This section will introduce earned value management using an illustrative example for a further project, the construction of a powerboat.

Drawing on project planning techniques presented earlier in this chapter, Figure 16.12 provides a brief summary of the planned project by showing a *task list*, including the duration and budget of each activity, and a simplified Gantt chart covering the planned eleven-month (rounded up) duration of the project. In the project, it is assumed that, if not indicated otherwise, work on each activity proceeds evenly. For example, after one month in an activity with a two-month duration, 50% of the work on the activity will have been completed. It is also assumed that costs are incurred evenly, again, if not indicated otherwise. For example, after one month in a four-month activity, it is assumed that 25% of the costs have been incurred. Note that the final activity is assumed to be instantaneous, with a negligible duration. To add additional structure, milestones have been placed at the completion of each activity. In the remainder of this chapter, however, the milestones will be ignored for simplicity.

The task list on the left side of Figure 16.12 shows the predicted, or *budgeted*, cost of each activity and its planned duration. This forms the *original budget* of the project. In project management, the budgeted cost is also known as *planned value*. To get a first overview of the expenditure on the project, or *project spend*, it is possible to plot the budgeted cost at convenient points, in this case, monthly, as done in Figure 16.13.

An important characteristic of project monitoring is that monitoring is always undertaken from a particular point in time, referred to as the *time of review* or *status date*. In the presented example, the time of review is at the end of August, after the completion of the eighth month of work on the project.

Budgeted Cost of Work Scheduled

The first step in earned value management is to investigate the estimated cumulative budgeted spend at each stage. In the case of the example project, this is assumed to occur at the end of each month. Figure 16.14 summarizes the planned state of the project at the end of August.

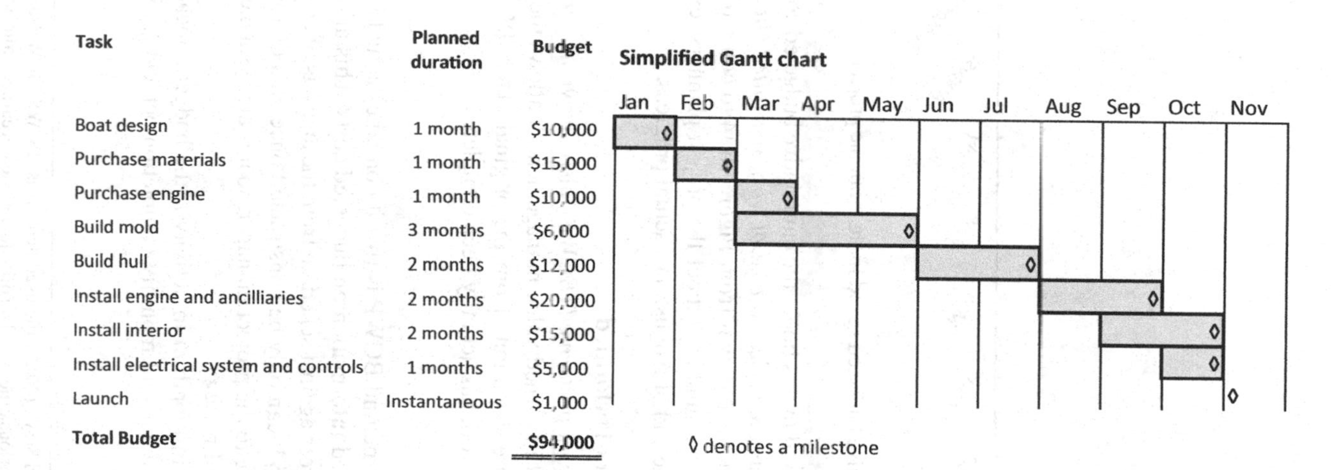

Task	Planned duration	Budget
Boat design	1 month	$10,000
Purchase materials	1 month	$15,000
Purchase engine	1 month	$10,000
Build mold	3 months	$6,000
Build hull	2 months	$12,000
Install engine and ancilliaries	2 months	$20,000
Install interior	2 months	$15,000
Install electrical system and controls	1 months	$5,000
Launch	Instantaneous	$1,000
Total Budget		**$94,000**

FIGURE 16.12 Task list and simplified Gantt chart for the example project

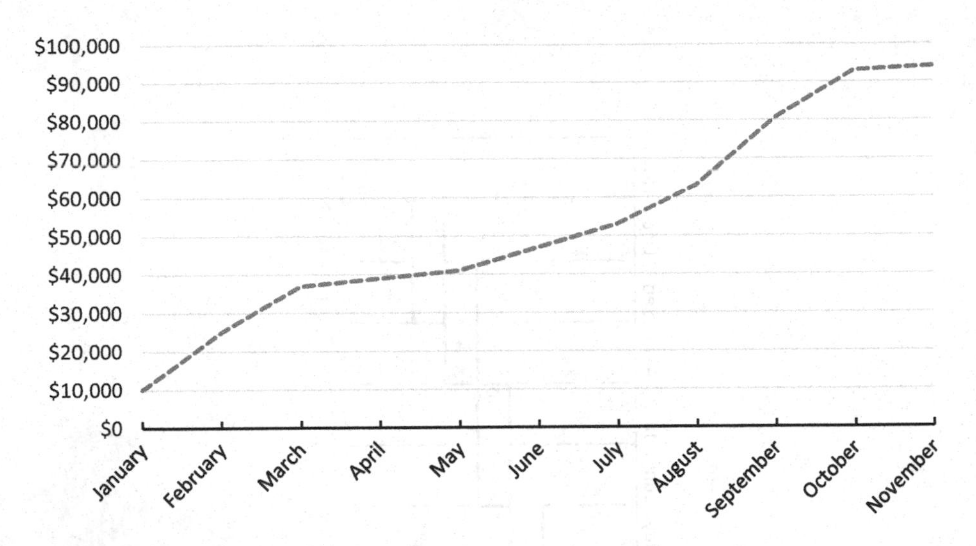

FIGURE 16.13 Cumulative budgeted spend in the example project

In the terminology used in this book, the cumulative budgeted cost of the planned activities is referred to as the *budgeted cost of work scheduled* (*BCWS*).* Table 16.5 shows the data that have been drawn from Figure 16.14 to construct the BCWS for each month. Note that the numbers reflect the project planners' expectation of the state of the project at the end of August, not actual progress.

Budgeted Cost of Work Performed

The *budgeted cost of work performed* (*BCWP*), also known as *earned value*, is the most important concept in earned value management. It allows the project manager to systematically capture deviations from the original plans. BCWP reflects how much of the scheduled work has actually been completed.

The conceptual leap behind BCWP is that it combines actual information on progress with planned information on budgeted costs. This allows separating the analysis of progress against schedule from the analysis of cost against budget. Of course, BCWP can only be constructed once data on real project performance is available to the project manager, so it is always a metric that looks back from the time of review.

To show BCWP in month *m* in activity *i*, the budget allocated to activity *i* is multiplied with the actual monthly completion percentage in activity *i* in

* According to Nicholas and Steyn (2017), the used acronyms (BCWS, BCWP, and ACWP) are industry-standards. There are various other valid terminologies used in earned value management, however.

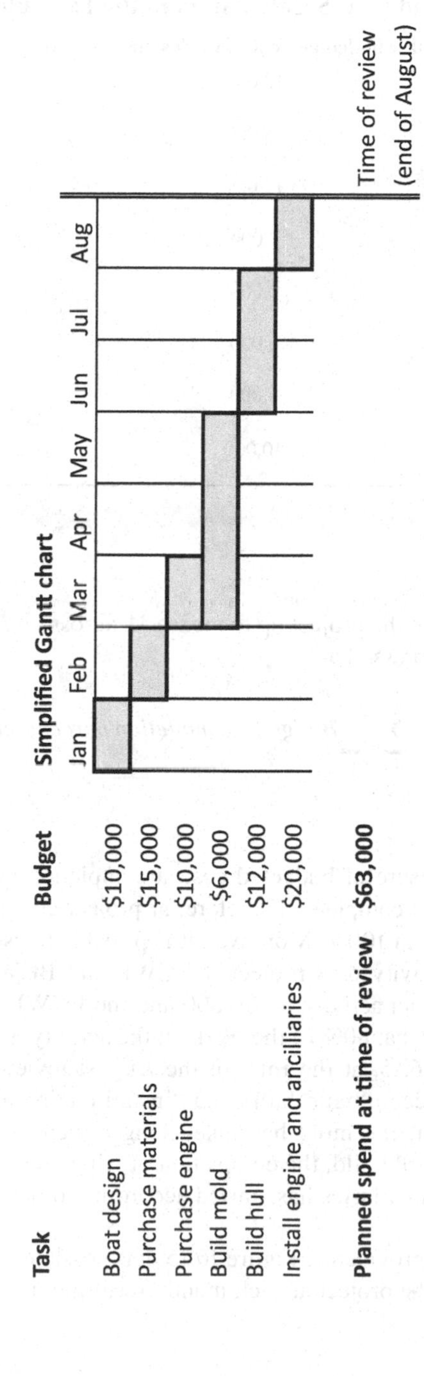

FIGURE 16.14 Planned progress at the time of review in the example project

TABLE 16.5
Monthly Data and BCWS Calculation in the Example Project

Month	Monthly budgeted cost of work scheduled ($)	BCWS ($)
January	10,000	10,000
February	15,000	25,000
March	12,000	37,000
April	2,000	39,000
May	2,000	41,000
June	6,000	47,000
July	6,000	53,000
August	10,000	63,000

month m. The BCWP of the project up to month M, across all I activities in the project, can thus be expressed as:

$$BCWP_M = \sum_{m=1}^{M}\sum_{i=1}^{I} Budget_i \times completion\ percentage_{im} \quad (16.2)$$

Since BCWP is a measure of budgeted cost, a completion percentage of 100% implies that the activity is complete. Therefore, at project completion, BCWP can never be greater or less than 100%. Moreover, it is possible to estimate the completion percentage of an activity or a project if BCWP and BCWS are known. For example, if the BCWS of an activity is $10,000 and the BCWP is $8,000, then the project manager can infer that 80% of the work on the activity is complete.

As shown in Figure 16.15, at the time of the August review in the powerboat example, all activities except "build hull" and "install engine and ancillaries" are 100% complete. However, the mold has taken longer than expected to produce, delaying the start of the hull build, the completion of which has slipped. The installation of the engine and ancillaries has started according to plan but is progressing far slower than expected.

Using the information provided in Figure 16.15, it is possible to expand Table 16.5 to include the BCWP in the project at each month, resulting in Table 16.6. It should

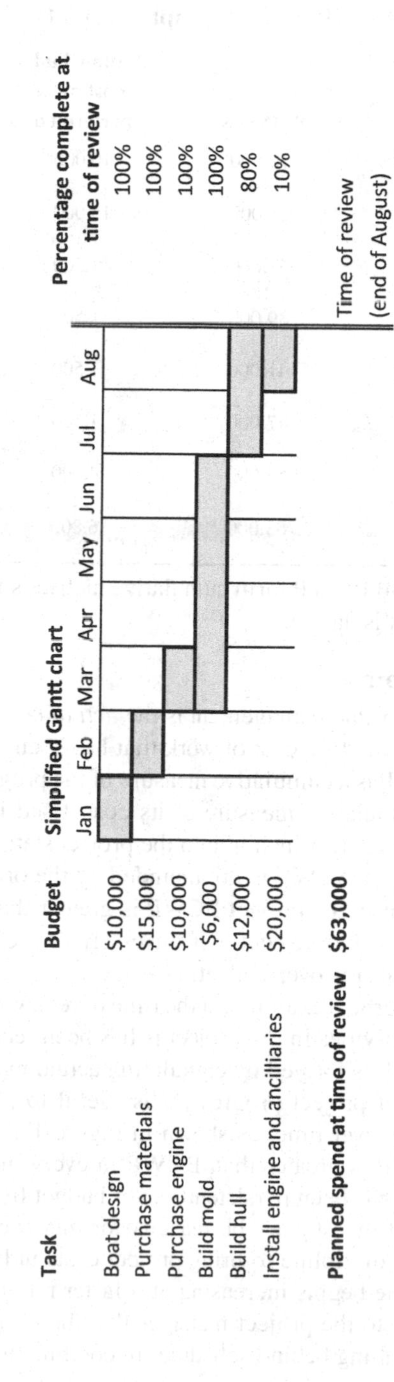

FIGURE 16.15 Budgeted costs and actual progress at the time of review in the example project

TABLE 16.6
Calculation of BCWS and BCWP in the Example Project

Month	Monthly budgeted cost of work scheduled ($)	BCWS ($)	Monthly budgeted cost of work performed ($)	BCWP ($)
January	10,000	10,000	10,000	10,000
February	15,000	25,000	15,000	25,000
March	12,000	37,000	11,500	36,500
April	2,000	39,000	1,500	38,000
May	2,000	41,000	1,500	39,500
June	6,000	47,000	1,500	41,000
July	6,000	53,000	4,800	45,800
August	10,000	63,000	6,800	52,600

be noted that both BCWS and BCWP form cumulative metrics so do not reflect project progress in any month in isolation.

Actual Cost of Work Performed

The final element of earned value management is the *actual cost of work performed (ACWP)*. ACWP represents the true cost of work that has been completed up to the time of review. While BCWP is a cumulative measure of the progress of work within the project, ACWP is a cumulative measure of its cost. Used in conjunction with BCWP, it gives the project manager insight into the project status.

Unlike BCWP, the value of ACWP is not bounded by the original budget of the project. If at any given point in the project BCWP is greater than ACWP, the work in the project is costing less than expected. Conversely, if ACWP is greater than BCWP, the work on the project is over budget.

Proceeding with the powerboat example, at the time of review in August the actual monthly spend across all activities in the project is has been recorded in Table 16.7, extending Table 16.6. ACWP is formed by cumulating actual monthly costs.

To obtain an overview of project progress, it is useful to plot the estimates of BCWS, BCWP, and ACWP over time, as shown in Figure 16.16 for the powerboat example. As evident, ACWP is greater than BCWP in every month of the project. This means that the project has been persistently over budget from the start. A competent project manager will investigate the reasons for this overspend and attempt to enforce greater spending discipline to bring project costs in line with the budget. Interestingly, the BCWS line begins increasing at a faster rate than the BCWP line from March. This indicates to the project manager that the progress in the work in the project is increasingly falling behind schedule. In combination, the cost overruns

TABLE 16.7
Actual Cost of Work Performed in the Example Project

Month	Monthly budgeted cost of work scheduled ($)	BCWS ($)	Monthly budgeted cost of work performed ($)	BCWP ($)	Monthly actual cost of work performed ($)	ACWP ($)
January	10,000	10,000	10,000	10,000	12,000	12,000
February	15,000	25,000	15,000	25,000	15,000	27,000
March	12,000	37,000	11,500	36,500	13,000	40,000
April	2,000	39,000	1,500	38,000	3,000	43,000
May	2,000	41,000	1,500	39,500	3,000	46,000
June	6,000	47,000	1,500	41,000	3,000	49,000
July	6,000	53,000	4,800	45,800	4,000	53,000
August	10,000	63,000	6,800	52,600	4,500	57,500

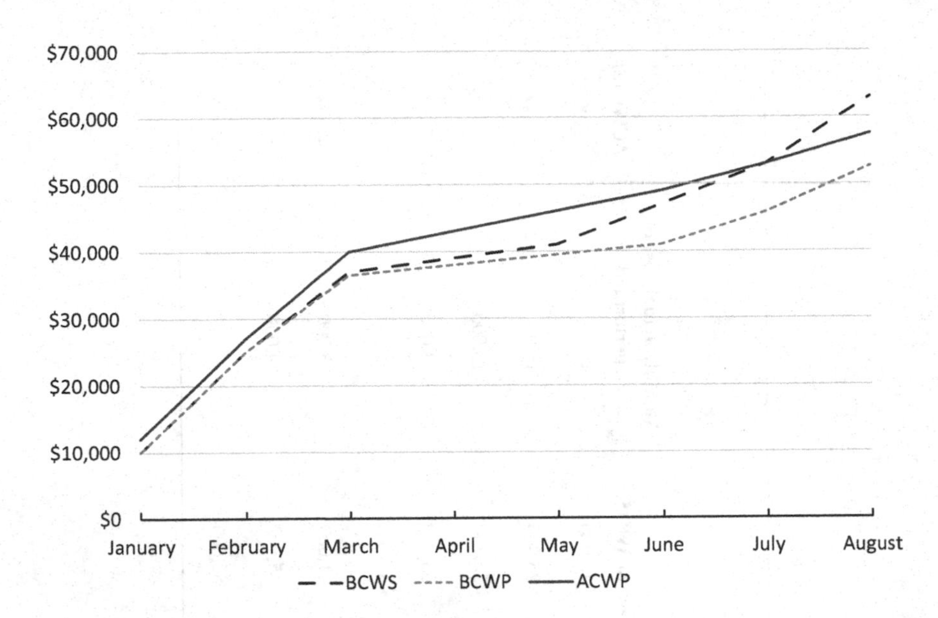

FIGURE 16.16 Plot of BCWS, BCWP, and ACWP in the example project

and slower-than-expected progress of the work carried out should give the project manager reason for concern.

Interpreting the Results of Earned Value Management

To allow an easier interpretation of the metrics generated through earned value management, variances and indices of schedule attainment and cost are frequently constructed. This section briefly discusses each.

Especially when presented in visual form, plots of the variances give an accessible impression of progress in the project. *Schedule Variance* (*SV*) reflects how well the project is meeting, or perhaps exceeding, its schedule. It is constructed as follows:

$$SV = BCWP - BCWS \tag{16.3}$$

A negative value for *SV* indicates that less work has been done than scheduled and the project is running late.

In an analogous fashion, the *Cost Variance* (*CV*) reflects how well the project is meeting its planned cost targets, with negative values for CV showing that the project work is costing more than planned. It is defined as follows:

$$CV = BCWP - ACWP \tag{16.4}$$

A drawback of using SV and CV is that these metrics are presented in money terms. This means that it is necessary to know the total value of the project to interpret their significance. The *Schedule Performance Index* (*SPI*) and the *Cost Performance Index* (*CPI*) overcome this limitation by being constructed as ratios, giving an immediate impression of the state of the program. SPI and CPI are constructed as follows:

$$SPI = \frac{BCWP}{BCWS} \tag{16.5}$$

$$CPI = \frac{BCWP}{ACWP} \tag{16.6}$$

For both indices, values smaller than 1 indicate that the work has progressed slower or, respectively, has cost more than planned. It is important to note that the interpretation in terms of "good" and "bad" is the reverse of SV and CV, such that values of SPI and CVI greater than one are beneficial and vice versa.

Table 16.8 presents the monthly values for SV, CV, SPI, and CPI calculated for the powerboat example. The same information is presented graphically in Figure 16.17, using two axes. As evident, the example project is not going well. Schedule attainment in the project has consistently declined starting around March. The project has never met the cost targets specified in its original budget, with cost performance deteriorating until June, after which it has recovered slightly.

While the presented metrics can be used by the project manager to give stakeholders, such as senior managers, an easily digestible snapshot of the project's performance at the time of review, it is important to keep in mind that the metrics used in earned value management are simplistic and need to be treated with a degree of caution. For example, positive values for CV may simply occur because payments to suppliers are still outstanding. Moreover, BCWP relies on an accurate reflection of

TABLE 16.8
Actual Cost of Work Performed in the Example Project

Month	Schedule variance ($)	Cost variance ($)	Schedule performance index	Cost performance Index
January	0	−2,000	1.00	0.83
February	0	−2,000	1.00	0.93
March	−500	−3,500	0.99	0.91
April	−1,000	−5,000	0.97	0.88
May	−1,500	−6,500	0.96	0.86
June	−6,000	−8,000	0.87	0.84
July	−7,200	−7,200	0.86	0.86
August	−10,400	−4,900	0.83	0.91

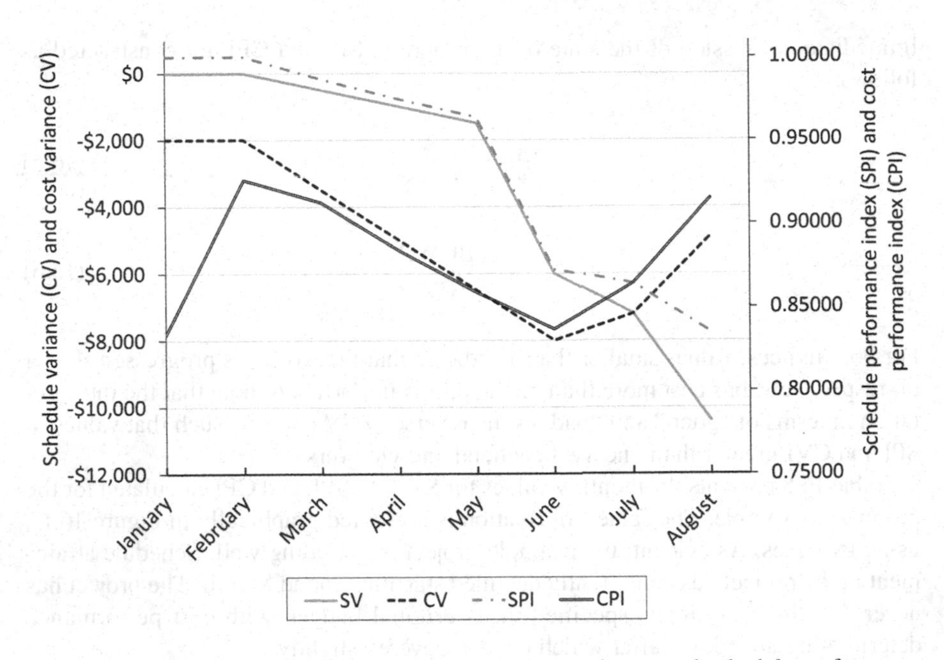

FIGURE 16.17 Combined plot of schedule and cost variance, and schedule performance and cost performance indices

the level of completeness of activities. This may be difficult or impossible to determine in reality, depending on the nature of the work involved. A reason for this may be, for example, that the design of machines or systems progresses unevenly or in increments.

Estimating Completion Time and Cost

Often, project managers will be required to generate forecasts estimating the time and money required to complete the project. There are many ways of doing this, some are very complex and some are very simple. A simple way of forecasting this is to use the SPI and the CPI at the time of review to predict the *estimated completion time* (*ECT*) and *estimated completion cost* (*ECC*). ECT can be specified by dividing the planned total project duration over the SPI:

$$\text{ECT} = \frac{\text{Planned project duration}}{\text{SPI}} \tag{16.7}$$

Given the available information, ECC can be constructed by dividing the total planned cost of the project $BCWS_{complete}$ over the CPI:

$$\text{ECC} = \frac{BCWS_{complete}}{\text{CPI}} \tag{16.8}$$

These metrics can be applied to the example of the powerboat project developed in this section. The ECT is generated at the time of review in August, at which the SPI stands at 0.83. Given the planned duration of the project of 11 months, the ECT is approximately 13.2 months. Also generated at the time of review, the relevant index for ECC is the CPI, standing at approximately 0.91 at the time of review. The project has an original total budget, or $BCWS_{complete}$ in the earned value terminology, of $94,000, as shown in Figure 16.12. On this basis, the ECC is estimated at $102,757.*

Since both SPI and CPI may change rapidly, it should be noted that both ECT and ECC are volatile and may not give a reliable outlook. As with all forms of forecasting, it is down to the judgment of the forecaster, in this case the project manager, to interpret the obtained values in the context of their own experience, general impressions of the project at the time of review, and knowledge of likely external influences.

Without wanting to discourage any budding project managers, in industrial practice, it is not uncommon for projects to run over time and cost. Generally, the project manager will specify contingency allowances of between 10% and 15% of the original budget. Experienced senior managers will be supportive of this.

CONCLUSION

After very briefly outlining the field of project management, this chapter has picked out two major topics in project management, project planning and project monitoring. As part of the section on project planning, which is aligned with the widely used techniques of PERT and CPM, the chapter has presented the distinction of activity-on-node networks and activity-on-arrow networks, alongside the concept of float, as core ideas of project planning. Without understanding this, it is difficult to fully grasp the meaning of modern project planning techniques and tools. The following section on project monitoring has concentrated on the technique of earned value management, which forms the conceptual basis for most current project monitoring tools. In this context, the chapter has highlighted the core idea of combining information on budgeted costs with data on actual work progress in the form of the BCWP metric. Again, without understanding this, it will be difficult to engage with professional project monitoring activity.

Of course, the presented project management techniques are only a very small part of the overall field of project management and many important aspects were omitted. Many useful textbooks are available on project management. A useful and up-to-date textbook is *Project Management for Engineering, Business and Technology* by John M. Nicholas and Herman Steyn (2017). Further treatments of the presented techniques can be found in the following textbooks: *Management for Engineers, Scientists and Technologists* by John Chelsom, Andrew Payne, and Lawrence Revill (2005), *Project Management* by Mike Field and Laurie Keller (1998), and *Engineering Project Management* edited by Nigel Smith and colleagues (2008).

---

* Note that the estimation of ECT and ECC is sensitive to rounding errors. To replicate the results presented in this chapter, the use of a spreadsheet tool is recommended.

REVIEW QUESTIONS

1. Which of the following are not valid criteria for an activity to be classified as a project?

   (Question type: Multiple response)

   ☐ Rewards must be great enough to begin it
   ☐ A clear purpose and well-defined goals
   ☐ Using careful analysis, risk can be eliminated
   ☐ Novel and unique in a sense that it has not been carried out before
   ☐ Engages people and resources from other organizations or functional units
   ☐ Temporary activity that ends after it has achieved its goals or failed
   ☐ Headed by a project manager

2. In project management, deliverables are tangible or intangible objects or outcomes that the team is contractually obliged to produce.

   (Question type: True/false)

   Is this statement true or false?

   O True O False

3. Which of the following statements about activity-on-node networks and activity-on-arrow networks are true?

   (Question type: Dichotomous)

True	False	
O	O	Both show activities as nodes connected by arrows
O	O	Activity-on-node networks show activities as nodes connected by arrows
O	O	In activity-on-arrow networks activities start and finish at nodes
O	O	Events have zero time duration
O	O	In activity-on-arrow networks, activities have zero time duration
O	O	Insight on project structure can be extracted from these networks

4. You are planning a project and have identified the following events network for your project:

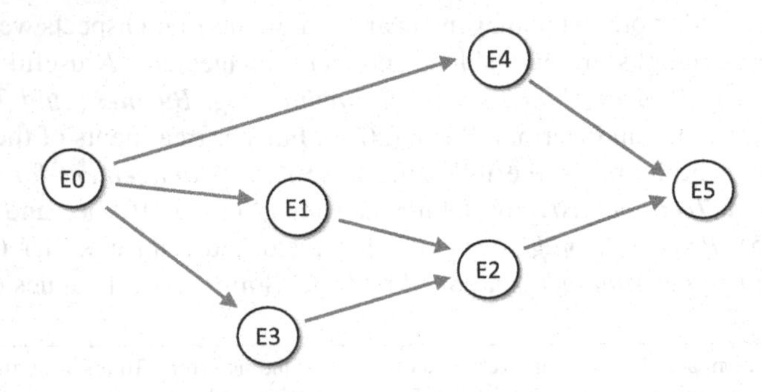

FIGURE 16.18 Events network

The duration of the activities associated with the events are as follows:

- E0,1: 1 week
- E1,2: 3.5 weeks
- E2,5: 1.5 weeks
- E0,3: 1.5 weeks
- E3,2: 11 weeks
- E0,4: 2 weeks
- E4,5: 4 weeks

(Question type: Calculation)

Using the backward pass, determine the latest time of occurrence of event E4.

5. In network analysis, the earliest activity start (EAS) of an activity is 7 and the latest activity finish (LAF) is 15. Which of the following is correct?

   (Question type: Multiple choice)

   O The duration of the activity is 8
   O The duration of the activity is 15
   O Duration cannot be stated using the information given
   O EAS has been identified using a backward pass

6. Complete the following figure showing the activity module for the activity-on-node diagram by inserting the correct labels.

   (Question type: Labeling)

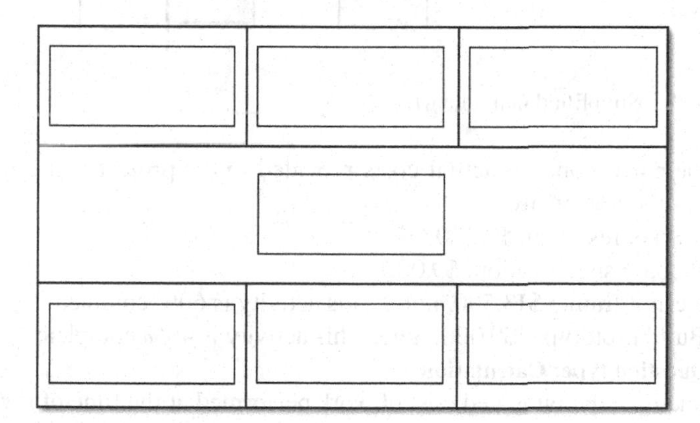

FIGURE 16.19 Insert the appropriate labels

7. The budgeted cost of work performed is a measure of how much of the scheduled work has actually been completed.

   (Question type: True/false)

   Is this statement true or false?

   O True O False

8. A project manager is monitoring a project to develop a bicycle frame. The following table is a simplified Gantt chart describing the project plan:

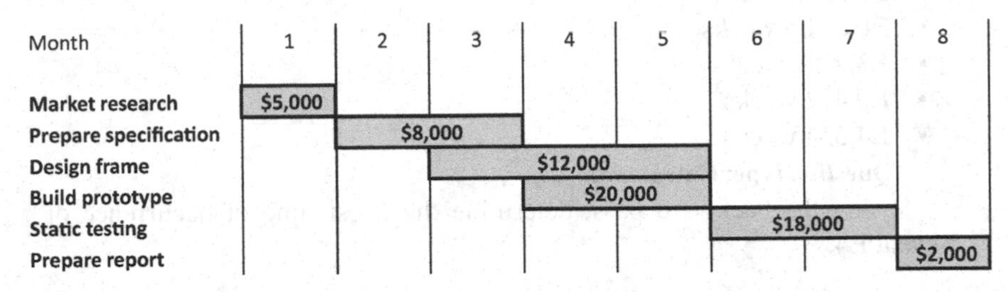

FIGURE 16.20 Simplified Gantt chart

A project review has taken place at the end of the fifth month and the project manager is given the following Gantt chart showing project progress:

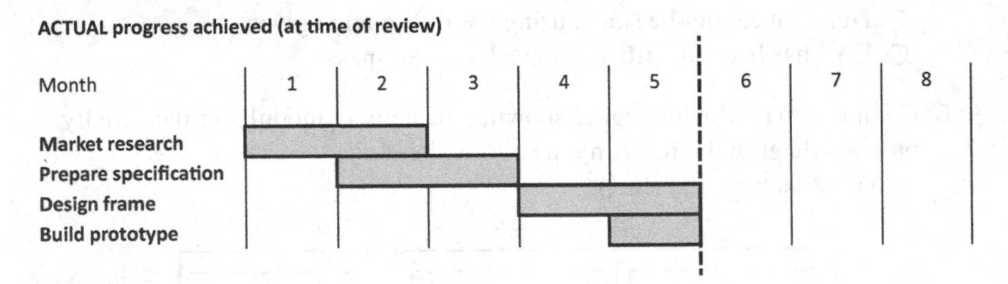

FIGURE 16.21 Simplified Gantt chart

The corresponding actual costs provided to the project manager at the time of the review are:

- Market research: $7,000
- Prepare specification: $9,000
- Design frame: $13,500, note: This activity is 60% complete
- Build prototype: $9,000, note: This activity is 40% complete

(Question type: Calculation)
Calculate the budgeted cost of work performed at the time of review.

9. The project manager has another question about the project outlined in the previous question.
   (Question type: Calculation)
   Calculate the actual cost of work performed at the end of month 4.

10. For a project you are managing you have estimated a budgeted cost of work performed of $57,500. According to your plans, the total expenditure at this point should be $94,000.

    (Question type: Calculation)

    Calculate the schedule performance index at the time of review.

REFERENCES AND FURTHER READING

Chelsom, J.V., Payne, A.C., and Reavill, L.R.P., 2005. *Management for engineers, scientists and technologists*. 2nd ed. Chichester: Wiley.

Field, M. and Keller, L., 1998. *Project management*. London: Thomson Learning.

Nicholas, J.M. and Steyn, H., 2017. *Project management for engineering, business and technology*. 5th ed. Abingdon: Routledge.

Smith, N.J. ed., 2008. *Engineering project management*. 3rd ed. Ames, IA: Blackwell Science.

Glossary

A

Account: In accounting and bookkeeping, an account is a record that tracks the financial activities and transactions involving a specific asset, liability, equity, revenue, or expense.

Accounting: Process of capturing information on the financial transactions of a business.

Accounting cycle: A model used to introduce the process of accounting, usually presented as an eight-stage cycle that is completed in each accounting period.

Accounting equation: The conceptual core of financial accounting. It is an equation that states that the assets in the business equate to the liabilities plus the owners' equity.

Actual cost of work performed (ACWP): The true cost of work that has been completed in a project up to the time of review.

Added value: Difference of sales revenue and input costs.

Agile development: Product development methodology designed to shorten the development time of products and to allow the continuous collection of new information to guide the developmental process.

Amortization (accounting): Decrease of the value of an intangible asset over time.

Amortization (finance): Process of repaying debt through principal and interest payments over time, lowering the balance of the debt.

Artificial intelligence (AI): The science and engineering of making intelligent machines, capable of accomplishing acts of reasoning and intelligence.

Assemble-to-order (ATO): Supply chain structure in which the assembly of the final product is delayed until a customer order is placed.

Asset: Any item of property, tangible or intangible, owned by a person or business that has value and can be sold or otherwise made available to meet commitments, debts, or obligations.

Automation: Any kind of technology by which a process, procedure, task, or activity is performed with minimal human assistance or intervention.

Average cost: The average cost associated with each unit of output at a specific quantity of units.

Average revenue: The total revenue obtained divided by the quantity supplied.

Axiom of non-satiety: The assumption that more of a good is always better than less.

B

Backward pass: Technique used in project management to determine the latest activity finish for each activity.

Balance sheet: Financial statement showing the assets and liabilities of a business and from which sources finance has been raised. The balance sheet reflects this information for a particular point in time.

Bargaining power: Ability to exert influence over other parties in a transaction in order to maximize own benefits.

Barrier to entry: Advantage that benefits established businesses in an industry but does not benefit businesses seeking to enter the industry. It thereby deters or prevents entry.

Batch operations system: Production system that uses special tooling, often employing flexible manufacturing systems. Generates higher volumes of products than job-shops.

Behavioral models: Set of models explaining the process of influencing others through the behavior exhibited by managers.

Benchmarking: Process used to identify and examine best practices or current standards of performance.

Branding: Practice of distinguishing a product or service offering from competing offerings through its overall features and characteristics as perceived by the customer.

Breakeven: Point of intersection of functions of interest to managers, especially of a cost function and a revenue function.

Budgeted cost of work performed (BCWP): Cumulative budgeted cost of the work that has been performed in a project.

Budgeted cost of work scheduled (BCWS): Cumulative budgeted cost of the planned activities in a project.

Burn rate: Negative cash flow, normally measured monthly, before a new business starts generating revenue.

Business angel: Wealthy individual who invests in new businesses by providing their own funds, usually in exchange for equity.

Business case: A description of the projected earnings and costs upon which the success of a distinct project, which may well be a business, depends.

Business model: Representation of the rationale of how an organization creates, delivers, and captures value in economic, social, cultural, or other contexts.

Business model canvas: Framework widely used by managers and entrepreneurs to develop new business models or to analyze existing ones.

Business model fit: A kind of fit that can be demonstrated when there is evidence that the adopted business model is profitable and can be scaled.

Business plan: Formal document describing a number of aspects in a new business, including its aims, the methods that will be adopted to attain these aims, and the timescale of the formation of the new business.

Business-to-business (B2B): Activity of selling products and services from a business to businesses, such as from manufacturers to wholesalers or from manufacturers to retailers.

Business-to-consumer (B2C): Activity of selling products and services from a business to consumers.

Business-to-government (B2G): Activity of selling products or services from a business to governments and public institutions.

C

Capacity: Upper limit for the processing or production of outputs, which can be products or services.

Capital: Any kind of financial or valuable object, physical or intangible, that is owned by a business or an individual and is useful in generating profit or progressing development otherwise.

Cash flow forecast: Plan that shows the amounts of money a business expects to receive as revenue and to pay out as expenses over a given period of time.

Cash flow shape: Pattern of cash flows associated with commercial initiatives.

Cash flow statement: Financial statement reporting how changes in the balance sheet entries and financial flows in and out of the business over a period of time change the cash and cash equivalents present in the business.

Central limit theorem (CLT): Statistical concept stating that the distribution of sample means will approach a normal distribution for samples of a reasonably large size. This is the case even if the underlying variable is not normally distributed.

Closed-end lease: Way of acquiring an object or asset in which leasing fees are paid over a period of time followed by the option of purchasing the object or asset for an additional sum.

Coercive power: Ability to make others comply by instilling fear of negative consequences, punishment, or harm.

Coherence: State that can be achieved by creating a value proposition which is well-aligned to the internal processes and goals of a business, developing a system of capabilities that is able to deliver the value proposition, and having a matching portfolio of products and services.

Collateral: Object or sum that acts as a security for the repayment of a loan or debt, will be forfeited in the event of a default.

Complementary good: Good whose appeal increases if the popularity of another related good also increases. In other words, the demand for a complement increases if the price of another good decreases.

Compound interest: Interest earned on the principal sum and all previously accumulated interest.

Conglomerate: Business that is characterized by the ownership of business units in seemingly unrelated lines of activity.

Consumer-to-consumer (C2C): Activity of selling products or services from a consumer to another consumer.

Continuous operations system: Production system processing a flow of bulk products that can only be measured in terms of a dimension, such as weight or volume. Geared for high production volumes, utilizing dedicated equipment and a high degree of automation.

Contribution: The additional profit obtained from the sale of an additional unit.

Control chart: Graphical tool showing how quality-relevant process variables fluctuate over time.

Copyright: Form of intellectual property granting authors rights that only they can exercise, such as the right to adapt, publish, broadcast, or prevent distorted copies of their work. Copyright relates to artistic and literary works, including music, paintings, books, sculptures, sound recordings, films, broadcasts, computer programs, and electronic databases.

Corporate social responsibility: Set of mechanisms integrated into the business by which managers evaluate and ensure compliance with ethical standards, the spirit of the law, and national or international norms.

Cost accounting: Process that records and reports costs. This includes the activities of classifying, grouping, and summarizing actual cost data and comparing these data with expected or standard cost data.

Cost-benefit analysis: The systematic assessment of the advantages and disadvantages of alternative courses of action to identify the option achieving the greatest benefits while protecting value for all affected parties.

Cost-volume-profit analysis: Group of techniques used to compare relevant functional relationships sharing a set of variables.

Cost of goods sold (COGS): Sum of input costs and process costs.

Critical path: The sequence of activities with the greatest collective duration in a project plan that must be carried out for the project to complete.

Cross price elasticity of demand (XED): The percentage change of demand for one good following a change in the price of another good.

Crowdfunding: A source of funding in which a new business raises money from a large number of amateur investors, normally through a service provided by an internet platform.

Current liability: Liability that is expected to be settled within one year. This normally includes salaries, fees, taxes, interest payments due within a year, and purchase expenses.

Customer creation: The stage in the customer development process in which the new business demonstrates that it can successfully convince new customers in sufficient numbers.

Customer demand: Customer needs and wants that coincide with a customers' willingness to pay.

Customer development: Commercial innovation approach aiming to expose customers as quickly and cheaply as possible to a product or service under development.

Customer discovery: Stage in the customer development process in which the concept of a business and its business model are formulated for the first time and the founders' original vision is translated into hypotheses about the business.

Customer engagement: Repeated interaction between a customer and a business through online and offline communication channels. Its objective is to strengthen the relationship a customer has with a business through mental, emotional, and social factors as well as through the objects the customer owns.

Customer lifetime value: (CLV) The total net benefit arising from a customer or a group of customers to a business over the duration of their relationship with the business.

Customer need: State of felt deprivation that a customer tries to satisfy. Customer needs include physical, material, social, emotional, and material needs.

Customer needs analysis: Technique used to identify what requirements customers have regarding a product or a service.

Customer relationship management (CRM): The systematic management of customer relationships using digital technologies.

Customer survey: Qualitative forecasting method in which customers are surveyed regarding their assessment of the future.

Customer validation: Stage in the customer development process in which the founders attempt to validate the business idea, testing whether the business and the product or service are viable.

Customer want: Customer need shaped by customer traits such as their personality if the customer is an individual, or their organizational culture, if the customer is an organization or a business.

D

Deliverable: Any measurable product, service, information, or result that must be completed to finish a project.

Delphi method: Qualitative forecasting method using a panel of experts for a survey, asking them to speculate about changes to their specialist area in the long-term future.

Demand: Quantity of a good that a group of buyers is willing and able to buy at a particular price.

Depreciation: Actual decrease of the value of a tangible asset over the time it has been in use.

Diffusion: Process by which products or services spread within a potential market. Only products and services that have undergone an innovation process can enter into diffusion. There is no guarantee that diffusion will occur.

Digital manufacturing operations system: Production system combining digital technologies with production operations systems to enable both high volumes and high variety, thereby escaping the traditional trade-off associated with the product-process matrix.

Digital marketing: Pursuit of marketing goals through the use of digital media, digital data, and information technology.

Digitalization: Process of using and leveraging digitization for various purposes and activities.

Digitization: Process of converting relevant information and procedures into digital form so they can be acted upon by computers, including their transmission through computer networks.

Direct material costs: The costs incurred for purchases of the materials processed in the business's activities.

Discount rate: Interest rate used to determine the present value of future cash flows or other objects.

Diseconomies of scale: Cost increases that are realized when the scale of an operation or process increases.

Disruptive innovation: Process of creating radically different new products or services that can be sold in a market. In this process, the new products or services destroy the existing market and the advantages held by the businesses in these markets.

Diversification: Activity of entering into multiple markets involving substantially different skills, technology, and knowledge or acquiring different investments in order to reduce the exposure to any particular risk.

Diversity: The different attributes of people and any characteristic that lead to another person being identified as different from the self.

Dividends: Net profit generated and paid out to shareholders.

Define-measure-analyze-improve-control (DMAIC) cycle: Improvement cycle forming the methodological core of Six Sigma.

E

Earliest activity start (EAS): The earliest time an activity in a project can start subject to precedence constraints.

Earned media: Exposure created through viral and social marketing, including any communications, ratings, and reviews published by customers.

Earned value management (EVM): Technique used to monitor the progress of work and project spend against original plans in terms of a set of specific project monitoring metrics.

Earnings before interest and taxes (EBIT): Sum of net profit, interest payments, and tax expenses.

Earnings before interest and taxes, depreciation and amortization (EBITDA): Sum of net profit, interest payments, tax payments, depreciation, and amortization.

Economies of scale: Cost reductions that are realized when the scale of an operation or process increases.

Economies of scope: Cost reductions that arise when the joint production of multiple products or product variants exhibits a lower cost than the production of each output separately. Economies of scope form a special kind of economies of scale.

Efficiency: State in which an individual, a business, or society gets the most out of the available resources or goods and services. Efficiency is the result of an ability to avoid wasting raw materials, energy, effort, money, and time or creating undesirable outcomes such as pollution.

Elevator pitch: Short verbal presentation of the founder's vision normally based on the executive summary contained in the business plan.

Engineer-to-order (ETO): Supply chain structure in which product development and design activities are triggered by the receipt of a customer order.

Enterprise resource planning (ERP) system: System providing functions for planning, scheduling, operation, and control. Usually in the form of integrated software systems.

Equity: Ownership of assets that may or may not have debts or other liabilities attached. Equity can apply to an individual asset or to an entire business. If the owners of a business own shares, their ownership stake is referred to as shareholders' equity.

Events: Artificial points of time defined in project management with a zero time duration and occurring only if all necessary activities leading to it are completed.

Expert power: Power flowing from mastery of indispensable abilities, knowledge, or expertise.

Externality: Cost or benefit resulting from an activity that affects a third party which did not choose to incur that cost or benefit. An externality is therefore an unintended by-product of an activity.

F

Failure modes and effects analysis (FMEA): Technique developed to analyze how products, services, or aspects thereof, can fail.

Financial accounting: The process of collecting and recording information on past transactions resulting from the activities of the business over a defined period of time.

Financial ratio: Method of analyzing the financial information presented in the financial statements issued by a business.

Five forces analysis: Method to analyze the effects of existing and potential competition on the profitability of an industry or market. Closely associated with the work of Michael Porter.

Fixed costs: Costs that are not linked to the total quantity of goods and services produced.

Flexible manufacturing systems: Manufacturing systems that have some degree of flexibility enabled by digital control.

Float: Variable in project management representing the spare time around an activity.

Forward pass: Technique used in project management to determine the earliest activity start for each activity.

Franchise: Agreement in which a new business enters into an agreement with an existing, normally large, business, known as a franchisor. This arrangement gives the new business, referred to as the franchisee, access to resources, including information, processes, and intellectual property.

Funding round: Event that gives outside investors the opportunity to invest cash in a new business in exchange for shares in the same business.

G

Gantt chart: Chart showing the required start and finish times for each activity in the project in an intuitive way and using a time axis.

General journal: Accounting document recording every relevant business transaction, for example, relating to sales, receipts of money, changes in inventory, and purchases. It records the transactions in chronological order.

General ledger: Main and most important information repository in any accounting system. It forms the collection of all accounts maintained by the accounting system.

Gig economy: Businesses relying on low-skilled workers flexibly hired on a contractual basis. Enabled by advanced software systems, normally by businesses running O2O business models.

Good: General expression for a beneficial material, or tangible, object, or item. Sometimes applied to services as well.

Goodwill: Accounting term for an asset that arises when an existing business, or a part of it, is sold to a new owner. Goodwill is the difference between the purchase price and the value of the assets owned by the acquired business minus other obligations and debts.

Grant: Amount of money that is given by a government or other grant-making organization to a business or individual for a specific purpose, normally linked to public benefit. Unlike loans, grants are not paid back.

Gross profit: Sales revenue minus the cost of goods sold.

H

Hierarchical organizational structure: A structure in which every person or group, except a single individual at the top, is subordinate to the authority of another person or group.

Hierarchy of needs: Theory that people are subject to a range of needs and will try to fulfill whichever need is most pressing in a given situation, prioritizing lower-order needs until they are at least partially fulfilled. Closely associated with the work of Abraham Maslow.

Hostile takeover: Act in which an outside bidder acquires ownership of a business whose managers are unwilling to agree to the acquisition.

Human capital: Overall explicit and implicit knowledge, habits, skills, abilities, behaviors, social attributes, and personality traits embodied by people and used in the attainment of business goals.

Hygiene factor: Factor surrounding work that causes dissatisfaction, thereby negatively affecting motivation. Closely associated with the work of Frederick Herzberg.

I

Incorporation: Act of adopting an incorporated legal form by a business, thus receiving the status of being a legal person.

Income: Financial inflows, often in the form of sales revenue.

Industrial design: Form of intellectual property that does not relate to an invention but to a unique appearance.

Inflation: Decrease of the value of money over time, captured by the annual inflation rate.

Information good: Good that is made of binary digits, referred to as "bits", rather than of physical materials. This means that anything that can be encoded in a sequence of bits can be an information good.

Innovation: Process of transforming an idea that is new or perceived as new into products or services that can be sold in a market.

Input cost: Financial outflow from the business due to bought-in materials and components.

Intangible asset: Non-physical object of value that can be used to meet the commitments, debts, or obligations of the business. Important forms of intangible assets are various forms of intellectual property and valuable information.

Integrated circuit (IC): A circuit in which all or some of the circuit elements are inseparably associated and electrically interconnected.

Intellectual property (IP): Form of property that results from original creative thought. As a legal object, the characteristics of intellectual property are defined by the laws of a country.

Intensive technologies: Technologies that produce changes and transformations through an intensive interaction with an object or a person in a specific situation.

Interest rate: The proportion of a principal sum that is charged as interest to the recipient of the principal sum, typically expressed as an annual percentage.

Internal rate of return (IRR): Rate of return at which the net present value of a cash flow profile is zero.

Inventory: Objects held by a business ready for use or processing or sale, including raw materials, components, assemblies, intermediate products, and finished products. Also referred to as "stock" in some circumstances.

Investor: Shareholder in a limited company who seeks to profit through the long-term ownership of shares through capital gains, dividends, or interest payments.

Ishikawa diagram: Graphical method used to relate possible events to their outcomes.

J

Job-shop operations system: Production system that is flexible enough to allow a wide variety of tasks while producing a greater volume of output than a project operations system. Overall, the volume of production is still relatively low.

Just-In-Time operations: (JIT) Processes in which the required inputs, materials, or components are delivered or made available immediately before they are used.

K

Kanban: Pull signal used in Lean operations.

L

Latest activity finish (LAF): Latest time an activity in a project can finish without causing delays to project completion.

Lean: Learning-based management philosophy focused on continuous improvement, which means continuously challenging oneself and learning by taking small steps and respect, which means supporting workers and making the best possible use of the available resources.

Legal form of ownership: Legal structure adopted by a business.

Legal personhood: Concept that under certain circumstances a business can do things normally reserved for human beings. This includes the ability to enter contracts, incur debt, own property, sue, and be sued.

Legitimate power: Power that flows from a manager's formal position of seniority.

Liability: Obligation of a business or person arising from a past event that is expected to result in the transfer of tangible or intangible assets, a provision of services or another benefit in the future.

Life cycle assessment (LCA): Study of the environmental aspects and potential impacts throughout an object's life from raw material acquisition through production, use, and disposal. The general categories of environmental impacts needing consideration include resource use, human health, and ecological impacts.

Life cycle (business): Series of distinct stages through which an object, such as a business, a product, or a service offering, passes during its existence.

Line operations system: Production system with dedicated manufacturing equipment and high levels of automation. Enables very high production volumes with small numbers of workers, mainly employing an unskilled workforce.

Liquidity: Ease with which an asset can be converted into cash without leading to changes in its value. Cash is, by definition, the most liquid asset.

Loan: Sum of money that is borrowed, normally expected to be paid back with interest.

Logic network: A simple network diagram used in project management capturing the sequence of precedence and parallelity inherent to project activities.

Long-linked technologies: Technologies that feature step-by-step processes of hand-off between distinct elements. Dominant in manufacturing, especially in mass-production, long-linked technologies exhibit an activity flow that can range from raw material generation to the assembly and packaging of the finished product.

Long-term liability: Liability that is not expected to be settled within a year. Long-term liabilities include long-term bonds, long-term debts, leases, pension obligations, and warranties and guarantees.

M

Macroeconomics: Branch of economics interested in the performance, behavior, structure, and decisions of an economy as a whole.

Make-to-order (MTO): Supply chain structure in which fully developed products, or components thereof, are fabricated, assembled, and shipped to the customer upon receipt of an order.

Make-to-stock (MTS): Supply chain structure in which all activities feed into an inventory of finished products from which the customer, who forms the end of the supply chain, is served.

Management: Group of functions in an organization that concerns itself with the direction of various activities to attain the organization's objectives. In doing so, management deals with the active direction of human effort.

Management by objectives: Practice of defining and using specific objectives within an organization so that managers can act to achieve each objective. This approach allows managers to assess the success of processes within businesses more clearly. Closely associated with the work of Peter Drucker.

Managerial accounting: Provision of information on a broad range of financial aspects required to make sound decisions in line with the goals of a business.

Margin of safety: Metric indicating the overall strength of a current or planned position of the business, allowing managers to identify what amount is or can be gained over or below the point of breakeven.

Marginal cost: Cost associated with the last, additional unit of output.

Marginal revenue: Additional revenue obtained by increasing the output quantity by one unit.

Market: Any structure or procedure that allows buyers and sellers to exchange any type of good, service, or information. A market may be specific to a certain good, service, or information.

Market equilibrium: Intersection of the supply and demand curves. Defines the quantity and the price at which the buyers' preferences and the sellers' preferences are in balance.

Market offer: Embodiment of the combined products or services the business intends to supply to the customer to satisfy an identified want.

Market power: Ability of a business to raise the price charged for a good or service over its cost of production without losing all of its customers.

Market pull innovation: Innovation process that creates new marketable products and services only in response to identified market forces and customer needs.

Market segmentation: Practice of dividing a market comprised of diverse customers into relatively more homogenous groups on the basis of demographic, geographic, socioeconomic, and behavioral factors.

Marketing: Management process responsible for identifying, anticipating, and satisfying customer requirements profitably.

Mass-production: Manufacturing approach combining systematic factory practices, advanced management techniques, and the expression of workers' interests.

Mass service system: Operations system delivering a standardized service in very large volumes. This mode of service provision features very limited interactions with the customers and very little, if any, customization.

Maximum investment: Maximum level of investment in a project or commercial initiative. It is the highest level of cumulative negative cash flow experienced in a project or initiative.

Mediating technologies: Technologies that facilitate interactions between individuals or businesses.

Metal–oxide–semiconductor field-effect transistor (MOSFET): Most important and widespread transistor technology.

Microeconomics: Branch of economics that is interested in the behavior of individuals and firms in making decisions about the allocation of resources and the interactions between individuals and firms.

Milestone: Specific point in time during the course of a project that can be used to measure the progress of the project toward its ultimate goal.

Minimum viable product (MVP): Object that can function as a product with the smallest possible number of features. An MVP embodies the hypotheses held by the managers about the customer needs and how they can be satisfied.

Monopolistic competition: Market in which there are many buyers and many sellers of a good, but each seller offers a slightly different version of the good.

Monopoly: Market in which there is only one seller of a product and multiple buyers.

Moore's law: Regularity that the power of computers relative to their cost approximately doubles each year.

Motivator factor: Factor that motivates people to invest effort in their work.

N

Net income: Alternative expression for net profit, sometimes also referring to other forms of profit.

Net present value (NPV): Cumulative present value of a cash flow profile subject to a given discount rate.

Net profit: Sales revenue minus the total costs incurred by the business.

Networking: Activities directed at building relationships with others to support work goals or career ambitions.

New business: Organization attempting to establish a viable and profitable business model under conditions of extreme uncertainty. New businesses are temporary because they either transition to being an established business or fail and cease to exist.

Nominal capacity: Level of capacity a system or operation is designed for. Utilization can normally exceed nominal capacity, at least temporarily.

O

Oligopoly: Market in which there is a small group of sellers and multiple buyers.

Online-to-offline (O2O): Businesses combining online elements with the transaction of physical goods or services.

Openness, conscientiousness, extraversion, agreeableness, and neuroticism (OCEAN) model: Framework explaining the process of influencing others through the personality traits of managers.

Operating costs: Financial outflows related to the operation of the business, reflecting the costs of resources employed to maintain its existence over time. Operating costs include labor, marketing, human resources management, and cleaners.

Operating profit: Sales revenue minus cost of goods sold minus operating costs.

Operations: Part of a business that takes inputs, such as capital, equipment, materials, labor, and information, and processes these to generate outputs, such as products and services, for customers.

Operations management: Set of activities, decisions, and responsibilities relating to the management of the creation and delivery of products and services.

Opportunity cost: Loss of not enjoying the benefit attached to the best alternative choice or course of action when one alternative is chosen.

Organizational lag model: Theory describing the general inability of management methods to adapt in step with rapidly evolving technologies.

Outright purchase: Way of acquiring an object or asset in which the full price is paid upfront and ownership is transferred immediately.

Overall equipment effectiveness (OEE): Important and versatile metric used to assess equipment performance in Lean.

Own-price elasticity of demand: Percentage change of the quantity demanded as a result of the percentage change of the price of a good.

Own-price elasticity of supply: Percentage change of the quantity supplied as a result of the percentage change of the price of a good.

Owned media: Channels owned by a business through which it can reach a target audience, including websites, blogs, social media accounts, and email.

Owners' equity: Value that would be returned to the shareholders if all assets in the businesses were sold after all of the debts owed by the business were paid off.

P

Paid media: Marketing channels requiring the business to pay for visitors or to display advertisements.

Partnership: Legal form of ownership in which the business is owned and controlled by a group of individuals. As a partnership, the business does not have legal personhood.

Patent: Form of intellectual property granting the legal right to prevent any other business or person from commercially using an invention for a limited duration, which is 20 years in most countries.

Perfect competition: Market in which there are many sellers and many buyers.

Persevere: Decision to retain a planned course of action in a new business.

Pitch deck: Brief digital slide show and oral presentation based on the executive summary of the business plan.

Pivot: Modification or reconfiguration of a planned business in light of new information.

Plan-do-check-act (PDCA) cycle: Well-known improvement cycle aiming to achieve continuous improvement over time.

Platform: Business providing an infrastructure service enabling value-creating interactions between other actors. These actors can be suppliers or customers of products and services, including information goods.

Political, Economic, Socio-cultural, Technological, Environmental, and Legal (PESTEL) analysis: Framework of environmental factors that can be used to evaluate the strategic context of a business.

Present value (PV): Present value of a cash flow subject to a given discount rate.

Principal sum: Original amount of money loaned, borrowed, or deposited in an investment.

Private enterprise: Organization owned and managed by individuals that freely decide to do so and for their own financial benefit.

Private limited company: Legal form of ownership in which the business is owned by shareholders and has legal personhood. This means that the shareholders are in principle not liable for the debts of the business.

Private venture: Commercial initiative producing a pattern of cash flow in which costs incurred in the short term are charged against the revenues generated by the same activity or project in the future.

Problem-solution fit: Kind of fit that is created when a value proposition is able to address the customers' needs or requirements.

Process costs: Financial outflows related to the transformation of inputs into outputs by the business.

Profit and loss budget: Financial plan setting out the amount of revenue the business will generate, the types and levels of cost that will be incurred, and the amount of profit or loss.

Product life cycle management (PLM): Activity of managing the full life cycle of a product or service offering from invention, design, and growth to in-life support, decline, and replacement.

Product-market fit: Kind of fit that is created when there is evidence that the product or service is creating value for customers by addressing the pains and gains.

Product-process matrix: Model suggesting that production systems cannot be set up to simultaneously deliver products in high volumes and with many varieties.

Professional service system: Service system in which customers interact intensively with the business, usually over a prolonged period of time. Designed to deliver high degrees of customization and adaptation to the customer.

Profit: A financial gain obtained by a business, resulting from the difference between an amount earned and an amount spent in buying, processing, or producing something. Also referred to as "net income" or "return".

Profit center: Segment in a business that achieves inflows of money and incurs costs and that has a positive effect of on financial performance of the business.

Profit and loss statement: Financial statement reporting the total sales revenues achieved by the business and showing how this financial inflow is transformed into net profit.

Profitability: An ability to generate income and grow in the short term and long term.

Project operations system: Production system producing goods in small numbers, down to a single unit. Usually organized to carry out individual tasks and processes in parallel, normally over a prolonged period of time.

Public limited company: Business with limited liability that is permitted to offer shares for sale to the wider public, for example, on the stock exchange.

Pull signal: Signal passed between the stages of a pull system.

Pull system: Operations system in which the pace of operations is determined by customer demand, arising as orders.

Pure online: Business focused on the provision of information goods through the internet.

Push system: Operations system in which work is progressed according to plans, schedules, and demand forecasts.

Q

Qualitative forecasting: A group of forecasting techniques based on information, knowledge, experience, and subjective opinions from decision-makers, specialists, and other experts.

Quality: Satisfaction enjoyed by customers as the outcome of the short-term and long-term performances of a product or service.

Quality loss function (QLF): Specification of the total cost resulting from a deviation in a specific quality-relevant target value in the long run.

Quality management: Activity of overseeing and controlling all processes and tasks needed to maintain a desired level of quality.

Quality management system: Set of processes aimed at consistently meeting quality standards.

Quantitative forecasting: Use of numerical and statistical techniques to forecast the future by using data and building models of relevant relationships.

R

Rate of return: Quantitative expression of the benefits received from cash flows, normally arising annually as a percentage of some initial investment.

Referent power: Power that is derived from the personal attractiveness and charisma of an individual.

Reservation price: Highest price customers are willing and able to pay for a good or a service.

Residual control rights: Rights to make decisions regarding the use of objects that are not explicitly assigned to another party in a contract.

Residual value: Value remaining in an object or asset after use or depreciation.

Return on capital employed (ROCE): A group of financial ratios comparing the profit earned by a business against the investment employed to earn this profit.

Return: Alternative expression for profit.

Revenue: Alternative expression for sales revenue.

Reward power: A manager's ability to issue rewards and compensation to incentivize others.

Risk: Possibility of an event which, should it occur, will have an effect on the achievement of objectives. A risk involves the likelihood that the event will occur and the magnitude of its impact on objectives.

Risk choice: Responses available for the treatment of an identified and prioritized risk.

Risk management: Identification, assessment, and control of financial and non-financial risks to stakeholders, together with the identification of procedures to avoid or minimize their impact.

Risk register: Risk management document that names and briefly describes each identified relevant risk in context. Allocates a category under which the risk falls, estimates the likelihood of the risk occurring, and anticipates its expected impact should it occur.

Robust quality: Quality of the products in use, including unexpected environments, atypical patterns of use, mild misuse, deviation from maintenance schedules, and so on.

Runway: Time a new business has before it runs out of money. Calculated using the business's overall initial funding and the burn rate.

S

Sales Revenue: Financial inflow received by a business from the sale of goods or services.

Sales budget: Itemized list of the sales projected in the sales plan, usually aggregating sales according to product types or geographic markets.

Sales force survey: Forecasting method in which sales data and the opinions of salespeople are used to construct forecasts of demand.

Sales plan: Forecast of the level of sales, measured in terms of the volume of products sold, that the business expects to achieve over a period of time, often measured in monthly or quarterly periods.

Sales revenue: Financial inflow to the business due to the sale of products or services.

Scalability: Characteristic of a business or business function that describes its ability to tolerate or benefit from an increased scale in activity.

Scarcity: Limited availability of objects, which may be required by individuals or organizations. Scarcity includes the limited availability of financial resources to buy other objects.

Scheduling: Activity of coordinating the operational resources available to a business across time and location.

Scientific management: Group of methods aiming to improve industrial efficiency by establishing a science of factory production. Closely associated with Frederick W. Taylor.

Seasonality: Pattern of regular fluctuation due to recurring seasons or other cycles.

Service-dominant logic: Marketing theory that builds on the idea that organizations, businesses, and markets are based on the exchange of services. This theory argues that the universal purpose of businesses is to apply their competencies to serve customers.

Service factory system: Operations system delivering a high volume and often requiring that an element of the service is generated away from the customers.

Service shop system: Operations system designed for intermediate levels of interaction with customers and customization. This allows such systems to efficiently serve relatively large number of customers.

Shannon's information theory: Symbolic representation of information as binary digits, or bits, allowing the error-free processing of symbols in imperfect systems, up to a threshold known as the Shannon limit.

Share: Unit of ownership in an incorporated business. Owning a share entitles its holder to a portion of the profits of the business and confers a range of responsibilities. A shareholder normally has the right to vote on certain issues pertaining to the business. Shares are also referred to as "stock" in some circumstances.

Shareholder: Person who owns shares in a business.

Short selling: Investment strategy aiming to benefit from decreases in the value of shares. Short selling requires the services of specialist financial companies who make available shares to rent in exchange for a fee.

Single-minute exchange of dies (SMED): Lean approach facilitating rapid changeover to meet varied production schedules.

Situational models: Set of models explaining the process of influencing others through the behavior exhibited by managers which is adapted to circumstances.

Six Sigma: Widely adopted set of techniques based on statistical process control focusing on eliminating defects and reducing variation.

Skin in the game: Demonstrating commitment to a business by investing a significant amount of own money into it.

Sole trader: Legal form of ownership in which the business is owned and controlled by a single individual. As a sole trader, the business does not have legal personhood.

Solvency: Degree to which the current assets of a business exceed the current liabilities of a business.

Sort-set-sweep-standardize-sustain (5S): Basic workplace organization method in Lean.

Span of operations: Overall extent of different operations performed by a business.

Specialized business: Business concentrating on the production of a narrow range of products or services in order to bring to bear special expertise, productivity, and other advantages.

Speculator: Shareholder to whom the characteristics of the business are largely irrelevant, instead attempting to profit from short-term fluctuations in the market value of shares.

Stability: Ability to continue operating in the long term and to survive temporary problems such as periods of low sales, lower-than ideal levels of funding, the unavailability of important equipment, or the loss of important employees.

Stakeholder: Individual, group, or organization that has a legitimate involvement with a project or organization. This legitimacy can be the result of ownership, legal rights, moral rights, or by being affected.

Statistical process control (SPC): Use of statistical techniques to analyze available quality data and graphical techniques to extract meaning from such information.

Stock (predominantly accounting): The supply of objects held by the business ready to sell to a customer. Sometimes stock is referred to as "inventory".

Stock (finance) shares: The shares (of various types) issued by a business.

Stock keeping unit (SKU): Distinct type of item for sale and all attributes associated with the item that distinguish it from other item types.

Stock turnover ratio: Financial ratio showing how many times inventory is being sold and purchased over a given time period.

Strengths, weaknesses, opportunities, threats (SWOT) analysis: Analysis aiming to provide a systematic overview, distinguishing between internal and external environments of a business.

Substitute: Good whose appeal increases if the popularity of another related good decreases. In other words, the demand for a substitute increases if the price of another good increases.

Supply: The quantity of a good that a group of sellers is willing and able to sell at a particular price.

Supply chain management: Set of activities, decisions, and responsibilities that coordinate suppliers, manufacturers, and distributors so that products or services are produced and distributed efficiently, while minimizing costs and satisfying the requirements of the members of the supply chain.

Sustainability: Ability of an entity or process to endure over time.

T

Tangible asset: Asset that is a physical object. Cash and cash equivalents, inventory, equipment, buildings, vehicles, and many forms of investments constitute tangible assets.

Tax accounting: Activity of accounting for taxation purposes.

Technology push innovation: Innovation process that deliberately selects an underlying idea for processing into a marketable product or service on the basis of the perceived quality of that idea.

Theory of the firm: Theory in microeconomics that is based on the idea that a firm exists and makes decisions to maximize profit.

Theory X: Belief that the typical worker has little ambition and motivation to do their work and sees their employment as a transaction in which commitment, effort, or time is exchanged against pay. Closely associated with the work of Douglas McGregor.

Theory Y: Belief that employees are motivated internally, see work as a natural activity, and take responsibility for successfully executing their work. Closely associated with the work of Douglas McGregor.

Time preference: Current relative value assigned by a person to receiving a good or an object at an earlier time compared with receiving it at a later time.

Total cost: Sum of all costs faced by the business.

Total revenue: Sum of all financial inflows due to the sale of products or services.

Trademark: Form of intellectual property associated with or attached to a good or a service to distinguish from the goods or services offered by other businesses.

Transaction cost economics: Branch of economics interested in how business transactions are structured in specific circumstances. Normally seen as part of the economic theory of the firm.

Transistor: Small solid-state semiconductor device employing quantum effects to act as logical switches.

Trial balance: Accounting report that summarizes the account balances from the general ledger in two columns, debits and credits, and shows if the debits and credits are equal.

U

Utility model: Form of intellectual property offering similar protections to patents but normally covering technically less complex inventions and those with a shorter life span.

V

Value: Desirable property of an object defined as either the amount of money that can be received for it or the importance or worth of it for someone.

Value chain: Set of activities that a business performs with the objective of delivering a valuable product, which can be a good, a service, or both, to its customers. Closely associated with the work of Michael Porter.

Value proposition: Statement reflecting the entirety of benefits associated with a product or service which a supplier promises a customer will obtain in return for a payment.

Value stream: Series of activities that are necessary to transform raw materials or other inputs into valuable products or services.
Value stream mapping (VSM): A method used to understand value streams.
Variable cost: Cost that changes in proportion to the total quantity of goods and services produced.
Venture capital: A form of financing that is provided by venture capital firms or funds to new companies in exchange for equity.
Vertical integration: Process by which a business acquires ownership over more than one element of its supply chain or a configuration in which a business owns more than one element of its supply chain.
Von Neumann architecture: Computer architecture containing a central processing unit, memory, input devices, and output devices that makes use of the same memory for both data and program instructions.

W

Whole-life cost: Total cost of owning an object over its life cycle as determined by financial analysis. It includes costs from purchase to disposal.
Work breakdown structure: Hierarchical decomposition of the work to be performed in a project team to reach project objectives and produce the required deliverables.
Work package: A unit of related tasks, products, or functions in a project. Work packages are defined in a work breakdown structure to break projects up into smaller sections of work.

X

X-bar: Sample mean of a quality-relevant variable in statistical process control.

Z

Zero-hours contract: Employment arrangement that does not require the business to state minimum working hours. Zero-hours contracts should not oblige the worker to take on the work offered.

Index

D

E

J

K

L

M

N

T

U

V

W

X

Z

Printed in the United States
by Baker & Taylor Publisher Services